Pelican Books
The Computerized Society

James Martin graduated from Oxford with an M.A.
in Physics, joined I.B.M. after two years' research on
rocket motors, and has been involved in the design of
many of the world's most advanced computer systems.
He is now on the staff of the I.B.M. Systems Research
Institute in New York, and is the author of *Programming
Real-Time Computer Systems*, *Design of Man-Computer
Dialogues*, *Telecommunications and the Computer*,
*Security, Accuracy, and Privacy in Data-Processing
Systems*, and many other books. James Martin was a
member of the first Russian–American committee to
study possible exchanges in computer knowledge, and
has done many radio and television broadcasts on both
sides of the Atlantic. He is married to an American.

Adrian Norman was born in 1938 and graduated
from Cambridge where he read Mathematics and
Physics. After several years of research in the
Atomic Energy Authority, he moved to I.B.M.'s
Systems Engineering team in the City. He then went
to Columbia University's Graduate School of
Business where he took an M.B.A., and
subsequently worked on the plans for I.B.M.'s
world-wide internal information systems. In 1969,
he returned to the City to develop computer
systems for investment analysis. He is currently
working for the Inter-Bank Research Organization
on studies of the impact of computers in banking.
He has lectured at home and abroad, broadcast on
radio and television, and done political research for
the Bow Group. He is married with two young
children.

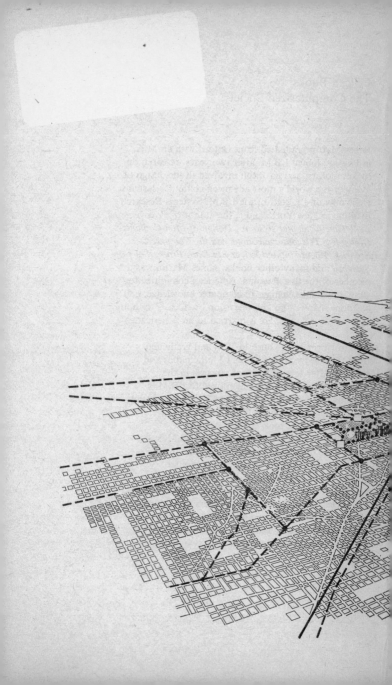

PREFACE

Arthur C. Clarke, who acquired a substantial reputation for the accuracy of his earlier scientific forecasts, claims that man's role as the dominant species on this planet is near its end.

Soon, he says, our species will be surpassed by ultraintelligent computers. As today's computers become more intelligent, eventually a critical point will be reached and a kind of chain reaction will occur because the machines will become capable of rapidly improving themselves. In a very few generations (*computer generations* which, when this happens, may be very short), there will be an explosion in the machine's capability; the merely intelligent machine will swiftly give way to the *ultra*intelligent machine.

This book looks ahead fifteen years, and we assume that in that time Mr Clark's critical point will not be reached. Indeed, in attempting to evaluate this future, we have deliberately avoided extrapolating much of today's work on 'artificial intelligence'. We think that this may be slow to bear fruit.

Nevertheless, we believe that a critical point is at hand. In the years immediately ahead, there will be a sudden, massive spread of computer usage that will affect the lives of almost everyone. Several factors will cause this: first, there will be mass production of the machines with a very sharp drop in cost. Second, there will be many areas of standardization that will have a snowball effect as programs used in one installation spread to others. Programs can be reproduced almost as cheaply as newsprint. Third, staggeringly large computer files are being developed in which every item stored can be retrieved in a fraction of a second. Fourth, and this may be the most important factor, the machines are becoming linked to the telephone network so that computers will be able to communicate with other computers. Small devices with a screen like a television set are enabling us to 'converse' with computers, and these devices will become cheap enough to allow many people to have one in their home.

Indeed, when the machines do eventually become much more capable we may someday talk not about separate computers, but rather about a vast organism interconnected by telecommunication links.

The man in the street has many misconceptions about the computer and the effect it will have on his world. The majority of persons outside computer circles totally underestimate its potential and the speed at which the changes are coming upon us. On the other hand, many of the current ideas are over-sensationalized and often highly alarmist. The computer is thought of in terms overly anthropomorphic. In this book, we, two computer systems analysts, have set ourselves the task of objectively explaining to the man in the street, in language as readable as possible, what is happening in the computer industry and its laboratories, and what impact this is likely to have upon society.

The story is an exciting one because the impact will be great. The end of our fifteen-year period is 1984, and a surprising amount of what George Orwell imagined now looks plausible.

Nevertheless, we believe that man can make the society of 1984 a better one than we have today. The potential wealth that the good use of computers opens up to us is staggering.

But like most of the greatest inventions, the computer has a potential for good that is matched by its potential for harm. Many factors are highly disturbing. To live with the computer of the future, new laws, new attitudes, and many forms of social action are needed. Some of these, the education of young children, for example, have a long lead time before they become effective. Because of this, it is important that the public, sociologists, teachers, law makers, and all levels of government officials, should understand *now* what is the likely course of computer technology in the next decade or two.

In an age so rich in innovation, we have a wide range of possible futures theoretically open to us. We say 'theoretically' because the institutions of our society have a high inertia, and the public has little understanding of the choices possible for the future. On the other hand, the technology that moulds our world is changing fast. It is changing much faster than it did when the automobile transformed our cities, and the rate of change is increasing. If devastating changes in technology occur at a faster rate than changes in our society's institutions or in the public's understanding, then society, to a major extent, is in the grip of the machine. As a new and powerful technique becomes economic, it is used *because it is economic*. The social implications are worked out afterwards, but this may be too late. The automobile has done much damage to some of our cities, especially old and beautiful cities in Europe. We know now how to build new cities in which the automobile and pedestrian amenities are separated, parking is adequate, and traffic jams are largely avoidable, but it will probably be several decades before such cities will flourish. The impact of the computer and new telecommunication facilities is going to be more sweeping than the impact of the automobile.

The question this book asks is: will we anticipate and plan for the new machines or will we let information technology race ahead undirected, leaving us to sort out the mess afterwards, as we are now doing with traffic in cities. If we permit the latter, then we have reason to be apprehensive.

ACKNOWLEDGEMENTS

The material in this book has been culled from a very large number of sources. The authors are grateful to all of the systems designers, politicians, engineers, sociologists, physicists, lawyers, and bar-room philosophers who made this book possible.

Part of the material was used in a study for the Federal Reserve Bank of Minneapolis entitled 'Developments in Telecommunications and Data Processing that will Impact the Possible Operations of the Federal Reserve Bank by the Year 1990' by James Martin.

Some of the material is used in six half-hour films called 'People and Computers' produced by John Cain for the BBC.

The authors are indebted to the following persons who read and commented on the manuscript: Mr Robert Baker of BP, Mr J. W. Greenwood of IBM, Rev. B. Shipman of the clergy, Miss Charity Anders, Mr R. MacLeod and others on the staff of a leading London stockbroker (whom we are not permitted to publicize), and Mr Kenneth Baker, British MP who read part of the book when drafting the Data Surveillance Bill. The authors also wish to thank Dean Courtney C. Brown and the faculty of Columbia Business School whose emphasis on the social responsibility of business led to the thesis from which this book was in part derived.

The authors feel particularly indebted to Mrs Kate Norman who managed to produce not only several drafts of their manuscript but two children as well.

He that will not apply new remedies must expect new evils for time is the greatest innovator – FRANCIS BACON

Part 1: Euphoria

1

AN EXPLODING TECHNOLOGY

After growing wildly for years, the field of computing now appears to be approaching its infancy. – *Opening sentence of the report of the US President's Science Advisory Committee on Computers in Higher Education, 1967.*

Popular writers and broadcasters tend to attach labels to the decades of the twentieth century. A momentous set of events, a political climate, or a way of thinking is often associated with the time period that it affected. The authors believe that the 1970s will be the decade when computer technology comes to the masses. Whether it comes in a way that will enrich lives and improve living conditions, or whether its coming will weave a bewildering mesh of controls and complexities, invading personal privacies, lessening freedom, wrecking social patterns, and making jobs ulcerously difficult, will be decided in the near future by the attitudes we create and the new laws we establish.

As yet, the 'man in the street' has felt the impact of computers only indirectly. His standard of living may have been raised marginally by their use. He may own a camera lens designed with their aid. His bank statement and telephone bill have taken on a neat if somewhat impersonal appearance, and he watches the most spectacular space circuses on television knowing that these would have been impossible without computers. In general, however, the new machines are not part of his life. He sees cartoons of strangely antiquated computers in *Punch* or the *New Yorker*, but they are largely a mystery to him. He does not, and he need not, understand them.

This, however, is only Round One. Computer technology is poised for a new and sweeping advance. This time the attack is going to be enormously more widespread because a fundamental change has occurred. The computer can now use our telephone lines.

A vast network of telecommunications links spans the industrialized countries of the world, carrying telephone, telegraph, and television signals, news photographs, and radio programmes. This network is constantly being expanded and new inventions are increasing its capacity, as we shall see on later pages, at a breathtaking rate. As yet, computer systems have made little use of this immense network. For a decade or so, data has been transmitted over communication links in a more or less pioneering, experimental manner. At first there were many problems, but slowly solutions have been found. We have learned to use the links more efficiently and the costs have started to come down. It is now practical for one computer to dial up another computer, just like a human being, and transmit information to it – at enormous speed if necessary. Even with the speed restrictions of a conventional telephone line, the computer can often code its information so that it sends hundreds or thousands of times as much in a given time as human speakers.

Perhaps more important, *we* shall be able to dial the computers and communicate with them. In offices, shops, factories, and probably in individual homes, there will be small machines designed to enable men to communicate with distant computers. We shall be able to ask them questions, to interrogate enormous banks of stored information, to perform calculations and to enter data which the computers will store, process, and act upon. In more advanced applications, we are seeing a new type of thinking in which the creative ability of the human user interacts with the enormous logic power of the machine, and has access to its vast store of data. This can produce results that neither man nor machine could achieve alone.

Systems described as *on-line* and *real-time* are now installed in many organizations. In these, data may be entered directly into the computer system from the environment it works with, and results relayed back. The word *real-time* implies that the response comes back very quickly – usually within two seconds or so if the response is to a man, and sometimes in a fraction of a second if it is to a machine. A wide variety of devices has been built for feeding data to and receiving replies from a computer – even from a distant location. Such devices are referred to in this book as *terminals*.

Terminals may be in the computer room or may be far away, connected to the computer by telephone line or other form of long-distance link. They may be on the shop floor of a factory, on the counter of a bank, in a warehouse or a supermarket, in the general manager's office – in fact, anywhere within an organization. In the future we envisage that individuals will have terminals installed in their homes. There are many different types of terminals, and they can be designed to fit as naturally as possible into their environment.

The operator of a terminal, whoever he may be, sends data to the computer. The data may be an inquiry or may be information to be processed. The computer deals with it – sometimes immediately, sometimes not – and a reply may or may not be sent back to the operator. Sometimes a computer sends data 'unsolicited' to a terminal. Items of data transmitted are referred to throughout this book as *messages* or *transactions*. An on-line system may transmit batches of data as well as single items.

Messages arriving at the computer from its terminals may be processed immediately, or they may be stored for later processing. Often a message is processed and a reply returned to the terminal within seconds. Transactions often update records immediately rather than being stored for later sorting and processing in batches. In many of the applications we shall discuss, the reply comes back to the terminal fast enough for the individual to carry on a high-speed 'conversation' with the computer.

A computer handling real-time transactions can usually carry on a 'conversation' with many terminal operators at the same time. The computer is much faster than the human users, so it shares its time between them like a chess champion playing twelve opponents at once.

Terminals may be designed for human operation or they may use instruments to collect data automatically. They may, for example, read thermocouples, strain-gauges, or other instrumentation. They may send a signal to the computer whenever an item is finished on a machine tool. They may enable the computer to count objects passing detectors on a production line. A very wide variety of devices is in use for collecting data at their source and for delivering the results of the computation where they are needed.

A few years ago almost all men using computing terminals were highly intelligent and trained in programming. Now exceptional intelligence is not a prerequisite and the reaction of the terminal is designed for men of lesser brilliance, and with no knowledge of computing. Indeed, some of the most exciting results have been obtained when backward children communicate with computers. Here the computer becomes a *teaching* machine, programmed to respond with infinite patience to the reactions of its pupils. The potential for computer-assisted instruction of children, and computer-learning programs available to adults, is one of the exciting parts of this story.

We are thus beginning to see persons from many different walks of life working with distant computers by means of terminals. An airline booking agent may talk to a machine on the other side of the Atlantic and to his customer at the same time in order to provide him with the best flight. Police in California pursuing a car down the freeway can radio a computer centre and be given information about that car. Secretaries can use a remote computer for storing and editing text, and what they type can be read or modified by someone in another city. Scientists and engineers carrying on much more complex conversations with their terminals can obtain results in a day that may previously have taken months. In some systems many different people contribute to the same real-time process in a multiple man–machine interaction.

In many fields the rate of scientific change increases every year; in computer technology it is truly breathtaking. In the late 1940s many informed and otherwise forward-looking scientists considered the computer to be a practical impossibility. Building a useful computer required very large numbers of logic circuits, logic circuits needed thermionic vacuum tubes, and vacuum tubes had a fairly high failure rate. A radio set with five vacuum tubes might fail only once a year, but a computer with 20,000 would fail once an hour. To find the failure would take more than an hour – with the machine running. So by the time one failure had been located, the chances were that another would have occurred. The repair man would have been chasing his own tail.

Nevertheless, the scientists' pessimism proved to be wrong. The reliability of vacuum tubes improved and they were used very conservatively to give them a long life. Less than two decades after the above ideas were accepted, the United States alone had 10,000 computers at work.

Vacuum-tube computers were later replaced by transistor machines. From transistors, 'solid-logic' circuits were developed with an entire logic unit built of pinhead-sized components on a wafer half an inch square. This type of circuit was replaced by minute 'monolithic' circuits, which are now to be replaced by 'larger-scale integration' in which many logic circuits are etched photographically on the single tiny face of a crystal of silicon.

An indication of this amazing rate of change can be obtained from Fig. 1.1, which shows the number of circuits that can be packed into one cubic inch. In 1950 almost twenty cubic inches were required for one circuit. A decade later the transistor logic could be packaged with one circuit per cubic inch, and five years after that almost ten circuits could be packed into a cubic inch. Developments still in the laboratory are exceeding densities of 1,000, although the user may not benefit till the mid 1970s. The time taken between operating a laboratory machine and having it sold, installed, and working appears relatively long; by the time it is installed it will be obsolete.

Because this book discusses what might happen in the 1970s, the reader might like to reflect upon how far the curve in Fig. 1.1 and the other curves illustrated in this chapter will have reached by 1980.

As the logic circuitry has become smaller in size, its speed*

---

* Electronic circuits process data by switching on and off according to the presence or absence of a bit of information. The faster they switch, the higher their processing *speed*.

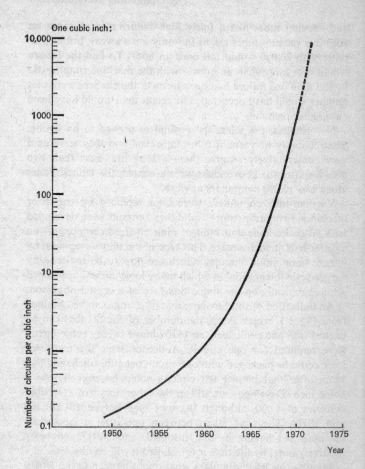

Fig. 1.1 The increase in number of logic circuits that can be packed into a given volume. (Note that the vertical scale does not go up in steps of equal magnitude. Each major division represents a tenfold increase.) *William A. Notz, Erwin Schischa, J. L. Smith, and Martin A. Smith, 'Benefitting the System Designer', from a special report on Large Scale Integration published by Electronics, 20 February 1967.*

has increased and its cost has decreased. The magnitudes of these changes have been equally dramatic. Figure 1.2 shows the circuitry cost in terms of millions of instructions per dollar. This relates only to the cost of logic circuits, not to such costs as programming, input/output devices, or printers. In 1955 about 100,000 program instructions could be executed for one dollar. In 1960 the same dollar bought 1 million, and in 1970 about 100 million. In other words, it is going up by a factor of 10 every five years.

The vertical scale in Fig. 1.2 and the scales in other figures in this chapter are *logarithmic* not linear. In other words, a straight line represents a constant percentage *rate* of growth, not a constant growth. The scale is deceptive insofar as it tends to give too low an idea of the growth. Suppose that we had drawn Fig. 1.3 with a *linear* scale, rather than its logarithmic one, and suppose that we had drawn it so that the entire scale prior to 1955 had occupied one inch of the page. Then in order to show the improvement between 1955 and 1970 we would have needed a page 250 feet in height! (Our publishers would not permit this and so we have had to use the logarithmic scale.)

The improvement in cost in Fig. 1.2 is a truly outstanding one by normal accounting standards. Lest the diagram should fail to illustrate this, let us translate it into terms of personal finances. Suppose that you were able to invest money so that it had the same growth rate as in Fig. 1.2 and that one dollar had been invested for you at birth. By the age of thirty you would have been a millionaire (before capital-gains tax). If you had lived to be seventy, then you would have overtaken the US Gross National Product!

Figure 1.3 shows the increase in speed of the computing circuitry. The first machines that the authors programmed, less than ten years ago, took several milliseconds to execute one instruction (and that seemed fast). Now the machines we normally use execute about 1 million such instructions per second, and they have big brothers 10 times as fast.

The capacity of computer storage has also increased phenomenally, and has made possible many of the uses of computers

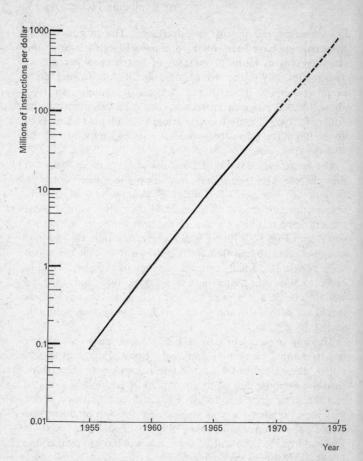

Fig. 1.2 The decrease in cost of logic circuitry in terms of millions of instructions per dollar. (Note that the vertical scale does not go up in steps of equal magnitude. Each major division represents a tenfold increase.) *William A. Notz, Erwin Schischa, J. L. Smith, and Martin A. Smith, 'Benefitting the System Designer', from a special report on Large Scale Integration by Electronics, 20 February 1967.*

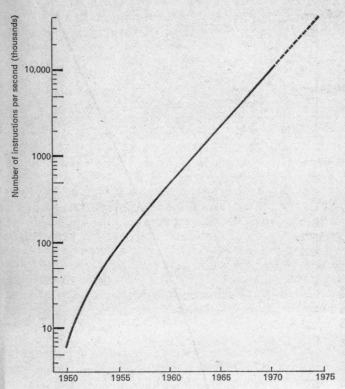

Fig. 1.3 Increase in speeds of high-speed computers, in terms of memory cycles per second. (Note that the vertical scale does not ascend in steps of equal magnitude. Each major division represents a tenfold increase.)

which we shall discuss. Applications which will affect our way of life will often need fast access to huge banks of electronically stored data.

There are two ways of storing huge files of machine readable information: they can be stored 'on-line' or 'off-line'. *On-line storage* enables the computer to read stored data without human intervention: the data are in the machine's own file units. It can read any piece of information at random, usually

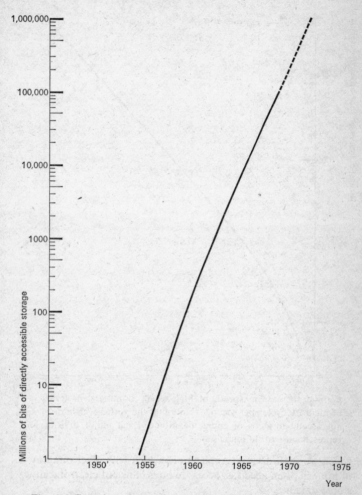

Fig. 1.4 Capacity of on-line data files directly accessible by the computer. The curve represents maximum amount of storage likely to be found on large systems in each area. (Note that the vertical scale does not ascend in steps of equal magnitude. Each major division represents a tenfold increase.)

in less than half a second. *Off-line storage* refers to a medium, such as magnetic tape, to which the machine does not have direct access. Here the operator must take a tape from the shelves of a storeroom and load it; then the computer scans through the tape sequentially – and thus relatively slowly.

Figure 1.4 indicates the volume of data the computer can store on-line – not the absolute maximum that *could* be stored on-line at any given time, but a reasonable practical maximum that might be found on an ambitious system at that time. Again the vertical scale is logarithmic. The increase in rate of growth is greater than in Fig. 1.2. Using the same illustration of investing a dollar at birth, with this growth rate you would over-take the Gross National Product by middle age, when you were still young enough to enjoy spending it!

As capacities go up, costs per 'character' stored come down as shown in Fig. 1.5. A *character* is a letter or digit, or a special symbol such as a comma, dash, or question mark. This book, for example, contains $1\frac{1}{2}$ million characters (not counting the pictures). If it were stored on-line in a large file using the redundant coding* that you are now reading, its filing cost would have been about $700 in 1955, $300 in 1960, and about $3 in 1970. As of 1970, systems existed in the laboratories which could store more than a million such books on-line (more than 20 miles of library shelving). With efficient nonredundant coding, vastly more information could be stored.

The reader might again like to extend the lines of Figs. 1.4 and 1.5 upward for a decade and reflect upon what could be achieved. It is staggeringly impressive – perhaps especially to a systems analyst who still thinks in terms of loading a file from punched cards! Is this exponential growth rate *likely* to con-tinue? We cannot say. Certainly there must be an upper limit somewhere. However, the developments now going on in the laboratories suggest that we are a *very* long way from it yet. Before long we shall be regarding today's billion-character storage systems as primitive and crude. And already the

---

* The written word uses a lot of letters to convey an idea. Less *redundant* codes than the English language use less characters to hold the same infor-mation.

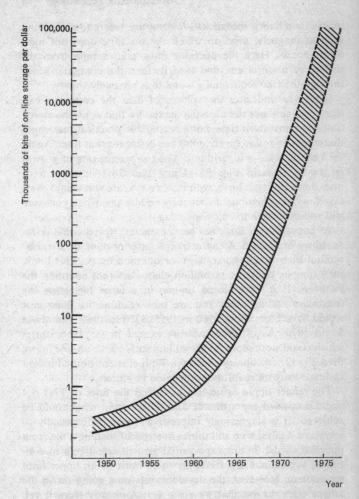

Fig. 1.5 Cost of on-line storage. The increase in number of
bits per dollar cost of on-line storage units. (Note that the
vertical scale does not ascend in steps of equal magnitude.
Each major division represents a tenfold increase.)

machines we are building are vastly outstripping our ability to use them to good effect.

To make good use of such vast storage capacities, telecommunication links are needed to make the data quickly available to their many users. Rapid progress is also taking place in telecommunications. A succession of inventions is increasing the amount of information that can be sent over one cross-country link, as will be discussed further in the next chapter. Figure 1.6 plots the maximum capacity of telecommunication links at different points in time, using as a measure of capacity the number of bits of data that could have been sent over the link in one second if the entire capacity of the link was used for data. Again, devices today in the laboratory suggest that this capacity will continue its constant and rapid rate of growth; one needs an inventive imagination to reflect what could be done with this increase if the straight line of Fig. 1.6 continued upward for another two inches or so.

The amount of data passing between one place and another is not directly related to the line in Fig. 1.6, because there can be many such channels. Figure 1.7 gives the American Telephone and Telegraph Company's 1965 forecast of the growth in their interstate telephone-grade links up to 1972. One physical link is likely to carry a large number of separate telephone channels. The lower curve in Fig. 1.7 relates to telephone channels used for human conversations. The difference between this and the upper curve represents the transmission of data, pictures, and other non-voice messages. This use of communication links is expanding much faster than that for conventional telephone conversations. Although AT&T has become the world's largest commercial organization mainly because of people talking to people, they recognize that greater revenue is likely to come in the future from machines talking to machines, and from men communicating with distant computers.

The graphs in Figs. 1.1 to 1.7 clearly tell one story. We are hurtling into the computer age at a pace which makes the Industrial Revolution look like a funeral procession.

The computers, their giant storage banks, and their telecommunication links constitute a force which is going to have

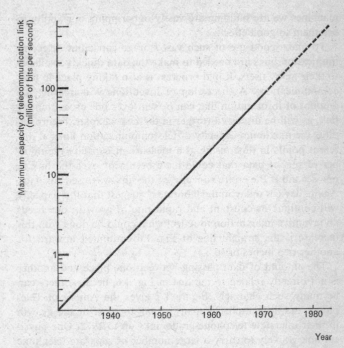

Fig. 1.6 Increase in capacity of single installed long-distance telecommunication link. (Note that the vertical scale does not ascend in steps of equal magnitude. Each major division represents a tenfold increase.)

the most enormous impact on society. Probably no other invention has greater potential for changing our lives. The machines are capable of benefiting mankind in many ways. However, they are also capable of causing very difficult new social problems. The Industrial Revolution changed men's lives and in so doing created great misery and loss of freedom for many people. Plunging into the computer era, can we avoid social chaos? Can we avoid creating with greater subtlety and intricacy some of the facets of Orwell's *1984*? Can we create a world that is more enjoyable than today's and worthwhile to live in? Or is it all happening too fast for us to control?

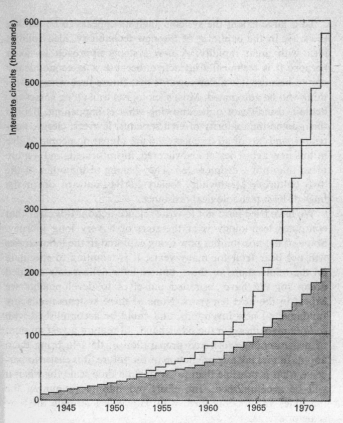

Fig. 1.7 Increase in number of long-line circuits in the Bell System. The upper curve shows thousands of inter-state voice channels in the USA. The lower curve shows channels used for conventional telephone conversations. The difference between the lower and upper curves represents equivalent voice circuits for data transmission, facsimile, television, and special services. *Reproduced from 'The Transmission of Computer Data' by J. R. Pierce in the Scientific American, September 1966.*

To a large extent the systems analysts appears to be leading the way. In the euphoria of the new technology, idea follows idea with great rapidity. A new systems approach is taken because it is technically attractive, because it is economically better, because it is a step towards that future time when everything will be automated. Most sociologists trail along some way behind, usually not quite knowing what is happening. Behind them come the majority of civil servants, lawyers, clergy, politicians, and last of all teachers who are preparing people to live in this new age. They are bewildered, misinformed, and – more often than not – disinterested. They belong to the other of the 'two cultures'. Meanwhile, society hurtles onward down the lines of least technological resistance.

We have tried hard not to write science-fiction. It is clear that computer technology is at the start of a very long journey. Some of the possibilities now being explored in the laboratories will not bear fruit for many years. It is tempting to speculate on the implications of these, but we have deliberately avoided doing so. We have restricted ourselves to developments we expect in the next ten years. None of these systems needs any fundamental new inventions, and could be accomplished with *today's* machines. On the other hand, all require a vast amount of programming work and preparation of data to bring them about. In making any prediction of the future, it is easier to perceive what is possible than to lay down a time scale for when it will be accomplished. The effort put into programming the machines and loading them with data will determine the time scale of events.

Compared with probable achievements in the next few decades, we have hardly started to use the facilities we have invented. The computers of today have a potential far beyond the way we use them today. In fact, if all hardware development were to cease now and we were to spend the next fifty years learning how to program and use today's machines and building up banks for them, we could accomplish what now would seem like miracles.

Certainly our way of life is going to change drastically as a result of computers. Equally certain, we need new laws, new education, and new attitudes to cope with this revolution. Can we acquire these in time? Or is the changing environment coming too quickly? Is it inevitable that, in a revolution such as this, the new technology explodes into action and we sort out the mess afterward?

2

THE NEW MACHINES

Scientists look forward to the fulfillment rather than the sacrifice of what the Anglican prayerbook calls 'the devices and desires of our own hearts'. – *Kenneth E. Boulding, The Image.*

Before computers, data processing installations were created by assembling separate machines, such as calculators, tabulators, card sorters, and collating machines. Large installations had hosts of clanking machines with girls pushing trolleys of cards between them.

This proliferation was cleaned up considerably with the advent of the computer. One general-purpose machine replaced many punched-card devices. However, the computer steadily became more complicated. Its peripheral devices became more numerous, and again we became faced with a kit of construction-set parts that could be connected together in many configurations. At present almost every month more devices are added to the collection, especially since telecommunication facilities can now link the machines. It is steadily being discovered that anything from a typewriter to a blast furnace can be on the other end of a telephone line leading to a computer. An installation today consists of many machines connected together with cables under the floor, and often with telephone lines bringing in data from distant places and conveying back the results.

1. *The Computers*

The computer is the heart of the system, performing the calculations and the logic and data manipulation, and to a large extent controlling the other equipment in the system.

Fig. 2.1 How to read this chapter. (A programmer prepares instructions for a computer in a form somewhat similar to this.)

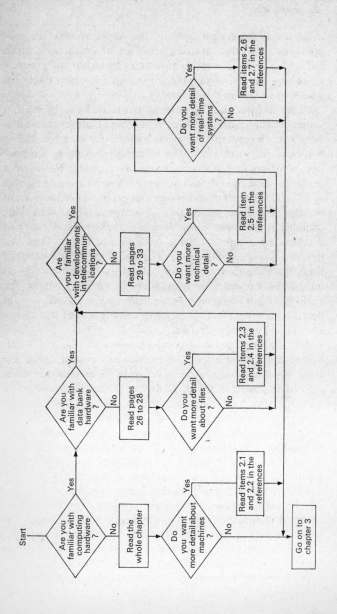

To make the system operate, the computer must be *programmed* – in other words, given a set of instructions which tell it what to do. The instructions must be logical and precise – something like the block diagram in Fig. 2.1, which might tell you what to read in this chapter.

Each 'decision' the computer is told to make is, by itself, a simple decision: for example, 'Is number A bigger than number B?' Each operation, by itself, is small: for example, adding two numbers, producing a line of print, or changing the position of letters in a piece of data in its memory. However, it performs these operations and decisions at an immensely high speed, and when it has completed a large number of them it can sometimes give the appearance of having achieved something complex. A computer which designs a transmission network or produces an optimum routing for a fleet of tankers does this by executing a lengthy series of simple steps. It goes through these exactly as its programmer instructed it to, perhaps repeating groups of instructions many times with minor changes. The result it obtains however, is better than any human could obtain because no person has the patience or time to perform such a lengthy and precise operation.

A typical medium-sized computer today can execute 1 million instructions in about one second, at a cost of about five cents. It will normally make no errors in the execution of these steps. The power of the machine lies in their immense speed and accuracy.

Putting such a computer to work requires a large programming effort. Fortunately, most programs are very highly repetitive, calling for the execution of sequences of steps over and over again. Many programs or segments of programs can be purchased complete from manufacturers, consultants, or other computer users. It is highly desirable that a program that works on one computer will also work on many (or all) others, although complete compatibility between different makes of computers has not yet been achieved. An industry is growing up to provide programs for computer users, and will probably become one of the major industries of the 1970s. Package programs are often referred to as *software* (in contrast to *hard-*

ware, which refers to the machines). A computer user purchases both hardware and software, and in addition usually writes programs of his own design.

The uses of computers discussed in the following chapters, will need a massive amount of programming work. As we note later, in the decades to come there is likely to be a severe shortage of programmers and of systems analysts who decide what should be programmed and what hardware and software is needed.

2. *Peripheral Devices*

In a typical computer installation there are many machines other than the computers themselves: high-speed printers, machines for reading punched cards and for punching them, machines for storing a massive quantity of information on magnetic tape or disks, and so on.

Figure 2.2 categorizes the main types of hardware that are attached to computers. First, there are *input units.* The commonest way to give information to a computer is by way of punched cards. Most installations have a room full of girls punching data into cards for the machine to read. The system itself also punches cards which may be fed back to it at a later time. There are various other types of input units which enable data on other media to be fed to the computer. For example, there are machines for reading punched paper tape, machines which read typewritten characters, and machines which read the magnetic-ink characters on our bank checks.

Second, there are *output units.* The usual way for a computer to give information to us is by means of a high-speed printer, which prints invoices, cheques, reports, and other documents at about 1,000 lines per minute. Sometimes a typewriter is used for printing occasional comments. These machines, like other 'peripheral devices', operate under the contral of the *central processing unit* (or CPU) of the computer system.

Third, there are *storage media.* The data-processing system must store a large amount of information. It may be stored in such a way that the computer can read it when required with-

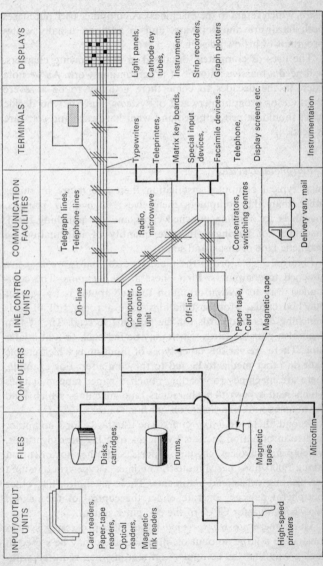

Fig. 2.2 The main types of devices that are attached to computers.

out human intervention, or it may be stored on shelves on a medium such as magnetic tape or magnetic cartridges. Ideally, we would like every record to be accessible almost instantly to the computer, so that appropriate files could be read or updated while a transaction was being processed. In practice, this would be too expensive, and so we compromise by storing some of the records on magnetic drums and 'disks', quickly accessible (typically in about 5 to 500 milliseconds), while others we store on reels of magnetic tape. Magnetic tape provides a cheap form of storage and any number of reels can be stacked away on shelves. Drums or disk units provide almost immediate access but they are more expensive. The facility to store massive quantities of data is an important part of the story we have to tell, and we shall describe the storage in more detail below. An equally significant part is the use of telecommunication links to connect the computer centre to distant terminals and displays.

3. *Data Banks*

The new technology of information storage and processing has turned up just in time to serve the needs of scientific progress. The history of technology records a number of fundamental inventions which arrived in one field of research just when they were needed to permit progress elsewhere. The thermionic vacuum tube, for example, arrived in the nick of time to allow development in telecommunications. The moon shots depended on a variety of recently developed technologies, and if any one of them had been developed a few decades later, the shots would probably have been impossible. We have reached a state now in man's learning when the quantity of information being generated in industry, in government, and in the academic world is reaching alarming proportions. The press euphemistically call it the 'information explosion'. But that is not a good term, because the violent growth of an explosion quickly ends – the growth of man's information has no end in prospect, only greater growth.

The sum total of human knowledge changed very slowly

prior to the relatively recent beginnings of scientific thought. It has been estimated that by 1800 it was doubling every 50 years; by 1950, doubling every 10 years; and that presently it is doubling every five years.[8] One wonders how long this acceleration can continue. Computer technology may make a frighteningly high rate of increase possible for centuries.

New knowledge must be recorded for others to use and as a result the quantity of stored information is increasing at an enormous rate. In the computer field itself, the volume of manuals and other documentation is far surpassing people's ability to keep track of it. The weight of the drawings required to build a jet plane is greater than the weight of the plane. Even in a field of research as old as medicine, more papers have been published since the Second World War than in all prior history.

Clearly, automated means of filing and indexing are becoming essential and must be associated with means of searching for and retrieving the required information. Computer storage coupled with computer-processing ability provides a large part of the answer. Telecommunication links will enable people to search for information with computer assistance.

Collections of automated files are now known as *data banks* – storehouses of information that can be added to or looked through on request. Economic considerations determine the storage media and access methods used. A library is a data bank with cheap storage (printed paper). It permits random access at slow speed via a series of indexes to the required data, and is not automated. More sophisticated systems might have microfilms of articles, and perhaps some measure of automation. A computerized data bank consists of a collection of high-capacity random-access files attached to a computer. In such a data bank the computer keeps the index to the data – in some cases much like the library index – and this may also be stored on the files.

Because these files are random-access, a large quantity of information is quickly available to the computer. It can read, or update, any item on the files at random in a fraction of a second. A different approach is used to store the data on magnetic tapes. Before an item can be read, the appropriate tape must be loaded onto one of the computer's tape units; then the machine will search through the tape sequentially, looking for the item it needs. To use a musical analogy, a jukebox is a random-access file. It will go directly to a record of the Beatles when you instruct it to from your 'terminal' at a cafe table. On the other hand, an ordinary tape recorder is not random-access. You take a tape from the shelf and load it on your tape recorder; then, because the tape contains many songs, you must scan sequentially through it to find the one you want. The juke-box could conceivably have been on the other end of a tele-phone line and you could have selected and listened to the Beatles over the telephone. The tape recorder could not, because there is no automatic mechanism for selecting and loading the tape. Similarly, data stored on magnetic tapes are not available in 'real-time', nor automatically accessible by telecommunication link.

Storage of data on magnetic tape has the advantage of economy. Storage of data on disks or other random-access devices has the advantage of permitting automatic access to any item of data *quickly*. Because of the long time it takes to scan tape files, requests for access to these files are normally grouped together sorted into the same order as the tape records, and processed in a batch.

There are also hybrids of the above two approaches. Tape files are being developed in which the tapes *can* be selected and loaded under computer control and scanned very quickly. Also, sets of disks from random-access files can be removed from their drive units and stored, like tape, on shelves. The computer then has access only to that part of the information which is loaded on the drive at the time.

The technology of random-access files is developing fast. As was indicated in Fig. 1.4, current inventions in this area make it likely that *very* large quantities of data will be immediately accessible to computers in the next decade. The entire Bible – 1,245 pages – can be reproduced on the surface of a two-inch square of plastic which can be read with a special viewer able to magnify the pages 50,000 times. A computer file in the future might have fast random access to millions of such squares of plastic.

Banks of data on many subjects will grow in the years to come. After an enormous quantity of coding and classification work has been done, the biochemist will be able to check what reports have been written on a particular topic, the lawyer will have the mass of literature he needs electronically available to him, and the patent attorney's agent will carry out a search by conversing with the computer.

Because the quantity of data in computer-scannable form will grow indefinitely, it should become standard practice *now* to give all technical reports not only a title and an abstract, but also a *data index* for information-retrieval purposes. A standard dictionary of content words could be used, and constantly be extended as required. For instance, a major subject category might be 'physics', with subdivisions such as 'superconductivity' and sub-subdivisions such as 'Meissner effect'. A number of words relating to the contents of the report would be in its data index and relationships between the words might be indicated.

In searching for reports on a given topic the user would carry on a two-way conversation with the distant computer. The computer might suggest more precise categories of what the user is seeking. He might *browse* through many data indices, titles, and abstracts before he found what he needed.

More interesting than files of such literature are the banks of data useful to the public, industry, government, or other organizations. The public can already use specialized data banks to find houses, obtain theatre tickets or hotel rooms, and make train reservations. In industry a wide variety of different files is going to be set up to provide information to management – details of customers, products, orders taken, personnel, machine-tool loadings, sales forecasts, and so on. The following chapters will discuss many more examples.

The information content of the large files in many computer systems is constantly altered. A major reason for installing these systems is to provide a large number of users with up-to-date information about a changing situation. Some files change their content only very slowly, whereas others change fast. A file to which new records are being added and old ones deleted at a rapid rate is sometimes referred to as a *volatile* file. A city telephone directory, for instance, is not a very volatile file because the rate at which changes are made is not high. On the other hand, a centralized file of reservations on an airline would be volatile, for bookings and cancellations on the many flights flood constantly into the system.

4. *Telecommunications*

Perhaps the most exciting recent development involving computer hardware is the ability to transmit computer data over a country's telephone network. By itself either the computer industry or the telecommunication industry is capable of bringing about changes in the working habits and the government of people that will alter their ways of life throughout the world. But the two technologies complement each other. In combination they add power to each other. Telecommunication links will bring the capabilities of the computers to the millions of locations where they can be used, and computers in return will control the immense switching centres and help divide the enormous capacity of the new linkages into usable channels.

In England in the eighteenth century the spinning jenny and a variety of weaving inventions portended a revolution in cloth-making. About the same time the steam engine was developed. These two inventions also complemented each other. Either by itself would have caused changes, but in combination they brought about an upheaval that was to alter drastically the lives of all the people involved. The attractive villages with their cottage industry gave way to the dark satanic mills of the early Industrial Revolution. The pounding new machines dominated men's lives. Today, as then, the new technologies are sweeping across our society too fast for the sociologists and politicians to plan the type of world they want to build.

The first Industrial Revolution depended primarily on man's building machines – such as the steam engine – to harness physical power and applying them in many different ways. The computer era has been referred to as 'the second Industrial Revolution', and depends upon harnessing mechanized logic and stored information for a wide variety of applications. Man first built extensions of his muscles which gave him power enormously greater than his own frail limbs. He then built extensions of his brain which are giving him mechanized memory of unimaginable magnitude and the logical abilities to process the data in this memory with absolute accuracy and at speeds millions of times greater than his own head-scratching thought. When this second Industrial Revolution has run its course it will have brought a change in mankind's environment even greater than the first.

The sources of physical energy in the first Industrial Revolution, however, required transportation. Canals were dug and weary horses spent their lives lugging barges of coal from the new mines to the factories. Many factories clustered round the coalfields, and regiments of the population moved into gloomy jerry-built rows of houses near by. It took many years before the means of transportation were adequate for the needs of the new industry. Later the railways were built; and when electricity was invented, overhead wires carried the necessary power across hill and dale.

Distribution, likewise, is of great importance in the second Industrial Revolution. We cannot afford to have a giant set of computer files duplicating data at every location where men need the information. We want a single 'data base' serving hundreds or thousands of locations. Similarly, we cannot afford to have a powerful computer at every place where men want to use one. Perhaps small computers will ultimately become as commonplace as adding machines, but there will always be a need to use the *million-dollar* computers, and these may be a long way away. Above all, many of the applications that are now envisaged require *fast* access to a machine or to data. A managing director wants a *quick* answer to his question so that he can make a decision. A doctor wants a *quick* analysis of an electrocardiogram. A factory shop floor has to be controlled in *real-time* as events there are taking place. There is no time to write a letter to the computer centre, or even to catch a train to it.

Distribution of information in the computer world will be done largely by telecommunication. Data are sent over transmission lines often using the same telephone links that we now use to communicate with each other. In the first Industrial Revolution the means of distribution lagged seriously behind the need for it. Today there is such a lag with computing. In the first decade of widely-accepted commercial usage of computers data transmission was restricted to a handful of pioneering experimental installations. Now, data transmission has become accepted and problems initially associated with it have been overcome. The data are being transmitted over conventional telephone and telegraph lines. In the 1970s we can expect to have a vast network, as complex as the physical distribution networks of commerce, making computer power and data widely available.

The telephone lines and probably other, newer cables will connect our homes and offices to the computers. One will be able to dial machines using particular programs or data, just as today one can dial a friend or, in England, a machine that gives the cricket score. This will change our working patterns just as thoroughly as the first Industrial Revolution changed them.

Although much work can now be done from individual homes, using telephone links to computers, vast quantities of programming are needed. In teaching, for example, computer-assisted instruction can be enormously effective, but only if the programs used for it are very elaborate and are prepared with great care. If computer assisted teaching is to be widely accepted, hundreds of thousands of man-years of programming will be needed. *Enormous* quantities of intelligent human work are needed to carry us forward into this new era – not the work of geniuses, but of ordinary intelligent step-by-step construction and testing of the multitude of programs needed. In general, it is creative, enjoyable work; work that wives and children can do; work that disabled and in some cases blind people are doing. Above all, it is work that can be done at home, provided the home has a telecommunication link to a computer. And so we may see a return to cottage industry, with the spinning wheel replaced by the computer terminal.

TODAY'S TRANSMISSION LINKS

Almost any telecommunication medium can be used by the computer. Open-wire pairs singing in the wind between telegraph poles are becoming an increasingly rare sight in many of the world's industrialized countries. One still sees them in France, stretched between green-glass insulators that look suspiciously like wine bottles, but in most areas they are becoming part of the romantic past. Instead, thicker cables, capable of carrying many telephone calls together, are buried underground or hung from poles with the power cables. These in turn are connected to more powerful transmission links which form the main telecommunication highways of our society, connecting towns and spanning long cross-country distances. The major links can carry many hundreds or thousands of telephone conversations at once.

A high-capacity coaxial-cable link can carry 17,000 voice conversations simultaneously. If all of this capacity were in use for transmitting data, the entire text of this book could be sent thousands of miles in one-tenth of a second. Alternatively, a signal may travel in the form of a radio beam from the dish-shaped antenna on a micro-wave radio tower to another on a tower within sight. Like the cable link, microwave links carry many thousands of telephone channels simultaneously, any of which can be used equally well for computer data. A group of such channels can be used for high-capacity data links. Such microwave links are in widespread use for television transmission. The towers with the microwave dishes are typically spaced about thirty miles apart, forming cross-country chains and are usually built with plenty of spare capacity, recognizing that tremendous expansion in data transmission is likely to occur in the 1970s.

Many cities of the world now have their skylines dominated

by a tower carrying microwave antennas. Tokyo has a tower like the Eiffel Tower but 40 feet higher. East Berlin is building one 1,185 feet high. One of London's most expensive dinners can be eaten in a revolving restaurant just above the microwave antennas, and Moscow, outdoing the rest, has one 250 feet higher than the Empire State Building. These may come to be regarded as the symbols of the computerized society. The gothic spires have been dwarfed by monuments to electronics.

It is particularly important that the telecommunication facilities should permit *dial-up* between data-handling machines, just as there is dial-up between domestic telephones. One terminal should be permitted to dial *any* appropriate computer that is connected to the public network. This is possible today in America on telephone and telegraph lines. As of this writing, telecommunication authorities in some other countries have not yet permitted a free use of the telephone networks in this fashion, and this is impeding progress there.

5. Terminals

A vast and ever-growing array of devices can be attached to communication lines for transmitting data to a computer. These can be devices into which data are entered by human operators, or which collect data automatically from instruments. Similarly, the remote output devices may present information for human use, or they may control machines, valves, or other components.

Terminals designed for human use may permit a fast two-way 'conversation' with the computer, or may be a remote equivalent of the computer-room input/output devices. Paper-tape readers and punched-card readers may provide input over communication lines. Printers may provide the output.

Some computer systems respond with a voice to their distant users. A number of spoken words or phrases are stored in the computing equipment and the computer is programmed to assemble appropriate words to respond to a user. Voice response can make the terminal very inexpensive, since a conventional telephones can be used for communicating with the computer. Input is keyed in, for example, on the keys of a touch-tone telephone and the response is obtained verbally. A representative can thus communicate with his firm's computer from any public telephone.

6. *Future Telecommunication Developments*

Important new inventions in telecommunications will bring about major changes in the cost of communication links and in the way we use them:

(a) *Picturephone.* With a picturephone you can see the person you telephone as well as hear him. AT&T are already installing picturephone sets – at present on a limited basis. The 1970s, however, will probably witness a mass sales campaign for picturephones and a decrease in their cost.

The picturephone should reduce some of the need for business travel if only because of its ability to show products and let technical staff draw diagrams as they talk. By the end of the 1970s AT&T forecasts that many business locations and 1 per cent of homes will have picturephone sets.

From the viewpoint of this book, perhaps the most significant aspect of the picturephone is that it requires more than 100 times the transmission capacity (bandwidth) of the telephone. Over a picturephone line, we can therefore transmit 100 times as much as over the telephone. (Over a leased telephone line we transmit today at speeds of about 5,000 bits per second. When picturephone sets are installed, we can expect to have links which can carry 500,000 bits per second or more – a very high speed by the standards of today's computer world.)

(b) *Waveguides.* The transmission capacity of the main telecommunication arteries is likely to increase greatly with the introduction of waveguides. A waveguide is, in essence, a metal tube down which travel radio waves of very high frequency. The 'helical' waveguide is a pipe about two inches in diameter with a fine enamelled-copper wire wound tightly round the inside in a spiral. This is surrounded by a layer of thin glass fibres, and then by a carbon layer. The whole is encased in a strong steel case, and bonded to it with epoxy resin.

A waveguide system is more difficult to construct and to lay down than coaxial cable or microwave links. The waveguide pipe, for example, can have only the most gentle bends in it. However, a waveguide system has been fully developed by Bell Laboratories which could span the United States coast-to-coast with sufficiently low noise and distortion, and this could transmit 100,000 simultaneous voice channels, and possibly more. This is probably the next step up the dotted line in Fig. 1.6.

(c) *Lasers.* Probably the step after this will be the *laser*, which portends a revolution in telecommunications as fundamental as the invention of radio. 'Laser' stands for Light Amplification by Stimulated Emission of Radiation. A laser produces a narrow beam of light which is sharply *monochromatic* – that is, it contains a single colour or frequency (in fact, several such single frequencies); and is *coherent* – that is, all of the waves travel in unison like the waves travelling away from a stone dropped in a pond. Normal light, even that of one colour, comprises a small range of frequencies and waves which are incoherent, bearing a random position relative to one another.

An analogy with sound waves, though somewhat inexact, will help the reader to visualize the difference between a laser beam and an ordinary light beam. The sound from a tuning fork consists of waves which are of one frequency, and which are reasonably coherent. On the other hand, if he puts a hammer through his window, the sound waves would be neither monochromatic nor coherent. The former may be compared with a laser beam, the latter with ordinary light. The laser beam is formed by forcing molecules to oscillate with a fixed frequency in much the same manner as the tuning fork.

A laser beam, because it is so sharply monochromatic, is not spread out by a prism; optical arrangements can be built for it so precisely that a beam of laser light can be shone onto the moon and illuminate only a small portion of its surface. A beam can be concentrated with a lens into so minute an area that the intense concentration of energy causes highly localized heating. The result is a cutting or welding tool of miniature precision beyond the dreams of Swiss watchmakers. The surgeon has a microscopic scalpel; the general a potential death-ray.

For telecommunications we have a beam of great intensity which is highly controllable, which can be amplified, and which can be made to carry information. It could be transmitted through the atmosphere as a pencil-thin beam of light. However, it is more likely to be sent through a pipe of optically transparent fibres, in which it cannot be interfered with by fog or snow. One laser beam has a potential information-carrying capacity many thousand times that of a microwave beam. (Its bandwidth is almost 100,000 times greater.) One laser pipe may carry more than one such beam. We have not yet learned how to make a laser beam carry this much information, but results of preliminary research are promising.

Systems already operating in the laboratory carry very high information rates, although probably far below the rate that will eventually be achieved.

(d) *Communication Satellites*. While commercial use of laser communications is still some way in the future, an invention that is to change our world perhaps even more dramatically has already been launched. On 6 April 1965, the world's first commercial satellite, Early Bird, rocketed into the evening sky at Cape Kennedy. The success of the transmission experiments that followed was spectacular. Before long, Earth stations were being built around the world and new, much more powerful satellites were on the drawing boards.

Originally the satellites were intended to span the oceans. The laying of subocean telephone cables is expensive and these cables have a bandwidth too low to carry live television. Early Bird alone increased trans-Atlantic telephone capacity by more than one-third. The television pictures were excellent. Pope Paul VI was soon seen live from the Vatican in America and President Johnson all over Europe. A Pacific satellite followed in January 1967 and President Johnson was seen 'live' in Asia.

The satellites were soon used with computer systems. Pan American Airlines successfully hooked up their spectacular reservation system to Early Bird, so that reservation agents in

Rome or London could smile at their customers while conversing with a computer in New York via satellite. IBM linked its laboratories on opposite sides of the Atlantic, and was soon talking about a far-flung information system using satellite-transmitted data responses to oil the wheels of its World Trade operations.

Early Bird is demonstrating a lifetime in orbit far beyond its original design estimates. The Intelsat II, and then III, satellites followed Early Bird and were found in practice to be capable of handling considerably more traffic than their design capacity. New generations of satellites are now being planned with much greater capacity and probably much longer life than the early ones. Since life and capacity govern the economics of satellites once they have been launched, they now contend well in the United States *domestic* market for long-distance transmission, as well as the international market. It will almost certainly become cheaper to transmit from New York to Chicago via satellite than via coaxial cables or microwave links.

This has interesting implications for international communications. If we transmit from New York to Chicago by satellite, the cost is about the same as from New York to Australia by satellite. If a firm has a tie-line from New York to Chicago, it could probably have one to Sydney as cheaply. If you see a relative in Chicago by calling on a picturephone in New York, you could presumably afford to call friends in Tokyo or Tel Aviv on a picturephone too. The links between nations will indeed have shrunk.

In data-processing systems, too, a firm that can today contemplate an on-line network for its domestic organization may before long be able to hook up its international organization for much the same cost.

The power of telecommunication satellites is expected to increase greatly in the years ahead. The rockets used for such launchings so far have been relatively small compared with Saturn V, the moon-shot rocket, and the componentry of the satellite will be further miniaturized.

A rapid expansion in telecommunication channels, then, is envisaged for the next decade. All of the various media we have described can be used to send information in a variety of different forms – voice, television, picturephone, diagrams, computer data, Xerox machines transmitting to Xerox machines, and so on. The many types of information will all be mixed together to travel over the main high-capacity highways. We can expect to have cables which can carry a thousand times more information than today's telephone cables linking our homes and offices.

Transmission between countries and over very long distances is likely to become no more expensive than transmission over a few hundred miles within one country, when future generations of satellites are put to work.

The story that unfolds in the rest of this book will be greatly influenced by these facilities. Our expectations for their use will emerge in the chapters that follow.

REFERENCES

1. Thomas Bartee, *Digital Computer Fundamentals*, Harvard University Press, Cambridge, Mass., 1966.

2. A. M. Weinberg and H. D. Leeds, *Computer Programming Fundamentals*, McGraw-Hill, New York, 1966.

3. David Lefkovitz, *File Structures for On-Line Systems*, Spartan/Macmillan & Co., London, 1969.

4. 'Introduction of IBM Direct Access Storage Devices and Organization Methods', IBM Corporation, Manual No. C20–1649, White Plains, N.Y.

5. James Martin, *Telecommunications and the Computer*, Prentice-Hall, Englewood Cliffs, New Jersey, 1969.

6. James Martin, *Programming Real-Time Computer Systems*, Prentice-Hall, Englewood Cliffs, New Jersey, 1965.

7. James Martin, *Design of Real-Time Computer Systems*, Prentice-Hall, Englewood Cliffs, New Jersey, 1967.

8. Edgar C. Gentle, Jr, *Data Communication in Business*, American Telephone and Telegraph Company, New York, 1965.

3

THE SYMBIOTIC AGE

The extension of man's intellect by machine, and the partnership of man and machine in handling information may well be the technological advance dominating this century. – Simon Ramo of Thompson Ramo Wooldridge.

The enormous data banks, the logic power of the computer – millions of times faster than man's own logic and free from errors – and the ability to transmit data at high speed over the world's telephone networks give us tools that are truly awesome in their potential. We are far from grasping as yet all that can be accomplished with the machines we have invented.

The word 'symbiosis' is being used in a new way by the computer technologists. Webster's International Dictionary defines 'symbiosis' as 'the living together in more or less intimate association or even close union of two dissimilar organisms'. In the new use of the word, one of the 'organisms' becomes a computer with its associated equipment. Man communicates with the machine by means of whatever 'terminal' device gives him the closest relationship. This might be a machine with a screen whereby the computer flashes diagrams, text, equations, or numbers on a screen something like a domestic television. The man gives the machine information with a keyboard, or with a 'light pen' with which he can, in effect, draw on the face of the screen, or can in other ways indicate his wishes to the computer. The potential of human intelligence combined with the best capabilities of machines will be explored in the rest of this book. However it will take years to fully understand, let alone exploit this power.

ARTIFICIAL INTELLIGENCE

Ingenious programs which suggest the eventual capabilities of the machines have received wide publicity. Computer programs have been written which compose passable pop music, write bad poetry, think up new names for detergents, and write scripts for Westerns that are hardly worse than TV's daily fare. One machine has been programmed to recognize human faces, and another to identify a human speaker by analysing his voice sounds when speaking a given phrase.

'Artificial intelligence' is a subject hotly pursued in Ph.D. theses. Thomas Evans at MIT programmed a machine to exhibit reasoning by analogy and hence answer questions in college admission intelligence tests involving the juxtaposition of geometrical shapes. Daniel Bobrow, also at MIT, endeavoured to make a computer understand questions posed in ordinary English. He programmed it to interpret high-school algebra questions such as the following: 'Mary is twice as old as Ann was when Mary was as old as Ann is now. If Mary is 24 years old, how old is Ann?'[1] The difficulty lies in understanding the loose format of English and translating it into easily solved equations. With some wording, the computer has to make intelligent assumptions (as we do) about the precise meaning. Another example is concerned with a computer set to work to prove theorems in geometry. The machine generated a new proof of a Euclidean theorem which was neater than Euclid's own proof.

There have been several attempts to set up a learning program in a machine so that the machine modifies its own performance on the basis of experience gained. A computer has been programmed to learn from its successes and failures at draughts to improve its own play so that eventually it beat the person who programmed it. We may be many years yet from the time when computer-learning techniques have wide commercial application, and probably we need new technical developments to greatly improve the hardware's capability for pattern recognition. The human brain's ability to quickly associate complex and non-identical relationships is startlingly good. The recognition of a face in a crowd or the ability to see depth in a detailed stereoscopic photograph are illustrations of processes for which the mechanisms of the brain are vastly better suited than the simple serial-logic circuits of today's computers. However, when a machine is programmed to learn and take certain actions as a result of what it has learned, it can go on learning and learning mechanically until, in this one decision-making process, it becomes superior to the human beings who created it.

One of the fascinating aspects of working in the computer industry is that one is never sure what new direction it is going to take next. The technology has produced many surprises in the last twenty years, and the above glimmerings of what is possible suggest that it has many more surprises in store for the next twenty.

MAN—MACHINE CAPABILITIES

Whatever the *eventual* capabilities of machines, however, there are still many types of thinking that are best done by human beings. In this book we want to restrict our speculation to the next decade or so, and within that time span we do not feel justified in assuming that current research will quickly lead to extensive use of creative logic in computers, heuristic programs, computer self-learning, or elaborate pattern recognition (although computers will be used to read print and handwriting, and to recognize human voice words and identify human speakers). The best way to carry out certain difficult types of processing, and make certain decisions, is to use neither computer nor human means alone, but to use an optimum combination of the two. Language translation illustrates this: human translation alone is slow; machine translation alone is either very expensive or else inaccurate and hilariously nonsensical.* In an optimum combination of the two the machine produces a rough translation and a man scans it rapidly, correcting the machine's imperfections. The same is true in many real-time situations. The setting of booking limits on multiple-stop airline flights may be an operation in which the intervention of flight-

---

* Two much quoted classics are 'The spirit is willing, but the flesh is weak' rendered as 'The liquor is prepared, though the meat is not tough' and 'Out of sight, out of mind' which became 'Invisible idiot'.

controllers may be desirable. The flight controllers have experience of the booking patterns that can occur at various times on various routes, and this experience can be used by the machine to improve the loading of multiple-leg flights. The use of human intervention at given points in processing or decision-making operations recognizes that certain operations are still too difficult or too expensive to program. On the other hand, intelligent human thinking and skilled use of a computer in real-time is a very powerful combination.

In the ideal system machine and human functions would be divided according to their respective abilities. The computer and its peripheral equipment can carry out the following:

1. rapid calculations; routine processing; fast editing and logic operations; optimum selection between multiple alternatives;

2. fast reading of certain types of documents; fast reading of cards punched from documents; high-speed printing; hence fast preparation of invoices, orders, work tickets, cheques, management reports, and so on;

3. the filing and maintenance of a vast mass of information, and the retrieval of information from the files, in a fraction of a second if necessary;

4. scanning, sorting, and correlating large amounts of information; searching for specific facts or correlations;

5. collecting information very rapidly from many sources; accepting and checking manual input of information or documents from many sources – the sources may be at the other end of a telecommunication link;

6. fast distribution of information to many locations; distant printouts or displays; the direct operation of machinery, chemical plant, or other equipment;

7. giving immediate answers to routine inquiries made in any location; facilitating man–machine conversations;

8. surveillance; watching with an unfailing eye for exception conditions, potential causes of trouble, and situations needing human attention; the notification of correct authorities by way of distant terminals;

9. the retention without loss of memory of routines that can be used again; the collection of information on which future action may be based. A computer system can be steadily improved and 'tuned' as experience grows, to a degree not possible with human systems.

A tool that can do all that is of great value in many fields of human endeavour. Human beings will nearly always be necessary in dealing with human beings; and even in the handling of data the following tasks are still best done by man:

1. handling unforeseen events; handling events of low probability;
2. selecting goals and criteria;
3. selecting approaches to a problem;
4. recognizing patterns in events; detecting relevance;
5. formulating questions and hypotheses – if a man formulates a hypothesis, a computer might be used to test it;
6. producing ideas; planning new products, new techniques, new business.

Let us now review some of the new uses of computers that will have a major effect on society, taking into consideration the present limitations of the machines as well as their capabilities. Today's limitations may well be removed in the next decade or so. A vast research effort, perhaps government-sponsored, would almost certainly make major breakthroughs leading to more intelligent machines.

MILITARY SYSTEMS

A good place to look first is at the American military, which has spent more money on individual computer systems than any other single organization. The computer technology developed for the military percolates down to civilian users, and most of the systems are likely to have civilian counterparts.

To a large extent, the pioneering work for systems combining telecommunications and computers has been done in the military. One of the first and most spectacular was the US Air

Force SAGE (Semi-Automatic Ground Environment), designed in the early 1950s to protect the United States from surprise air attack. SAGE became operational with one computer centre in 1958 and has extended to many centres in the 1960s. A system to back up SAGE, called BUIC (Back-Up Interceptor Control) is also being implemented to provide some of the SAGE functions with much greater dispersion, so that if SAGE is knocked out BUIC will still give some protection.

SAGE is designed to maintain a constant watch on the air space over North America, provide early warning of airborne attack, and give its Air Force operators the information needed for conducting an air battle. Its input comes over data transmission links from a variety of radars which unceasingly sweep the skies of the continent. These include the CADIN line (Canadian Integration North) and many other land-based radar systems. Input also comes from observation aircraft, ground observer-corps stations, and ships on picket duty out at sea. The computers digest this constant stream of data and prepare displays for their output screens. In addition, the system contains aircraft flight plans, weapons and base status reports, weather data, and much other information. Figure 3.1 gives a highly simplified diagram of one sector of SAGE, and Fig. 3.2 shows the division of the United States into SAGE sectors, with the approximate locations of the radar sites and computer centres. These locations are all connected to the computers, and the computers interlinked by telegraph and telephone lines carrying digital data.

Air Force personnel watch the screens of console units designed for their different responsibilities. They communicate with the computer using a light pen – a forerunner of the light pen on today's commercial systems. They can request displays of particular situations stored in the computer, and can command the machines to compile certain special displays for their use. The computer will of its own initiative flash urgent messages on the screen. If an unidentified aircraft approaches, the system notifies command personnel, who will then use the screen to investigate and if necessary dispatch interceptor aircraft or activate weapons.

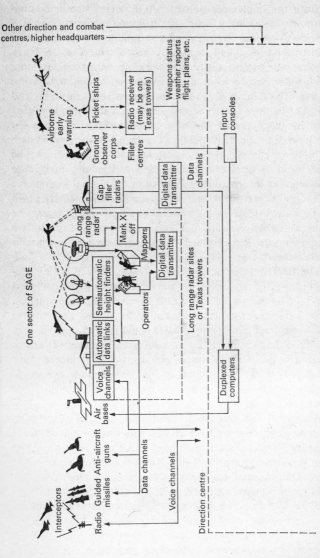

Fig. 3.1 One sector of SAGE. Redrawn from 'Sage – A data processing system for air defense', by R. R. Everett, C. A. Zroket, and H. B. Benington. Eastern Joint Computer Conferences 1957.

Since the implementation of SAGE, many command-and-control systems have been installed and planned. The US Marine Corps has what is in effect a mobile SAGE, transportable to an operational theatre by helicopters. The US Navy has computers and screens on board ships linked together on a worldwide basis to form a vast mobile command-and-control system. The Pentagon can obtain on its screens positions and status of its ships anywhere and can display data that are being used by the Navy on the other side of the world. The US Army, meanwhile, is planning a command-and-control system of immense complexity for operation in the 1970s. This will be used by every level of command and will handle intelligence and logistics information. Information on tactical operations, fire support, personnel, and other administrative matters will be kept up to date in the system. The output will, again, be available either on individual consoles or large group displays and, again, the Pentagon will be placed directly in touch with the theatres of war.

The BMEWS System (Ballistic Missile Early Warning System) uses giant radar units for the detection of potentially hostile missiles aimed at the West. Its computers in America, Greenland, and England scan the radar signals they receive ceaselessly and notify appropriate military personnel if they conclude that an attack might have been launched. It is this computer system that would give us twenty minutes grace prior to nuclear devastation, during which time we could launch our counterattack and do whatever else needed to be done in that interval.

Another system keeps up-to-the-minute watch on every object orbiting in space – a job which becomes more involved as the quantity of satellites and discarded 'space junk' increases. The objects observed have their orbits compared with those recorded for the known satellites, and any deviations are reported.

There are many other US military information systems in addition to these, each system having as integrated components computers, files, display and input units with highly trained operators, and a variety of telecommunication links. An all-embracing system is now under design, intended to tie together into one integrated complex the global systems of the Army, Air Force, Navy, and other government agencies. Using this system, the President of the United States or the Chiefs of Staff will be able to display details of all military and related affairs. No doubt the communist countries are also involved in massive military automation, and the prospect of these systems grappling in some future warfare is a systems analyst's nightmare. Chapter 20, on page 437, discusses the effects of computers on military power.

In setting up such systems, a massive use is made of telecommunications of all types. The military communication systems are, in general, kept separate from civilian ones. It is perhaps not surprising to find that, as of this writing, the US military had at least twenty-four synchronous telecommunication satellites, whereas civilian organizations have only four.

The US Defense Communications Agency has set up two vast telecommunication networks: Autovon (for voice), and Autodin (for data). Although initially these were domestic networks, they now function as an integral part of a rapidly developing global system. The networks are designed to give precedence to priority traffic – a general's telephone call pre-empts, if necessary, lower-priority calls. Also, the system is designed to handle the unusually high volume of traffic that may occur in times of emergency; and its facilities are distributed so as to present no unusually attractive target to potential attack. Thus it has sufficiently widespread alternate routings for the purpose of surviving massive nuclear attack. A configuration of trunk groups, in the form of multiple overlapping hexagons and diagonals, links the many switching centres in America by many alternate paths. Even if most of the United States were annihilated, the computers in the remaining parts would still be able to transmit to one another!

'WAR ROOMS' IN INDUSTRY

A 'war room' in which top military decisions are made – by now familiar to the public from the movies – may have display screens around the wall on which information can be displayed from the computers. Gone are the map tables and counters of the Second World War's RAF Bomber Command. The significant information is stored in a computer file and is electronically displayed.

Probably in a decade or so, industry also will have its 'war rooms' equipped to display all manner of facts from the computer data bank. Future board rooms or *management control rooms* in industry may be designed with wall-size computer display screens and individual consoles for communicating with the computer. It may even look something like today's computerized military war-room (see Chapter 9).

Already there are management control rooms in which a group of men assist the computer in controlling the flow of work through the organization. Such rooms have a number of functions. First, they are used to control errors in many systems – often errors in human input, or errors found in the filed data. Second, they handle exceptional conditions that arise and require human intervention. Third, they represent the recognition that it is not necessarily the best policy to make the computer do *all* of the required processing. Some transactions or situations still need human judgement; in such cases the computer requests help and transmits details to the man with the relevant experience or ability to judge. Four, they enable management to obtain needed information or to test the possible effects of contemplated actions. Thus, those members of management who do not have the ability to communicate with the computer directly may do so via the staff in the control room.

In industry the uses of data transmission are growing fast. Links not connected directly to a computer are used for sending data from one point to another, or for collecting information from factory work areas; and now there are developments in which the transmission links and the computer are inseparable parts of the same system. Most of these systems are tailor-made to carry out a specific set of operations in a given company.

The work of most data-processing installations of earlier

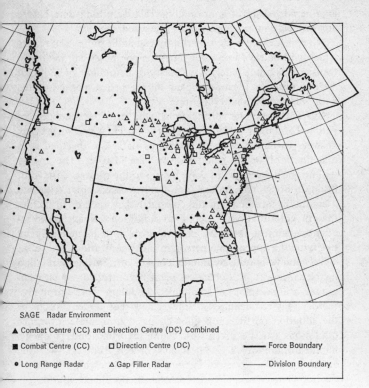

SAGE   Radar Environment

▲ Combat Centre (CC) and Direction Centre (DC) Combined

■ Combat Centre (CC)   □ Direction Centre (DC)   —— Force Boundary

● Long Range Radar   △ Gap Filler Radar   —— Division Boundary

Fig. 3.2 Courtesy SAGE.

generations of machines was organized exclusively to use 'batch processing', in which several hundred transactions would be grouped into a batch. In a punched-card installation, many trays of cards would be fed through a machine before its set-up was changed for its next function. The 'batch' would then have to wait for the next operation, possibly sorting or merging to ready it for another 'run'. Similarly, large magnetic tape files are processed with one program before the file is sorted ready for the next operation. It is economic to have large batches, so a cycle of operations – perhaps repeated weekly or monthly –

is adhered to, and it is normally a relatively long time before any one particular transaction is processed.

Now, instead of operating run-by-run and printing out results which will be read at a later time, the computer can deal with transactions immediately if necessary or in an hour or so if that is good enough. It can complete all the processing associated with one transaction at one time.

Current computer installations can provide management with information *when they ask for it*. They can give quick answers to inquiries about the status of a job, the load of a machine tool, or the amount of an item in stock. The answer will not be a week out of date, but up to the minute. Instead of filling in documents on the shop floor, workers enter data into a computer terminal and the computer then records in its files the current situation. Exceptional conditions, which arise frequently in most factories, can be dealt with immediately by notifying the appropriate person in the organization.

To achieve minute by minute control of a business situation in the same way as a chemical plant is controlled – such techniques are essential. A computer scheduling work through a factory may reschedule it whenever new requirements occur or the situation on the shop floor changes. Immediate insurance quotations can be given and claim reporting and payments handled more quickly. Salesmen's orders from all over the country can be processed in time for the following day's production. The computer can control a steel mill and optimize its efficiency. It can monitor a commercial process in the same way that it monitors a manned space flight. Traffic in a turbulent city has been speeded up by computers manoeuvering vehicles into groups and changing traffic lights at the best moment; in a similar way the steps in an industrial process can be regulated in an optimum manner. The benefits of such control are not necessarily the traditional reduction in manpower. Often they accrue from increased efficiency, better customer service, more profitable utilization of facilities, or increased business.

Many companies with widely dispersed facilities are employing data links to connect separate plants, warehouses, and large

numbers of sales offices or service centres to a central information system. As in other fields, paperwork is being replaced by
electronic storage and data transmission. The airline office
does away with its filing cabinets of cards for each passenger.
Industry branch offices will soon be able to enter and retrieve
data at a device resembling a television set with a keyboard,
and a computer will exercise centralized control over, and store
centralized information about, all that is happening.

Such systems in industry today appear to be more a tool of
middle management than of top management. They can handle
the logistics of production scheduling and inventory planning,
but they provide little help in top management's policy-making.
The potential, however, is as yet largely unexploited. In the
future the display screens, the data banks, and the telecommunication links will be indispensible for computer-assisted
decision-making in big companies. As with other computer
applications, the limitations of control systems are no longer in
the hardware, but in the programming and organization of the
vast quantities of information involved.

FINANCE SYSTEMS

The world of finance employs many files of constantly changing
information. Among the earliest real-time computer systems
were those employed by savings-bank tellers to update customers' passbooks. In this application a central computer keeps
records of all customers' accounts, which are updated by tellers
at the counter machines (see Chapter 4). The teller has the
information he needs about the account automatically transmitted to him. The same system often provides information to
management on request.

This small but successful beginning portends a sweeping
change in the financial community that may eventually revolutionize our entire monetary system. Many financial transactions
will be made without using cash or cheques, as are credit-card
transactions today. The bank customer will be required to
have some means of identifying himself to a telecommunica-

tion terminal, and details of his transaction will be keyed into the terminal. These data will reach computers holding the records of the payee and payer. The latter's record will be debited and the former's credited. In this way financial transactions will proceed by means of data transmission and without paperwork – at least, what paperwork there is will be external to the system, not internal as today, and will merely inform the users what transactions have taken place. Payrolls, for example, will be worked out by a firm's computer and transmitted to the appropriate bank computer; when an employee of that firm then buys something at a store, the store's computer will find whether the cash is in his account.

Some writers have envisaged a completely chequeless and others a cashless society. It seems unlikely that we shall ever do away with cash entirely. Cash will always be needed for such small transactions as buying a drink or tipping a porter. However, the use of large sums of cash and large cheques will probably decline rapidly as the concept of credit transfer gains acceptance and the different systems spread, interlinking, across the nation.

The development of such systems could be easier in a country like Great Britain than in the US, because in Britain one organization, the Post Office, owns and operates all telecommunications and the nationwide computer-run Giro banking system. The commercial banks are also largely nationwide and some are planning to interconnect over 2,000 branches into one giant on-line computer network. In the US, on the other hand, there is a patchwork collection of small banks and a set of restrictive state regulations. Nevertheless, one suspects that as in other areas the US will, in fact, lead the way. If only Britain could overcome its reluctance to change its systems and invest courageously in technological innovation, it could make spectacular use of the new computer techniques. The country is geographically small and data highways of massive capacity could interlink it at a fraction of the cost of those in America.

TIME-SHARING

Each of the military, commercial, and financial systems discussed above is tailor-made for one application. A type of system that is spreading rapidly is one which gives a user remote access to a computer and allows him both to program the machine himself and to use programs from the computer's files. The user dials up the computer on a conventional telephone or telegraph line (a private line could be used also). The responses he receives at the terminal are normally fast. Since with most systems he receives a reply in about three seconds (unless the computer is performing a lengthy computation for him), he gains the impression at the terminal that he is the only user of the system.

In fact, the computer has many such users and switches its attention from one to another at lightning speed. By chopping up its work into small slices of time and scanning from one user to another, it ensures that nobody is kept waiting unduly long at his terminal. A wide variety of possible scheduling arrangements and hardware configurations permits the computer to pay attention to many users simultaneously. This process is called *time-sharing*.

Time-sharing offers several benefits. First, it permits users who could not afford their own computer to have a part-share in one. Inasmuch as communication lines provide them with fast access to a computer and fast replies, they can behave much as though they had a machine on their own premises. Bulk input/output – such as high-speed printing – will be limited on their premises only by the speed of the communication lines used.

Secondly, work is received back from the computer more quickly. To obtain a run on a heavily loaded computer may take half a day. If the data have to be sent out and returned by station wagon, this may be a day; by mail, two or more days. This aspect of time-sharing is especially useful, as most programmers – particularly non-professionals – make small mistakes in their programs. Finding and correcting these can be frustratingly slow. If a program has twelve logic errors ('bugs') in it, it might require eight brief runs on a computer before all are located. If there is a day between each run, it will be eight days before the program is working. With a terminal on a time-sharing computer, the user could 'de-bug' it in an hour or so. An engineer or scientist can solve in one afternoon while sitting at the terminal problems that might otherwise consume days or weeks.

Thirdly, the user is in direct contact with the machine. He can use the power of the machine to supplement his own thinking as in the 'good old days', when he might have had a computer all to himself. He is even better off now because he can use modern computer languages, and the computer has a large random-access store in which many useful program routines are kept. He can 'experiment' with his programs at the terminal, testing the effect of modifications. He can change parameters or logic elements and observe the subsequent effects. Man–machine interaction is greatly expanding the creative power of skilled users. In some ways the most exciting results of early time-sharing systems have been the entirely new ways of using computers that have emerged from the real-time combination of a highly-skilled man and a powerful machine.

Fourthly, the users of time-sharing systems have access to each other's programs. The files holding programs and data are commonly divided into 'private' and 'public' areas. The private areas hold the programs and work that individuals retain for their exclusive use. The public areas contain those programs and data that are made available for *any* user.

One of the most surprising developments on the early university time-sharing systems has been the growth of the public files. Many users wrote programs or compiled data that were deemed valuable for other users. These were made available and very rapidly the public files became filled with useful items. When this happens on a nationwide or even global scale, as it will, time-sharing systems will provide a store-house of human thinking. The work of one programmer will be picked up and developed by interested parties elsewhere. Today we are confronted with a bewildering array of technical literature; tomorrow we shall have an equally bewildering array of on-line programs.

SPECIAL-PURPOSE TIME-SHARING SYSTEMS

In addition to general-purpose systems in which the user can program anything he wants, other time-sharing systems available for public lease carry out a specific operation. The user of such a system does not necessarily program the machine, he may merely use its services.

Some banks in the US, for example, offer professional billing services. A doctor or a dentist may have a small terminal in his office, and his nurse or secretary dials the bank computer to transmit details of the charges to patients. Or these may be punched into cards or transferred to tape and transmitted as a group later. The bank computer bills the patients, credits payments to the doctor's account, and sends out receipts. It also compiles complete tax information for the user; (perhaps for this reason the service is not quite so popular as it might have been otherwise).

The Keydata Corporation operates a time-sharing system in the Boston area offering business services to users. The system initially offered such functions as on-line invoicing, order entry, inventory checking and updating, and credit checking. In addition, many off-line services are available. Future plans for the system include expansion to provide functions for scientists and engineers. Thus it could grow into a general-purpose system which the user could program. Conversely, present general-purpose systems are making special-purpose programs available so that they can become useful to non-programmers.

The authors used a time-sharing system to help in the preparation of part of this book. This was IBM's ATS system, which consists of a typewriter similar to an ordinary office typewriter, linked to a distant computer. A secretary using the typewriter can store, on the computer's files, sentences, paragraphs, insertions, or entire documents, which she may use later for compiling letters or a report; she also uses the computer to help her in editing, correcting, and inserting items into documents.

The average secretary spends a large amount of time correcting, revising, and typing material that has been typed before. Where a perfect copy is required, she may have to retype a whole page because of a misplaced comma. To assemble a bulky document she may have to copy and recopy standardized paragraphs or items from previously prepared documents. In preparing a book like this, the manuscript becomes modified many times before it is fit for the printer. The computer inserts each modification into the text and will print out a clean copy on request.

MAN–MACHINE INTERACTION

With the logic power of the computer and mass-information retrieval available to it, the ability of the human brain to tackle certain types of problems becomes immensely greater. The human operator may formulate hypotheses and immediately test them on pertinent data. The computer can help formulate and evaluate the effects of management policies. It can assist in producing engineering drawings, planning charts, or simulation programs. It can help in planning complex schedules. It can help programmers write programs by keying them directly into a terminal and changing them where necessary as they go along. In this way, complex programs can be produced much more quickly than with conventional methods.

Much research is needed – (and is being done, now that time-sharing has made it economical) – on the combined use of man and computers. Among the possibilities already tried are the following:

Designers may use computers to assist them. A circuit designer, for example, may store details of all the components available to him and quickly make the computer investigate the effect of various arrangements or modifications. An *architect* or *bridge engineer* can use a display screen to produce plans, fitting together elements of the design and modifying the results as required. A linkage of steel girders in a building or on a bridge may be built up on a display screen and the computer programmed to place figures by each girder indicating the results of stress calculations. The designer may change his linkages and immediately see the stresses caused by different loadings, and so optimize his design efficiently. An *automobile designer* may sketch surfaces of body panels or fender designs on the graphic input device. The computer will fit a mathematical expression to the surface and then 'rotate' the surface on the screen so that the designer may observe its shape from all viewpoints. The computer can apply constraints that are forced by mechanical or manufacturing considerations, and the designer may then adjust his shapes to achieve a compromise within the given constraints.

If the calculations involved in these procedures were done by hand or with a batch-processing computer, they would take a long time – so long that the designer would be restricted in the full use of his imagination or inventive ability. Most design processes are 1 per cent inspiration or inventiveness and 99 per cent calculation and the laborious working out of detail. The object of using a real-time computer for design is to relieve the designer of as much tedious work as possible and to enable him to observe as quickly as possible the effects of his ideas. In a trial-and-error process, three months between the trial and the error can cause the designer to lose many of his original ideas. Where it is possible to explore the effect of these ideas in real-time, much more fruitful and exciting thinking can result.

Similarly, *simulation* or *mathematical 'models'* may be 'built' at a display terminal to investigate the passage of work through a factory, the flow of traffic through a city, or any other situation capable of being simulated. A standard simulation 'language' may be used for this, the model being tested as it is built. The effect of different actions may be investigated using such models. A model of the functioning of an organization may be permitted to 'grow' until it accurately represents the behavior of the organization. Sales forecasts may form inputs to a model used for planning. The effects of different policies may be investigated, and experiments carried out with the model. As in a design process, the work of building a simulation model today can be long and laborious. If somebody asks, 'What would be the effect of changing so-and-so?' the answer may come two weeks later. This is too slow for many purposes, especially in the turmoil of the commercial world. To demonstrate to management that quick and effective answers can be obtained from simulation, real-time methods may be needed.

The surprising effectiveness of computers used for *teaching* has already been demonstrated. Many pupils can be handled at once and different subjects taught simultaneously with a time-sharing system. The computer gives individual treatment to students and modifies its behaviour according to individual progress in a much more adaptable way than simpler machines.

Used in this way, the computer becomes a type of storehouse for human learning and thinking. An effective teaching program can be used throughout the world on many types of computers and should be constantly improved as student reactions are observed. Languages now exist that permit a professional educator with little knowledge of computers to write teaching programs. As the potential of this becomes developed and better understood, it is likely that many educators throughout the world will set to work writing and improving computerized instruction courses (see Chapter 6).

Doctors are beginning to make use of distant computers as a help in diagnosis. A patient's symptoms may be transmitted to the distant machine, which then makes suggestions to the doctor and perhaps asks for additional information. Already, electrocardiograph results have been transmitted from patients' bedsides over the telephone for immediate analysis by a distant computer. The computer will not in any way replace the human qualities of the doctor, but would add to his limited store of information. The computer in this application is also carrying on a type of information-retrieval work, but probably with more logical or analytical programming in it than straightforward information retrieval.

Another likely development is the storing and automatic retrieval of patients' case histories. The case history accessible to the computer would contain details of the patient's past diseases and symptoms, diagnostic tests, inoculations, drugs administered, the effectiveness of attempted treatments, side-effects, and so on. All this information could be retrieved by the doctor to help him in his diagnosis, and might even in the future be processed by the computer so that the accumulated experience of the medical profession might help in a given case. (See Chapter 10.)

PUBLIC USE OF COMPUTERS

It now becomes apparent that we are going to need to make computers available for public use. People everywhere must be able to dial up a computing facility appropriate to their needs. They may have a portion of a file reserved for them personally in a distant machine. The grocer on the street corner will transmit the paper tape from his cash register to a computer to do his accounting. Doctors' assistants will transmit details of charges to patients to their bank's computer, which will bill them and keep accounts. An engineer will perform routine calculations on a teleprinter or on a terminal that looks like a desk calculating machine. Computer programmers will develop programs at a remote input/output device.

There will be on-line files of data available of numerous types, with automated means for searching them. Commercial and economic data files will be interrogated for business planning. A private investor will be able to check his hypotheses on stock investment against files of data on past stock movements. One visualizes the clergyman of the future preparing his Sunday sermon at a computer terminal, with the ability to retrieve appropriate literature and perhaps past sermons. (The result will probably be better than some of the sermons we get today.) Architects, electronic designers, market researchers, and many other professional people are already having programs written for their specific needs.

The same types of computer could handle most of the applications in this vast spectrum. The files needed vary in size and speed of access, but are conventional computer files in each case. None of the above applications needs a unique file technology. The telecommunication links are the same, although some applications could benefit economically from faster links. There is more variation in the input/output devices envisaged, but although many special-purpose devices will undoubtedly be built, two or three types of machines *could* handle the majority of the applications. Only the programs will be widely different, and these will be as diverse as the applications themselves. The programs, however, will be stored, along with data, on the computer files and in each case the user will simply request the required program before he begins.

It is likely that before 1980 many persons will have terminals in their own homes. Some have them already, and sooner or later individuals will constitute a mass market for terminals. A wide variety of computer systems will be dialable from the home. This topic is explored further in Chapter 7.

A NATIONAL GRID OF COMPUTERS

We are moving into an age when intelligent men in all walks of life will need, and constantly use, their computer terminals – this will be a *symbiotic age* when the limited brain of man is supplemented by the vast data banks and logic power of distant machines. Probably all of the professions will have their own data banks and possibly their own languages. The non-professional man will use the terminals for performing calculations, working out his tax returns, computer dating, planning vacations, or just for sheer entertainment. In Great Britain and some other countries, the possibility of setting up a government-controlled 'national grid' of computers is being examined. Under private enterprise, telecommunication companies, computer manufacturers, and third parties are setting up schemes in which users lease an input/output device just as a telephone is leased, and are then billed for the computer time they use. The input/output devices include touch-tone telephones, teleprinters, and specially manufactured computer terminals.

Although some general-purpose systems which can be programmed to meet most of the needs of most types of users have been put into operation, it seems likely that in the beginning, at least, the majority of nationwide systems will be much more specialized than these. Part of the reason for this is machine efficiency. The ratio of productive to nonproductive computer utilization can be considerably higher on the special-purpose, or restricted, system. Secondly, the massive files that are envisaged for certain applications are more likely to reside on their own special system. Thirdly, special-purpose systems are, in general, easier to implement than the Big System which does everything for everybody. And last, we have not yet mastered the art of building the gigantic computers necessary for a general-purpose system for a large number of users.

The user, then, is likely to have a catalogue of remote systems which he could dial up for different operations. He will, it is hoped, use the same input/output equipment for all of the various computers he communicates with.

A variety of public time-sharing systems is now working successfully. The next generation of such schemes will improve upon them enormously, and before long it can be expected that development will reach that critical point in the growth of a technology when it achieves mass acceptance, and mass-produced machines will be sold. The systems will grow, multiply, and interlink. A country-wide, and probably a worldwide, network of computers that can be dialled up on the existing telecommunication facilities will be available to us before many years have passed. We shall find enthusiastic users in many fields adding programs and data to files for public use. Dedicated amateurs will probably begin to write and sell programs to the 'utilities'.

We must achieve *mass* participation in the enormous programming task to be performed if such schemes are to flower fully. Computer assisted instruction can only achieve maturity after thousands of talented and dedicated teachers have been persuaded to prepare the required programs. And this is equally true in other fields.

Technology is bringing the computer to the masses. People

everywhere will be able to participate in using and building up an enormous quantity of computerized information and logic. We now have a tool so powerful that it will take many decades for us to use it to its full potential. Today we cannot even imagine what that potential will be.

REFERENCE

1. Marvin L. Minsky, 'Artificial Intelligence,' *Scientific American*, September 1966.

4

GOODBYE TO MONEY?

Nothing is certain save death and taxes. – *Benjamin Franklin*

MONEY IS INFORMATION?

Primitive man learned the value of the division of labour when he became involved in bartering his product for those of his fellows. Later he devised systems of exchange in which one commodity became the standard of value against which all others were measured. Usually that commodity had intrinsic value, such as cattle, slaves, or wives. Usually, too, it had the disadvantage of indivisibility. (How do you obtain 'half a wife' of wheat?) But this did not apply to precious metals. Gold was found to be rare, infinitely divisible, easily assayed, unattacked by rust and lichen, and beautiful enough to inspire poets. For millennia it has been the world's standard of value, fought over, traded, ornamented, stolen, and worshipped. Only in our century has man gathered the effrontery to question the necessity of this 'symbol of pre-eminence ordered by the celestial will'. The gold standard has been relinquished – for trade purposes superseded by paper. Paper money once represented an equivalent amount of gold stored in some banker's vault, but gradually that ceased to be the case. The currency note is no longer convertible into gold. Canada's dollar bill states that the Bank of Canada 'will pay to the bearer on demand' one dollar. What it does not say is that it has nothing to pay with, except more dollar bills. Certainly, there is no gold for this purpose. Most US dollar bills simply say 'In God We Trust'.

Money – the paper which passes from pocket to pocket – has become merely a demonstration of a man's ability to pay. The same *information* could be passed with a cheque. The cheque system, one of the great inventions of commerce, gradually gained acceptance by the 'man in the street'. Although the process of eliminating tangible currency is still incomplete, we are heading in that general direction: a man can now use a credit card, and we hear talk of a 'chequeless society'.

A man need no longer display his wealth for us to accept it. A token suffices. *Money is information.* That is, in making a payment, two forms of information are required: a record of the transaction and certification that the payer has the requisite wealth to make the transaction. Since it is information, money should be firmly in the province of the computer, and be telephoned between machines like all the other information discussed in this book. Money to be accounted for must not become lost and it must be secure – we can make computers loss-proof and as secure as any bank. Since money is a *record* of a man's worth, it is easy for computers to keep each other informed about all the customers of various banks by transmitting to each other electronic 'cheques' recording changes in their 'worth'. This is what we mean by *automatic credit transfer.*

The replacement of gold by paper money and of paper money by cheques were each revolutionary in their day. Now we must become used to financial transfers occurring in the form of electronic pulses on a data link. The paperwork associated with the transaction will now merely inform us *about* the transaction, rather than represent the transaction itself. Nor will it have to be punched into cards and fed into a receiving computer.

A pilot scheme is operating at the Bank of Delaware at Wilmington.[1] Shop assistants in selected stores have been equipped with touch-tone telephones. Plastic ID cards are issued to selected customers who are also the bank's customers. When any of these makes a purchase, the assistant puts the card into a special slot in the phone and pushes the appropriate buttons. The telephone transmits data to the bank's computer. Its mechanism reads the customer's number from the card, and the shop assistant keys in the cost of the purchase. When the computer receives this information, it adjusts its data files. The sum of money involved is deducted from the customer's account and added to the store's. Neither cash nor a cheque is involved. Careful electronic error-detection mechanisms are employed to insure that transactions are not falsely transmitted and that the files are not incorrectly updated. Full details of the transaction are automatically printed in the bank statements of both store and customer.

AUTOMATIC CREDIT TRANSFER

The eventual consequences of the simple idea of automatic credit transfer will be enormous. Vast random-access computer files in banks will hold full details of all accounts. Some banks have operated this way since the early 1960s, but now (as in other parts of our story) the second phase begins and use is being made of telecommunications. As a transaction is entered into the system, transmitted data will cause the appropriate amount to be deducted from an account in one computer and added to an account in another. Eventually, the financial community will become one vast network of electronic files with data links carrying information between them.

Thomas J. Watson, Jr, the President of IBM, foresaw the revolution in banking as follows in 1965:

In banking ... the advances of yesterday are merely a faint prologue to the marvels of tomorrow. In our lifetime we may see electronic transactions virtually eliminate the need for cash. Giant computers in banks, with massive memories, will contain individual customer accounts. To draw from or add to his balance, the customer in a store, office, or filling station will do two things: insert an identification into the terminal located there; punch out the transaction figures on the terminal's keyboard. Instantaneously, the amount he punches out will move out of his account and enter another.

Consider this same process repeated thousands, hundreds of thousands, millions of times each day; billions upon billions of dollars changing hands without the use of one pen, one piece of paper, one check, or one green dollar bill. Finally, consider the extension of such a network of terminals and memories – an extension across city and state lines, spanning our whole country.[2]

There are two prerequisites for making such a system viable. First, it must be possible to identify a customer with reasonable certainty – at least the same level of certainty as today when he presents a cheque.

Second, it must be possible to ascertain that the customer has sufficient money in his account or, if he does not, that some organization will guarantee that the payee will indeed be paid. Problems exist with today's methods of paying bills but much has already been done to solve them. There are at present a variety of possible ways of identifying customers. If we put the computers to work, we could produce much better ways. Probably the simplest method is to use some form of identification card or credit card.

CREDIT CARDS

The Director of Automation of the American Bankers Association has stated that 'credit cards are going to be our keys to a future chequeless–cashless banking system'. Unfortunately, with a thousand banks issuing cards, as well as oil companies and travel agencies, confirming that the user and owner of the cards are identical is difficult, since the cards may be known only locally. Already, many Americans carry a bulging wallet filled with credit cards, and lost credit cards falling into felonious hands cause a mounting problem. Tens of thousands of credit cards are being falsely used. The safety and security of the cards is presently inadequate for a nationwide system of credit transfer.

Until a couple of years ago, the problems engendered by 'bent' cards were merely a major nuisance. Now they threaten the whole system. In 1969 annual losses reached $40 million as organized crime moved in. In the USA the number of cards in circulation exceeds the population, and is rising at 25 per cent per annum.

Stolen cards usually reach the racketeers from muggers or pickpockets, who get perhaps $100 apiece for them. Within a day or two, they are used to buy thousands of dollars worth of expensive goods, particularly airline tickets and fur coats. To get the card number into a blacklist which informs retailers of its invalidation takes at least three days following notification, and frequently as long as ten days. By then, the card will have been milked and destroyed. The blacklist is hardly helpful anyway, since the merchant seldom has time or inclination to consult it. The cardholder ceases to be responsible for purchases once he has informed the company, and the retailer gets his money either way.

American Express has now installed an Audio Response Computer System. When a card is presented to an airline or hotel clerk, for example, credit authorization is usually required for transactions above a certain limit. The clerk contacts the computer in New York by means of a touch-tone card-dialler telephone, putting in his employee's identity card and pushing the buttons to indicate the card number and amount. The computer checks its file of 2 million accounts for 'lost, stolen, or cancelled' cards and then checks credit rating. It also looks for any unusual spending pattern, since a spending spree on a stolen card usually involves the purchase of a number of expensive items within a short period. If the computer cannot itself give a 'spoken' approval immediately, it transfers the call to a 'credit authorizer' and displays the account record for him.

Another solution to identification problems would be to have each person carry a unique personal ID card. Like the ID cards issued in some large corporations, this card might carry the owner's photograph and be machine-readable. Persons using a stolen card would soon be in trouble as the computers could check card numbers against a single national list of cards reported missing.

Some countries may well adopt a national credit card. In others, where private enterprise produces a proliferation of cards, at least the same unique number could be used on all of an individual's cards. Such cards might contain two identifying numbers: one for the individual and one for the source of his credit. The American Bankers Association has been studying the idea of a single identifying number for every individual, and leans toward the use of the Social Security number, since it is already well established as an identifying number. Most of the 250 organizations surveyed by the ABA, ranging from credit-card concerns to hospitals, favoured the single-number rather than a different one for each card.

Whether the one-number card will lead to one card for every purpose is another matter. 'Looking at the matter dispassionately, however,' says one commentator, 'it appears certain that credit-card companies eventually will have to merge into the regular banking system or disappear'.[3] Banks do not have to invoice their cardholders – they simply deduct charges from the cardholder's account – and so can offer better terms to stores and restaurants, either in the form of quicker payment or lower charges. Whether stores and oil companies will be able to maintain their cards and their 'captive' clientele in the face of a national competitor will depend on the attraction of deferred-payment plans.

Since the single card would also lead to a national credit-rating system, another potential advantage would involve uniform ratings: a store would hardly rate a credit applicant differently from a bank, and the bank would be able to lend money to individuals more efficiently.

A sign of future innovations was the launching in mid 1967 by the Western States Bankcard Association of the Master Charge Plan. With an *Inter*bank Card Association and a *multi*bank plan, cooperation between traditional competitors arrived, and within one year achieved a level of business that took the 'independent' Bankamericard eight years to reach. The scheme spread in 1969 to the East Coast where several banks joined together to enrol thousands of merchants. Each bank then mailed Master Charge cards to its creditworthy (and sometimes uncreditworthy) customers, who were able to use them, not only at any participating merchant, but also at any participating bank for drawing cash. In four years prior to 1970, more than 100 million *unsolicited* credit cards were mailed to the unsuspecting American public, so it is hardly surprising that credit card crime is reaching high proportions. The unsolicited mailing, made possible by computers, has been heavily criticized but the mailers would argue that it is necessary to get the 'plastic' revolution to take off.

The Bankcard Association runs a clearing house that receives (for payment) sales slips prepared by merchants on their card imprinters, and which also maintains all cardholder accounts. The clearing house also authorizes sales above the 'floor limits' that the merchants can allow. The telephone wires so far only contribute to this scheme by carrying verbal authorizations for above 'floor limit' sales. When eventually the imprinters are replaced with touch-tone phones, as in Wilmington, and the Eastern and Western clearing houses are joined by microwave links, the cashless and chequeless society will be in sight.

VOICE RECOGNITION*

Research is now under way on more fundamental methods of verifying an individual's identification. Pattern-recognition techniques, for example, may enable the computer to recognize certain identifying characteristics of the human voice. It is within the reach of our technology to build a machine into which an individual reads his Socal Security number over the telephone and a computer interprets the digits and looks up a record containing details of that individual's voice characteristics. Just as the police use details of a fingerprint to identify a suspect, so computers can use details of voice sounds to determine whether the person speaking is the one with *that* identifying number. It is not yet known how accurately the machine could identify the speaker. After weeks of practice a mimic might fool the machine, just as a forger can falsify a signature. However, it seems likely that 'voice forging' would be more difficult than signature forging. Another problem is that the computer might not recognize the digits spoken by everybody. A Scotsman with a broad accent might attempt to buy goods in a store only to find that the machine cannot understand the digits he speaks. (He would have to go home and practise speaking computer-recognizable English or else pay by cheque!)

Perhaps eventually with such techniques the credit card itself may be eliminated.

*See also Chapter 5, 'Law Enforcement', p. 111.

CONSUMER CREDIT BUREAUX

Since many consumer durables are purchased by young people with an income but no savings, retailers generally need to extend credit to make sales. Organizations able to provide financing need to check the credit-worthiness of the buyer, particularly since the value of a repossessed article may be less than the outstanding debt.

Credit-worthiness naturally varies with the extent and nature of the loan in question; charge accounts, hire purchase, and mortgages have different criteria. But the data on which the assessment is made are the same. Credit bureaux in the US collate information provided by their subscribers, sometimes supplementing it with data gathered by investigators. Some are national or even international in scope, others cover just one trade or area. One individual may be on every possible file or just one of many files, most of which might have to be consulted before a decision can be made. Typically, each report costs from a few cents to a few dollars, and may take days by mail, or minutes by phone, for the subscriber to obtain. Department stores need quick responses but credit-card and travel-card issuers will be able to wait for reports without prejudice to customer relations.

Granting credit is an expensive operation. Even without instalment-loan delinquencies, the gathering, sifting, and provision of information about the borrower raises the cost of lending (and of borrowing). Computer systems have been enlisted in the attempt to reduce this cost.

Credit Data Corporation, a subsidiary of the TRW aerospace conglomerate, operates in Los Angeles a large computer that collects details of all credit granted by any of its subscribers (1,200 in mid 1969). The CDC subscribers made 300,000 telephone inquiries a day, at a cost of less than $1 a call.

The call is answered by one of 200 girls, each equipped with a terminal that has a screen on which she can display the payment record of anyone who has previously been granted credit by a CDC subscriber. (By microfilming and then transferring to tape all a new subscriber's credit records, histories predate the computer's instalation.) Not only does CDC check credit worthiness, it also provides a charge authorization service for some credit cards when their holders want to charge more than a small sum. If the sum would take the holder over his credit limit, a 'NO' appears on the terminal, and on occasion even 'PICK UP CARD' is flashed, indicating a stolen card or a delinquent holder.

Not only can the computer gather all the relevant information, it can also evaluate it automatically, an operation known as *credit scoring*. Now, instead of providing the undigested information by which an applicant can be judged, the bureau can give the inquirer a simple rating, based on standardized objective criteria, which is cheaper and simpler to use and to keep up to date. Credit scoring also makes possible direct inquiry by touch-tone telephone or teleprinter to the computer, since the reply is brief and standardized. (The programs by which the 'scoring' is done will vary from one geographical and sociological group to another; automobile ownership is generally significant in Texas, irrelevant in Manhattan. General Electric Credit Corporation has 35 different systems in operation in various parts of the US.)

There was a time when social status determined behaviour and mode of living. Present-day behavioural patterns are becoming more determined by the credit rating you want.

The ultimate, then, would be a single card issued by (or in conjunction with) a nationwide credit bureau utilizing a network of on-line computer systems programmed to evaluate a rating based on standardized objective criteria. More likely, however, a variety of credit-card and credit-checking organizations will persist.

FAMILY FINANCE

At a point such as this, authors are required by long tradition to create a hypothetical family and put them through an improbable series of daily events. Far be it from us to ignore the mores of crystal gazing.

Winston Smith, a respected member of the technostructure of the early 1980s, lives with his wife Julia in a flat with many of the latest devices in home electronics. (We have borrowed 'Winston Smith' from a notable literary work about the period, in which he also had electronic devices in his apartment. In our version, however, he is happily married and lives in a benevolent, though hard-selling, capitalist democracy.)

Winston works for a large manufacturing organization. Each day at 5 p.m. his company's computer transmits a message to his bank and deposits there his day's wages, less the cost of his lunch in the cafeteria. (These deposits were made weekly until the union proved that the company had no right to withhold the employee's earnings.) A proportion of his earnings are transmitted straight to the various tax departments by the company.

When the bank computer receives the credit, it checks his account and pays any standing orders due that day. The telephone bill is to be paid, since it is now exactly fourteen days after the telephone company's computer sent the account to the bank's and notified Winston by mail in the event he wanted to question the amount and prevent payment. The bank's salesman was enthusiastic about this kind of 'management by exception.' The rent on the flat is paid direct to the landlord's account, and the forty-third instalment on the home video-terminal is deducted.

Winston's wife has been on a deferred-spending spree and a list of bills against her credit card are due. The Central Comprehensive Computerized Credit Card Clearing Corporation (known as 'Seven Cs') has been on the line and debited Winston's account. The Smith's are now overdrawn by $70, so the computer looks at their savings account: at the end of last month it had cut their current account balance back to $500 and put the rest into their savings according to a long-standing arrangement. Since this brought the savings account to over $5,000 the sum of $1,000 had been automatically used to buy the bank's mutual-fund shares. The Smiths' credit rating was good when the computer looked it up; he had no bills due next week, and a loan of $100 for two weeks would cost less than a transfer from the savings account and back again later. So a loan was made immediately, and all the charges computed. Since tomorrow would be the 17th, when the Smiths get statements in the monthly cycle, no earlier statement was sent to advise of the overdraft.

When the statement arrives, Winston and Julia go through it. The local supermarket computer has her regular order on its files and telephones her each week to confirm it. The computer reads out her order over the telephone and then asks for her account number and 'YES' or 'NO', two of the very few words it understands. When the delivery is made, an order form for next week and advertisements for special offers come too. (This week's 'exotic' import is Scotch shortbread and the computer reminds her that she bought it last time.) Winston skips the supermarket bills and looks at the department store items. The last time that Julia was there they had a special sale of classic clothes from Carnaby Street. Since these traditional styles were hard to find, she let the salesman talk her into buying. After all, she only had to put her credit card in the touch-tone telephone and key in two numbers – she did not have to spend any money!

Suddenly the picturephone* rings and Winston's aunt appears on the screen in a state of some agitation. She is not wearing her usual picturephone make-up. Apparently her local credit-rating computer with a correct knowledge of her name, address, previous addresses, employment past and present, length of service, husband's name, standing debts, how long her account has been active, outstanding loans, name of bank, children's numbers and ages, telephone, salary, car, insurance and assurance, legal judgments, criminal record, and machine aptitude scores, has given her an *omega-minus* credit rating. They discuss whether they could sue the computer for libel (or possibly slander, as it uses a voice response unit).[4]

*Service mark of AT&T Co.

CONSUMER ACCEPTANCE

Response to new banking techniques and services has not always been universally favourable. Great Britain is still far short of paying all wages by cheque, a backwardness for which the redistribution industry shows its gratitude every Friday. Similarly, the acceptance of credit cards was rapid among businessmen and among more socially knowledgeable and affluent members of the community, but their use has spread only slowly among the mass public. Their use can be speeded up by banks discouraging the acceptance of cheques, and by other organizations expressing reluctance to take large sums of cash. In the United States one frequently sees posted notices to the effect that cheques will not be accepted, and in some places – especially in the West – one finds signs refusing to take payment in cash. As robbery in most areas is growing faster than other professions, the incentive for a cashless society is also growing. It is more difficult to rob a computer. Successful crime in the computerized society will probably require, as will most other activities, longer training and a higher IQ.

Cheques and some other documents now have numbers on them which can be read automatically by computers and sorting machines. The numbers – whose distinctive shape has become one of the symbols of the scientific 60s – are coded on cheques in a special magnetic ink. The acronym MICR (Magnetic Ink Character Recognition) is used to describe these numbers and the machines that read them. Optical machine recognition of non-magnetic characters is now also used.

When MICR first made automatic sorting of cheques possible, clerical handling was reduced from fifty or more times per cheque to perhaps one-half this. Local encoding of the amount of the cheque was needed to enable the computer to read the value as well as the account number. Using only the current-account number and value, however, the computer could not print out the recipient on monthly statements. This caused consternation among many depositors. Furthermore, the depositor must now use only cheques with his magnetic-ink number on it. He cannot borrow cheques from another person as he might have before. Some customers got the feeling that the banking system was reducing people to computerized numbers rather than giving them the personal treatment that they had before.

Some banks have started a service for small businessmen which provides a terminal in their home or place of work. The bank computer can be dialled on the telephone and details of transactions entered at the terminal. The bank computer then invoices customers and keeps the requisite accounts. In particular, a service for doctors and dentists has been introduced in the USA. The doctor's assistant enters the details of each patient's bill at the terminal and is relieved of much of the subsequent paperwork. Such schemes have not met with the wide and immediate acceptance that was hoped for. This is due partially to the lack of flexibility of the system, but probably more to lack of trust in, and acceptance of, the remote computer. Possibly the largest factor is the question of income tax. If all details are sent to the computer, is it not likely that the tax authorities will see everything?

Solely in terms of technology we can have the automated money system today. Psychologically and socially, we are not ready for it. The concept of automatic credit transfer has yet to be fully accepted. Credit cards tend to make people spend more. Few are capable of the budgeting necessary to handle a non-cash income, and to employ the computer to handle their savings and investments. Loans at high interest up to a credit limit represent a standing temptation to overdraw. The on-line terminal services, while accepted with delight by some, are regarded with distrust by others. There is still a fear, if not always consciously stated, that the remote machine will lose information, make mistakes, or print out one's financial details for other people to see. The computers *can* be made trustworthy, superbly accurate, and secure. The computer files *can* be made quite private, but it will take time for these facts to be understood by the public.

THE BANKING INDUSTRY

Goldsmiths became bankers when they discovered that their customers did not insist on being handed back the same gold they had deposited, but would accept any equivalent amount. Fractional-reserve banking is reputed to have begun when a thief removed 95 per cent of one banker's gold and the depositors did not find out for fifty years. The reason is that what one pays in, another can withdraw. Cash reserves need only be adequate for day-to-day fluctuations in net withdrawals.

Although the law of the land determines how large these reserves must be, statistics show that the amount to be kept ready need not increase in direct proportion to the total deposits, but in proportion to the square root thereof. A small American bank with deposits totalling $100 million and complying with the requirements of the Federal Reserve Board need have only $15 million in reserves and may have $85 million invested in securities or lent out. For equivalent security for its depositors, a bank with $1.6 billion in demand deposits would need only $60 million – or less than 4 per cent – in ready money. The government, however, must be in a position not only to guarantee that depositors will not lose their assets, but also to regulate the cash in circulation in order to counteract booms and depressions. By altering the requirement on reserves that banks must hold, the government alters what can be lent out by big and small banks alike.

It is an accident of constitutional history that the USA has many small banks, while the British have only a handful of large institutions. There are thousands of branches of Barclays in Britain, while banks in Florida are allowed but one head office. In Texas they can spread across a county; in California and Connecticut throughout the state. The Bank of America in California has nearly 1,000 branches and three times the deposits of any London bank. New York's largest bank has huge deposits but relatively few branches, although it, like many American banks, has set up offices abroad.

Between 1930 and 1933, 8,000 US banks collapsed. Their

average assets were only $6 million. The little ones went first: the 346 that failed in the first half-year had a mere $115 million deposited.[5] The legislation that gave rise to this plethora of tiny banks and the consequent weakness of the whole system also militates against efficient credit transfer.

Nationwide banks have for many years provided British bank-account holders with many of the services now being lauded as new in the US. The automatic payment of recurrent bills by 'banker's order', immediate overdrafts for small amounts, deposit of salary direct to a bank account, and simultaneous deposit and withdrawal are normal and easy when aided by a computer.

Since the eighteenth century London clearing banks have utilized a central clearing house to settle accounts with one another. Large cheques are handled in the afternoon if drawn on one of the ninety or more London 'town clearing' branches; all other cheques are handled next morning. When the 7 million cheques, totalling £3 thousand million, have been cleared, the eleven clearing banks add up what each owes the others and all their accounts with the Bank of England are then adjusted.

Most Americans have bank accounts today; together, they exchange 100 million cheques a day, fourteen times as many as the British. Sheer volume of transactions makes central cheque clearing and credit transfer by computer economically attractive.

As current accounts have spread to a wider clientele, the number of accounts has grown at a faster rate than total deposits. The amount of book-keeping, however, has increased in proportion to the increase in number of accounts, and so clerical costs per dollar deposited have risen. Few banks on either side of the Atlantic knew, before computers became available, what these clerical costs really were. Certainly, the ten-cents-per-cheque charge normally made is less than the average cost of the transaction; however, it may well be more than the marginal cost of processing each extra cheque. In the past bankers reckoned to make most of their profit on balances which they could lend out, and were happy to break even on book-keeping. Now, however, they can improve their cost accounting and have started to offer money-making–money-handling services independent of the customer's balance. Many extra services have a low marginal cost. If, however, banks are to start offering financial management as a service to their customers, they may need to increase their charges.

Again we see the change in business methods resulting in a change in business policy, which in turn alters an institution's relations with us, the general public. The Industrial Revolution stopped women from spinning cloth for the family's clothes; the data-processing revolution will force us to leave the handling of our money to the more efficient professionals. The new techniques and competitive drive to give the public a better service may bring about changes in banking laws, and perhaps in America the chartering of interstate banks.

THE BRITISH POST OFFICE GIRO

The Austrians first introduced a *Giro* in 1883. Basically it is a system of circular credit, in which money circulates from one holder's account to another. The word is derived from the Greek *guros*, meaning 'ring', which conveys the circular concept. However, to be of any use, money must also pass into and out of the Giro system in cash, and to and from the other banking systems.

Unlike accounts in conventional banking systems, Giro accounts are all held in one place, to which all instructions for credit or debit transfers must be sent. Transactions are then completed immediately. The British centre is at Bootle in Lancashire, close enough to the hub of communications for any mail to reach overnight. With millions of mailboxes and thousands of post offices, communicating with the system is very easy.

Parliament had been toying with the idea of a Giro for more than half a century, but it was not until 1959 that a committee investigating money-transfer services recommended that, in the absence of suitable action by the joint-stock banks, the

Post Office should provide a Giro. At that time, however, two factors were not auspicious.

The first was the comparatively low rates of interest in force at the time. A Giro relies on the interest on the money circulating within it to meet the costs of the services that it provides free. If the £125 million target balance can be invested at $6\frac{1}{2}$ per cent, the rate prevailing in 1968, there will be a profit of £3 million after costs have been cleared. At 4 per cent per annum, which was usual a decade ago, the scheme would not have broken even.

The other discouraging factor was the absence of suitable computers and peripheral machines for the reliable handling of a million transactions a day. Key-punching that lot into punched cards would take 2,000 full-time operators. Optical character readers which can handle typewritten documents are essential. Only third-generation computers can provide the requisite speed and reliability at reasonable cost, which meant waiting until 1965. Now the British national Giro provides facilities for transfers between Giro accounts and accounts outside the system. Every account-holder is supplied with printed forms for this purpose, along with postpaid envelopes. Unlike cheques, these forms have message spaces and thus serve as both instructions to pay and covering letters. If the payee has no Giro account, his name and address is filled in on the form, which then becomes a cheque, and cashable at any post office (subject to certain safeguards for large sums). Since these can be made out to 'Self', these Giro forms also constitute 'travellers' cheques'. Although British Giro links with the European systems are contemplated, exchange control regulations still apply; thus the use of travellers' cheques across the Continent will not be possible immediately.

If the Giro system was to compete with the banking system, it was decided, 24-hour clearance would be essential. This means that processing cannot wait for the preparation of all the input data, but must overlap with it. Great Britain is therefore divided into six regions, and the Giro stationery of each section is colour-coded for quick identification. Forms originating farthest from the Giro center are processed first so that the output mail can be on its way quickly. Only a small proportion of total transactions are conducted across regional boundaries. The computers therefore have only the files for one region accessible at any one time and build up an 'exceptions' list for subsequent processing against the regional files. When instructions from account-holders arrive at the centre, envelopes are opened, the contents sorted by colour, and the documents scrutinized for completeness and authenticity of signature. A large number of documents are pre-encoded, but those that are not encoded from manuscript to a machine-readable type-script, and all of them microfilmed before being transferred to magnetic tape. The document readers accept numerous type fonts, including letterpress and high-speed computer-printer character sets, on a great variety of sizes and qualities of paper. As the forms are read, they are spread with a bar of fluorescent ink which carries the code number for subsequent sorting, since the input forms are sent on to the other party to the transaction as proof of its completion.

Because some forms are unsuitable for character reading, one computer is fed directly from 150 keyboards by operators who read and type in from lists and non-standard documents. Another set of readers accepts the magnetic-ink characters used by the joint-stock banks, since the two systems must communicate with each other easily.

A large number of punched cards record the details of new accounts, standing orders, and direct-debiting arrangements. A large proportion of all money transactions are anticipated by those that make them. These include payments of rent, wages, fuel bills, premiums, subscriptions, and so on. If payer and payee agree in advance and notify the Giro, these can be directly debited to the payer's account on a specific date. The payee sends his list of payers' demands – in many cases on tape – to the Giro, and notifies the payers that the sums will be debited unless they register objections – which constitutes the exception, not the rule.

Charges to the British Giro user are very small. Individual account-holders can buy standard stationery at cost at a post office and all transactions within the system are free, including postage. Transactions across the post office counter cost four-pence which is therefore the cost of depositing or withdrawing cash. Firms wishing to use the Giro to collect their accounts may agree to pay the counter charge. A television rental company, for example, may issue the hirer with a printed book of payment slips listing the dates and amounts due in machine-readable form, adding its own account number and codes. The hirer goes to the nearest post office on the date due and passes the slip and cash across the counter and the operation is virtually finished. If the hirer had his own account, he would merely sign the slip, add his account number if it were not pre-printed, and drop it into the mailbox, without even needing to go to the post office.

The logic of such a scheme is overwhelming. Many of Britain's largest enterprises have followed the nationalized industries and government bodies in opening accounts, and the volume of cash is such that the Giro is considering an investment service for short-term funds built up by companies in their accounts. The real test of the system, however, is the support it gets from the 'man in the street'. Two thirds of British wage-earners have no bank accounts. Large quantities of cash are therefore moved to pay offices each week, which has made 'redistribution' a leading growth industry and payroll robberies a weekly occurrence. The hope is that ultimately every employee will have either a Giro account or a bank account, enabling employers merely to send in magnetic tapes of lists of numbers. The recipients would be notified by mail the day after their accounts had been credited. Once they became used to keeping their money at the post office instead of under the mattress British employees would soon realize the advantages of paying bills without effort or extra cost merely by signing a form and dropping it in a mailbox.

THE BANK GIRO

The British National Giro was set up because the joint-stock banks had failed to provide a suitable service. Now that the threat has been carried out, the banks have decided to provide a similar scheme – a 'bank Giro' – themselves. About a dozen banks (i.e., most of the British banks) are involved, and they have set up a centre in Lombard Street in the heart of the city of London. This Interbank City Computer Bureau receives magnetic tapes from each of the banks with details of interbank standing orders, direct debits, and bank Giro credits, which must be cleared from one bank's accounts to another's. Those which will accept debits from bank customers with their own computers will thereby be able to cut accounting and cheque-paying systems to a minimum.

The banks plan to be nationwide but decentralized, and therein lies their main advantage over the National Giro. Most bank customers expect the bank to provide them with money when they are short and to give personal attention to their financial problems. The personal touch matters, and so, too, does the ability to overdraw slightly on occasion, a facility not provided by the Post Office Giro.

The British business customer is being wooed in the same way as his American counterpart. Magnetic-tape interchange between banks and other banks, and between banks and their customers, will ultimately be extended to direct terminal-to-computer transmission. Direct debiting cuts bookkeeping costs and speeds up receipts of accounts payable, while credit transfer – the paying of many accounts with a single cheque – cuts the cost of paying accounts, since there is less paper to handle, fewer cheques to sign, and less postage.

The banks have also cut their costs, particularly on space at the branch office. One third of the space in each of 2,500 branches of each big bank had been devoted to accounting. Accounting staff are now rapidly being replaced by terminal links to the computer, freeing the premises for improved customer contact. Branch productivity typically shows a 20 per cent increase, 15 per cent fewer staff being needed to handle 5 per cent more business.

Only time will tell whether the efficiency and availability of the National Giro will prove more attractive than the personal contact of the bank. In any event, the handling of money will never be quite the same again.

POSTSCRIPT, 1972

We have deliberately chosen to leave the above discussion of the banks and the Giro as we wrote it in the fall of 1968, because it illustrates so effectively many of the difficulties facing the computerized society. A year after the system was opened (on 18 October 1968) the target number of accounts was cut by 75 per cent, and the break-even point was forecast for 1972 at best.

What went wrong? The public was bewildered by sales literature of daunting incomprehensibility; the banks fought the scheme, not only with their own Bank Giro, but with competitive advertising and non-cooperation (banks will transfer funds to accounts in rival banks but not to a Giro account); the system was not vigorously advertised since it was thought capable of gaining acceptance on its merits; such advertising as there was concentrated on the computers rather than the service itself; cash could only be drawn at a specified post office branch, and then only £20 on demand.

In fact, the National Giro is doing very much what the 1965 Government White Paper projected. Unfortunately, in the excitement culminating in its launch, the Giro's prospects were inflated by spokesmen drawing on later optimistic market research. The 'failure' of which Giro is accused lies with its public relations, not its performance.

The advertising in the autumn of 1969 was aimed at the big customers as well as the small. An experiment agreed upon with a large industrial company, its staff and unions, involved the payment of 27,000 men's wages through the Giro with direct payment of union dues and of contractual savings into holiday tour operators' accounts. Union and employer clearly gain; the experiment has to test whether the employees think they do too.

Following the change of government, there came an announcement in November 1970 of an inquiry into the future, if any, of the system; speculation about its survival did nothing to improve its health. A year later, the government accepted the report that with improved management and financial control, and new marketing policies and tariffs, it could provide a commercially viable service to public and government without a subsidy from the Post Office. But its prospects really depend on its ability to attract accounts.

Since most holders of Giro accounts seem happy with the system, success should in the end breed success. Meanwhile, the moral to be drawn by those who expect the early achievement of the computerized society, is that logic alone will not bring about change. Only people can do that.

INCOME TAX

Let us now turn to the question of income tax. The title of this chapter is perhaps more applicable here, because in banking, for the time being, we are clearly not going to say 'Goodbye' to money entirely.

In 1967 a computer was used for the first time to check all US Federal Income Tax returns. The system, introduced on a less comprehensive scale in 1962, now consists of the National Computing Center at Martinsburg, West Virginia, and seven regional centres across the country. The central master files keep magnetic-tape- records of all taxpayers and are supplied with current information from the regional centres.

Whenever something interesting is discovered by the juxtaposition of new and old information, the local tax officials follow it up. The records are all identified by Social Security number or by a business identification number. One man's expenditure is another's income, and the tax man now has files to prove it. Even before tax authorities could check, say, a company's dividend-payment records against all taxpayers' receipts, the threat to do so had resulted in 45 per cent more taxpayers reporting nearly $3 thousand million more interest and dividend income.

Tax returns are first checked by examiners for correct filling in of entries and then converted to punched cards for the regional computers, which verify the mathematics and do some initial calculations, or report the errors. Validated information, now on magnetic tape, goes to Martinsburg, where it is combined with master-file information to detect those who have failed to pay, have not filed a return, or should get a refund. The Treasury's Disbursing Office had to refund more than $6 million to 51,000 taxpayers in January 1967 alone.

Information concerning delinquents and others less fortunate is listed on tapes which are returned to the regional centres, resulting in communications to the parties about their defaults. The computers also notice any peculiarities worthy of audit by special inspectors.

Once the present system is fully proved, it seems likely that banks and employers will be asked to feed the tax computers with magnetic tapes directly, or even to transmit the information needed by communication line.

Having gone to the expense of collecting all this material, it makes sound economic sense to use it for policy-planning. A sample of tax-payers' records is already being used to simulate changes in taxation.

'PAYONS, ENFANTS DE LA PATRIE!'

All tax-collecting systems are complicated, but none has achieved the same level as the French system. All citizens in any country expect to arrange their affairs to attract the minimum tax; ten years ago the *citoyen* had so perfected the art that he could nearly avoid paying at all. De Gaulle changed all that when he needed money for his grandiose plans.

The Direction Generale des Impôts (DGI) collects its money annually, not as it is earned. Ten computers, a Honeywell 1200 and 200 at each of five centres, issue tax forms each February in accordance with any changes in the laws that might have been made in the previous months. The taxpayer completes his form, and the programmers adjust their programs, ready for processing in May. Meanwhile, in March, the DGI collects one third of the previous year's tax, another one third in May, and the rest – i.e., the 'correct' tax for the year ending December 31st preceding – in October. Stringent penalties are imposed for late payment and the tax forms are fantastically complicated. To help the taxpayer, therefore, the computer fills in a lot of basic information from its records. The local tax officer reads through the return and prepares it for the computer, which checks for payment on time and keeps the taxpayer's file up to date.

The DGI computers not only handle the 9 million demands to 3 million taxpayers, and thousands of reminders, but also provide government statistics, collect turnover and profits taxes, and collect rental taxes for half a million apartments – and are creating a land registry for every plot in France.

The information gathered is analysed for use in modelling the national economy and for making policy decisions. Future plans include utilizing the computers for rate-accounting and state property rental.

Schemes such as this will no doubt be extended and elaborated. The tax evader in the computerized society will have little hope of avoiding detection except in minor matters, and the government's planning of changes in taxation will become highly scientific, with vast banks of data for use in anticipating changes in revenue. As in other aspects of this story, the use of computers will mean a tight network of controls. Our society will become a more precise and intricate machine. The maverick tax evader will quickly be rounded up.

THE STOCK MARKET

At the time of writing not only taxation but also the stock market seemed that it should be included in a chapter called 'Goodbye to Money.' Wall Street is well up with bankers and Internal Revenue Service in its use of computers. The famous

Big Board of the New York Stock Exchange (NYSE) must be the most elaborate display terminal outside Houston's Space Center.

One of the most spectacular systems belongs to Hornblower & Weeks – Hemphill, Noyes. During 1968, their back office was in the same mess as everyone else's. To install the computers they had to revolutionize traditional procedures and automate all communications and book-keeping involved in a securities transaction. One of the twin machines accepts orders from the firm's salesmen around the country, checks them, passes the relevant details to the exchange floor, and matches the order with the report of the trade. All details are recorded, and any errors or confusions are tidied up on-line by clerks at visual display terminals. The second machine stands by lest the first should fail and fills up its time keeping the routine accounts. The customer confirmatory invoices are ready minutes after the deal, and Hornblower gets its money that much quicker.

Mere automation of present procedures, however, has been quite inadequate in dealing with the rising volume of trading. The paperwork problem that caused Wall Street to reduce trading hours centres on the archaic methods and systems by which the industry processes transactions and transfers the ownership of securities. The most obvious symptoms of the trouble is the broker who 'fails to deliver' security certificates to a buyer's broker. Even the big, efficient, automated firms cannot always meet their obligations if the previous deal had led to a 'fail'.

The exchanges themselves have been improving and radically changing their systems. The New York and American exchanges are cooperating in floor automation projects, a common message switching system and a computerized system to reduce fraudulent negotiation of stolen securities. 'Am-Ex' is joining 'NYSE's' existing Central Certificate Service (CCS) and Block Automation System (BAS). Meanwhile, the National Association of Securities Dealers (NASD) has set up a national quotation and clearance system for unlisted stocks.

The CCS is a depository for share certificates registered in brokers' names with a computer recording the current owners.

When ownership changes, the computer updates the records – but the certificate never moves in the vault, any more than cash physically moves when checking accounts are used.

The 'over-the-counter' market for unlisted securities is in effect located in the offices of brokers across the nation, a diffusion which adds further to delivery problems. The NASD scheme cuts out intermediate deliveries in a chain of transactions in the same stock, only the first seller having to make a delivery to the final buyer.

Ultimately, the share certificate may go the way of the banknote in the cash-less society. A central register will record the identity of share-issuing corporations and their owners and make the changes as trades occur. Certificates are only information, like so much else.

The market itself is only an information exchange and the traders would not need to be present if a substitute were available for voice communications. The *Wall Street Journal* (Tuesday, 13 May, 1969) revealed 'the existence of the automation report, made jointly by the Big Board and International Business Machines Corporation' which proposed to do away with the hordes of brokers and clerks clustering around trading posts. In fact, it is technically feasible to do without a physical trading floor at all and replace it with a computer system.

Under present trading methods, orders are sent to brokers' clerks in their booths by teleprinter or telephone. Some teleprinters are fed from computers in brokers' head and branch offices. The clerk has to summon the broker from the floor by posting his number on a signboard. The broker then hurries with the order to the 'crowd' by the appropriate 'trading post' and bargains to establish a price for the number of shares he wishes to buy or sell. Every trading post has 'specialists' who match buyers with sellers and if necessary trade on their own account to minimize the swings caused by a preponderance of either bids or offers.

The proposed Advanced Trading System will still require specialists and brokers, but there will be no vociferous crowd calling out bids by a trading post, and many fewer clerks. Instead, there would be 'Ordermats' around the floor accepting

orders from brokers' computers and paging the broker concerned automatically. The broker would then insert an identification card and the Ordermat would print the order or display
it on a screen. Using a console, the broker displays the current
auction information about the stock – such as bid and asked
prices, volume, recent transactions, and in particular the size
of individual orders bid and offered in the crowd and the
brokers responsible for them. The broker can alter his bid or
agree to trade on the best terms.

When the order is executed, the trade is 'locked in', all subsequent paperwork – culminating in the new owner receiving
his share certificate – being handled by computers. If the order-
delivery system forces all brokerage houses to provide machine-
readable orders, only the bid and offer decisions of the brokers
and specialists on the floor will be human. Machines will produce the ticker reports, confirmations, and book-keeping entries
for brokerage houses, and match buy and sell orders for the
clearing house.

The London Stock Exchange

Since the Second World War, investment business in London
has come more and more from institutions collecting small
savings, such as pension funds, insurance companies, investment and unit trusts, trade unions and so on, and less and less
from individual investors. The market machine, however, was
still rooted in its club-like traditions, and its commission rates
were geared to small deals for private clients. The institutions
baulked at paying substantial sums in commission when they
could make large deals with each other outside the market.
This not only kept money away from brokers, but also upset
the traditional stock jobbing system built up under different
circumstances a century ago. Jobbing firms merged, cutting
their number from 250 at the end of the war to less than 30 by
1970. This gave each more capital (though little enough
measured against the size of institutional deals) but also coupled
the fate of one kind of share with all the others. With big insti-

tutions and big jobbers thinking alike, prices tended to move together without the stability that comes from numerous small deals.

The Stock Exchange fixes prices for its members. Instead of reducing the prices of their services, brokers compete with each other for the lucrative institutional business by gift-wrapping their product with free (to the client) additional services, particularly investment research.

The merchant banks, some of whom pay £1 million in commissions each year, threatened at the end of 1971 to set up their own computer system to handle large deals after a study of American and European automated block trading systems. They suggested that the system would be open to any institution wishing to make a deal worth more than, say, £50,000. Such a system would take a great deal of business away from the Stock Exchange, which is already worried about competition likely to come from banks dealing in their own right on European stock exchanges following Britain's entry into the Common Market. Tele-communications could link all Europe's exchanges into a single trading floor.

The London exchange has had unhappy experience with computers. In 1970 it had to abandon its inter-firm accounting system at a cost to its members of £1 million or so. As in most professions, the men at the top are stockbrokers first and managers a long way second. The skills they bring to negotiating numerous deals at the best price as they arise are irrelevant to the problem of managing a large project over a long period. Nevertheless, although they hired an architect and contractors to build their new exchange, they felt quite capable of tackling the much more complicated task of installing a computer system themselves. The management structure they set up diffused authority through committees instead of concentrating authority, and the system that came near to being installed leant so far over backwards to accommodate the idiosyncracies of every member firm that it collapsed when first tested.

The exchange has now proposed major changes in the way that deals are processed and that holdings and records are transferred. These proposals recognize that the best way of running a system with computers is to design it for computers in the first place, not just to substitute machines for men where one can.

There is no logical reason why a stock exchange should be involved in a share deal at all, nor is a broker necessary if other means could be provided to find buyers or sellers. Institutional Networks Corporation has proposed a means for big institutional investors to eliminate brokerage commissions when they trade big blocks of shares with each other. Naturally, the brokers are concerned since the mutual, pension and endowment funds and the banks and insurance companies across the country who might join the scheme are responsible for one-third of their business.

'Instinet', as the scheme has been dubbed, allows the subscribers to bid and offer in total anonymity through terminals attached to a time-sharing computer. Code numbers identify others who are active, enabling terminal-to-terminal bargaining. Trades are posted to the computer, which closes the trade and copes with the record changes. For those who do not wish to bargain, the system will circulate to print-only terminals details of bids and offers, which can later be reviewed by interrogating the computer until they are withdrawn.

The New York Stock Exchange has a Block Automation System (BAS) using a network of cathode-ray tubes and keyboards. As buy or sell orders for blocks are put into the system, the computer matches potential buyers and sellers and tells the brokers concerned, who proceed to bargain in the normal way. A privately run competitor, AutEx, offering more than BAS, has different facilities for brokers and institutions and allows brokers to screen out their competitors or direct messages only to selected institutions.

The latest market prices are vital to all dealers in stocks. The New York Stock Exchange has a colossal display board and a ticker tape' containing the details of each deal as it is reported from the floor. Both are computer-operated and dealers are obliged to provide the input data. Several services feed the 'ticker' signals straight to their computers and then redistribute the figures to customers.

'Stockmaster' is an information network supplying prices from most major stock exchanges in the world to thousands of terminals from South America to Asia via most places in between. Outside the USA it is run by Reuters who collect nearly ten thousand prices from the various exchanges, a simple operation in America and Germany, but more difficult, of course, in London where there is no computer to tap.

The London Market Price display service is quite different from New York's. Because no central record or deals in the market is maintained, messengers have to walk round the floor asking the prices which they type into a terminal near by. The terminal is connected to a computer which checks and edits the prices and then converts them from digital records to analog signals for transmission to conventional television sets in subscribers' offices all over the City. The display is effective and cheap, but limited to active shares and unsuitable for computer-to-computer transmission. Those wanting price data for updating computer records have to wait till the Exchange Telegraph delivers during the night a magnetic tape of stock information, with price details which are collected by different messengers.

Brokers do a great deal more than just buying and selling shares for their clients. In the past, clients have paid for these other services by directing their business, and hence commissions, to those who help them with recommendations concerning their portfolios. If the customers take their business elsewhere, who will pay for expensive research departments to give advice away free?

Research is getting still more expensive as security analysts turn more and more to the computer to help them. Computerized research has yet to prove itself, and there is as yet no record of any analyst retiring and leaving a computer to manage his affairs.

Many millions of dollars have been spent by institutions as well as brokerage houses to improve the quality of their investment analysis. Reasons for doing so vary. William J. O'Neil, who runs the O'Neil Fund, told *Forbes Magazine*: 'They're going to put a man on the moon with computers. The stock market isn't that complicated.' The O'Neil fund is typical of many performance funds which buy only shares whose prices are moving fast and sell within months if the up-trend slows. His computer watches thousands of potentially rising shares to pick out the beginning of a pattern likely to develop well.

O'Neil's is an extreme view, but even conservative institutions are actively investigating how the computer can help. Opinion is sharply divided between *fundamentalists*, who believe the stock market is an economic phenomenon amenable to financial analysis, and the *technicians*, who hold that it is a behavioural phenomenon to be analyzed psychologically. Many brokers broadly use the former theory to determine what to buy and the latter to tell them when to do so.

Fundamentalists

The proper price to pay for a share depends on the income it is expected to provide for its owner, either as dividends or from its ultimate sale. Since, when a stock is sold, its price reflects its earning power in the same way, the sole determinant of a proper purchase price is the value to the purchase of all those dividends stretching way into the future. If the managers of a business have the best interests of its owners (i.e., shareholders) in mind, they will reinvest in the firm that part of corporate earnings they can use more profitably than the owners themselves, and distribute the rest in the form of dividends.

Fundamentalists consequently devote their efforts to the accurate prediction of earnings in the future, and take a glance at the market to see how much investors seem prepared to pay for such a stream of income. To determine the earnings, the analysts will try to estimate the future sales of the industry, the firm's share therein, and the margin between sales income and costs which provides the earnings. Past history of the firm and others like it provides the principal source of data for estimation. There is so much information that computers are necessary for analysing it.

Company financial statements and allied facts have been

available in computer-readable tapes from Standard and Poor's Corporation for some time. These immense 'Compustat' files going back over many years are widely used despite their limitations. Certain potentially useful facts are necessarily omitted, while others had to be adjusted to fit the format used on the tape. The adjustments necessarily involved human judgment, and many professional analysts doubt the ability of their colleagues at S&P to do the job quite as well as they could. Nonetheless the Compustat data form the basis on which many theories of investment are tested. These theories, with which Wall Street is rife, are designed to simplify the determination of future earnings or to predict future market price changes.

Most theories involve mathematical models which describe anything from the whole economy by way of a single industry to one product or firm. Burnham and Co., for instance, have an airline industry model designed to derive earnings of airline companies from considerations of – among other factors – traffic, route structures, load factors, and so on. At E. F. Hutton one man has been engaged for a few years on the development of a model to predict the share price of many hundreds of companies.

Theories and models are not everything. Much analytical work involves numerous small calculations; many time-sharing services are available to analysts who wish to call on more computing power than their slide rules provide. Some firms have their terminal on wheels so that it can be moved to where it is needed and then acoustically coupled to one of several computers dialled up from the nearest telephone.

Many time-sharing services provide not just computing routines but private and public data bases and even special languages. One of the best known started life as the private computer department of White Weld, the large New York brokers. Some years and millions of dollars later they decided that the department would never generate the extra commissions needed to pay its way. So they set up Interactive Data Services Corporation as an independent company and transferred the know-how and their computer chief to it. The latter, an ex-

MIT engineer, teamed up with a group of computer specialists from MIT's Lincoln Laboratory and together they developed a service for which each of their clients pays thousands of dollars per month. Each subscriber has a terminal with which he can interrogate a data base containing all the Compustat data, any private files they like to add, and a history of price movements. The computer communicates with the analyst through 'FFL' (First Financial Language), which uses terminology similar to the natural language of finance.

Security analysts with access to a computer soon start to use it for 'screening' the huge volume of data about the recent or distant-past performance of individual shares to discover those warranting further attention. They may feed the ticker tape straight from the exchange to the computer to watch for unusual volume or perhaps signs for a recovery in a particular share, alternatively, they might hunt for individual firms which over the years have increased their profit margins in a declining industry. Before computers, such broad surveys were quite impossible.

Technicians

Technicians and 'chartists' contend that the price history of a share drawn on a chart reveals something about the way it will move in the future. Chartists have been around for a long time, and they have developed numerous plausible theories about the patterns they see. Sceptics maintained for years that the only ones who made any money from analyzing trends were those who published books on the subject, and when they got computers they proceeded to try to prove it.

Opponents of charting maintain that stock-market prices take a 'random walk'. Readers with a taste for academic in-fighting and abstruse statistical theory can pursue the arguments for and against charting in learned financial journals and texts. Essentially, the 'random walk' theory maintains that prices have no memory, so that what they did yesterday has no bearing on tomorrow's movements. (This argument is similar to – though by no means so obvious as – the incontrovertible

statement that each toss of a coin is independent of the preceding toss and that even after a run of heads, both heads and
tails are equally likely to occur.) Numerous attempts have been
made by computer analysis of millions of price changes to
relate the size and direction of price changes to those that have
gone before. None of the theories makes any sense – which
does not worry the technician, pursuing his laughing way to
the bank: 'I make dollars so who needs sense!'

Some chartists will claim that although most prices do indeed
wander randomly, the exceptions are worth studying. Others
maintain that enough chart watchers exist for their common
action to fulfil the chart's prophecy. If the signs indicate a
price decline, the chartists will sell – and behold, the price declines as the chartists said it would.

The computer is a great boon to technical analysts, for not
only does it watch the charts with tireless accuracy, it can also
draw them in all shapes and sizes. The ultimate in sophistica-
tion is the system that is fed constantly with the ticker prices
which it compares with previous history recorded in its data
banks. Any development of exceptional interest causes it to
call in an analyst to whom it displays the price history and
trading volume on a screen; the analyst examines the results,
adjusts the scales or time period covered, and instructs the
computer to print the chart.

Volume of trading is important to technical analysts, who
look for unusual activity in association with price movements.
Heavy trading usually means that the institutions are getting
involved. Those small investors who trade their few shares in
'odd lots' are considered by the cognoscenti to be always wrong,
whereas the institutions, backed by armies of analysts and
platoons of computers, are considered to be right. Many a
computer is watching for high volume following the 'bottom-
ing out' of the market, expecting the purchasing power to the
funds to run the price up quickly. This is the moment to buy in
and go along for the ride.

One of the more popular technical tools is the 'principle of
relative strength'. This is based on the statistical finding that
shares in the strongest up-trend over a short period up to today

will outperform most others for some time from tomorrow. The computer watches for the strong short-term price trends (ignoring volume entirely) and for breaks in the trend, and manages the portfolio so as to cut losses short and let the gains run.

One brokerage house gained a large following for its advice service based on this method. Those who followed the advice faithfully owned shares in a certain small company when the computer decided it had seen an unmistakable 'sell' signal, with the price standing at $21\frac{1}{2}$. Obediently, nearly every client gave his broker instructions to sell, and when trading opened, the harassed specialist offered a large part of the corporation's stock, first marked his price down to $20\frac{1}{2}$, then refused to trade at all. Sometime later he re-opened with an offer of $16\frac{1}{2}$, which brokers had to take. The advice service closed, and its originators are prepared to try again only if they can control the buying and selling of portfolios as the mutual funds can. Unloading over a few days would have kept that price near 20.

Buying and selling depends as much on the investor as the stock. Individuals need portfolios which reflect their tax, family, and income circumstances, which naturally vary widely. Once the analysts have determined the risk and return associated with stocks, the computer can be left to choose a portfolio for a client and then to supervise all further purchases and sales.

The computer could even have a portfolio of its own. It might well outperform its human rivals, but Adam Smith, the pseudonymous author of *The Money Game* has reason to think it will not. One of his old friends, he reports, told him:

'Our computer is running one portfolio.'
'How is it doing?' I asked.
'When we first put the computer on the air, we asked it what it wanted to buy and we couldn't wait to see what it reached for. It said "Treasury Bills, Cash." We couldn't get it to buy anything. So we checked out the program again, and while we were checking out the program, the market went down. Then we asked it again. The computer insisted on staying in cash. The market went down some more. We begged it to buy something. "There must be one

stock somewhere that's a buy," we said. You see, even computer people are victims of these old atavistic instincts from the pre-computer days. The computer just folded its arms. It wouldn't buy anything. Then, just when we were worried that it never would buy anything, right at the bottom it stepped in and started buying. Pretty soon the computer was fully invested and the market was still going up.'

'What did it do then?' I asked.

'The market was still going up,' Irwin said, 'and then one day it came and asked us for margin. It wanted to keep buying. So we gave it some margin. After the market went up some more it sold out a bit and came back to being fully invested. Right now it's got buying power.'

REFERENCES

1. *New York Times*, 12 May 1967, Financial section, p. 13.

2. Thomas J. Watson, Jr, 'Man and Machines – The Dynamic Alliance', *Proceedings of the ABA National Automation Conference*, 1965.

3. 'Electronic Money,' *Forbes Magazine*, 1 April 1967. p. 42 (unsigned).

4. Ask A. P. Herbert, 'Reign of Error', in *Bardot MP and other Modern Misleading Cases*, Doubleday, Garden City, NY, 1965.

5. J. K. Galbraith, *The Great Crash, 1929*. Penguin, 1961, p. 197.

5

LAW ENFORCEMENT

When little old ladies have to wear tennis shoes so they can outleg
the criminals on the city streets, there's something wrong with the
way we're doing things. – *Spiro T. Agnew*.

The time was shortly before midnight. A burglar moved
quietly down the fire escape of a midtown apartment house
looking for a darkened room with an unlocked window. Finding one, he lifted it high enough to slip through, grabbed a
radio and an electric razor, and searched quickly for jewelry.
In the top drawer of the dressing table he found a gun, which
he pocketed, and a handful of rings, brooches, necklaces, and
earrings. The living room yielded a chequebook and three credit
cards, a camera, and a miniature television. Within three minutes he was on his way out, only to find the building superintendent at the foot of the fire escape. Drawing the gun, he told
the 'super' to face the wall, and then clubbed him behind the
ear before running off.

The occupant of the first-floor apartment heard a raised
voice and running footsteps. He snatched his telephone and
dialled the police emergency number – 911. As he did so, the
telephone company's computer identified his number, looked
up his address, and transmitted it to the police network computer.

A flashing light alerted the complaint clerk. He lifted the
telephone and on the screen in front of him viewed a projected
map showing the origin of the call and the name and address
of the telephone subscriber. In a brief conversation, the clerk
discovered that only one man was involved, and confirmed the
address. The conversation took 45 seconds – far less time than
with a manual system.

The clerk typed the name of the call-originator into his computer terminal, and specified that a two-man car was to go to

the scene. Other cars were to watch for a running man. With a light pen he pointed out where the crime occurred on an enlarged map on his computer screen. The nature of the incident was indicated and the whole display transferred to the dispatcher's screen.

A buzzer sounded in a police car prowling through the night. The dashboard teleprinter spat out details. The patrolman radioed the dispatcher that they were on their way. The computer had selected this car because it was the nearest one not involved in another incident. Confirmation was received from the car within fifteen seconds and so the computer did not alert a second car.

Ninety seconds after dialing 911, the caller hears the police car scream to a halt down the block.

The computer tracked the movements of the assigned car. Each car has a small transmitter which continuously puts out a signal indicating its number. These signals were picked up by receivers all over the city. The car's direction and identification number were sent straight to the computer in the form of a ten-digit code. These receivers are spaced about a quarter of a mile apart over the 100 square miles of the jurisdiction. At any time the central computer has the position of each car and a record of the job assigned to it.

The police quickly found the superintendent, alive but unconscious. Their computer arranged for an ambulance to be on the scene within minutes. The burgled apartment was discovered by the same technique the burglar used, going up the fire escape and looking for loose windows. Two hours later the tenants returned, to find the police in occupation. The missing property was listed and the serial numbers of the television set, camera, and radio recorded from the old warranty cards, and that of the pistol from the gun certificate.

Following the assassination of several prominent political figures, the state legislature had brought in a gun-registration law. Every commercial sale of a gun has to be accompanied by a form giving the purchaser's name, descriptions, and address. The forms must be taken to the precinct office, where the police take the buyer's fingerprints, check that his record is clean, and issue a certificate in duplicate, charging a few dollars for the cost of the clerical work. The buyer signs an undertaking not to dispose of the gun to anyone without a certificate and to report loss or theft. The seller then hands over the gun, and sends the police his own certificate of the weapon and the buyer's duplicate. All the details go straight into the computer, from which within seconds the history of the gun can be abstracted and its most recent owner identified. Possession of an unregistered gun is now an offence.

The first thing the following morning, the victims of the burglary telephoned the Credit Card Bureau maintained nationally and jointly by all credit-card companies for the supervision of lost or stolen cards. The police updated the pawnbrokers' lists, on which are recorded all serial numbers of stolen goods. This list is circulated locally each week, so that the pawnbroker can avoid receiving stolen goods. He also records all pledges with serial numbers on punched cards which must be submitted weekly for computer comparison with the central files. Most criminals dispose of stolen goods too quickly to be apprehended this way, but the recovery of property and its return to its rightful owner is facilitated. (We might anticipate that insurance companies would indemnify pawnbrokers against loss on any item restored in this way, thus making the whole exercise advantageous to everyone.)

In this case the crook moved fast. The household items and the valuable jewelry were pawned by 10 a.m. Four days later the computer discovered them from the pledge records and the police collected them for fingerprint checks. The clerk who took the goods in gave a partial description of the man, but the address and signature turned out to be false. Although the burglar had worn gloves, the man who pledged the stolen property had not. With the description and a few latent prints it would be easier to convict the man once he was arrested.

The routine data-processing did not find the burglar. Large files of data were searched for persons who met the description or operated in the same way and in the same district. Every time new material on the crime was found, the appropriate files were searched. In this instance too little data and too many similar crimes made the list too long. The police had to wait for the criminal to reveal himself.

A week later, a three-year-old Chevrolet pulled into a gas station in a town 500 miles away. The driver pulled out an oil company credit card, and the pump attendant inserted it into a touch-tone telephone, in contact with a credit-checking computer. The 'lost-or-stolen-card' light flashed on at once. The attendant went back to the car and asked the driver to identify himself. He did – by stepping on the accelerator and screeching off downtown.

The 'alert' routine went into action. The gas-station attendant telephoned the police and gave a description of the car, its driver, and the number of the stolen credit card left clutched in his hand. All near-by patrols were alerted by their prowl-car teleprinters. A rapid exchange between computers identified the card as the one stolen in the burglary, and listed the other items taken at the same time, among them a gun. Ten seconds after the alert, still only two minutes after the car had left the pump, the patrols knew they were after a man who might be armed. One minute later, the getaway car stopped at a red light, and a patrol car drew up across its front. Two armed police had the suspect covered before he could get his hand to his gun.

The above account is fictional, but it is probably only a few years ahead of its time.

Automation in law enforcement is badly needed and is proceeding fast. At the present time the police forces of most advanced countries throughout the world are losing ground in the battle against crime. Crime in most Western nations is increasing at a much faster rate than industry. At the present rate of growth in the US, it is doubling every five years. Figure 5.1 shows the growth since 1944. The latest statistics published for the percentages of crimes solved indicate a drop each year since 1965.

We suspect that as our technological society becomes more complex these appalling crime figures will become worse.

The most effective method of crime prevention is the threat

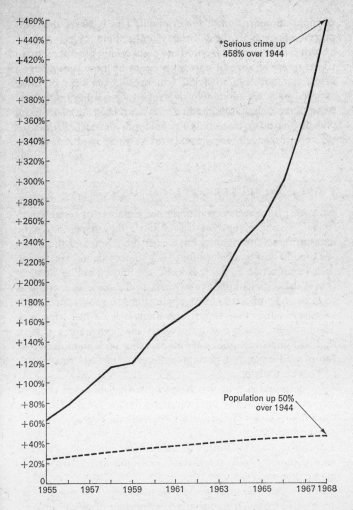

*Serious crime up 458% over 1944

Population up 50% over 1944

*Serious crimes = Murder, Forcible rape, Robbery, Aggravated assault, Burglary, Larceny $50 and over, and Auto theft.

Fig. 5.1 The rising crime rate in the United States. Can a massive use of computers in police work slow this down?

of rapid apprehension of the criminal. This is where the computers can help. In the US today, only about 25 per cent of known crimes lead to arrest. Many petty crimes are not reported to the police at all, and only 20 per cent of those for which an arrest is made lead to a jail sentence. 75 per cent of those arrested once are ultimately re-arrested, suggesting on the one hand that rehabilitation methods have not been very successful and on the other that computer methods of keeping track of persons who have been arrested are likely to pay off.

POLICE COMPUTER APPLICATIONS

No police force can recruit unlimited numbers of suitable personnel. To make the best use of those they have, automatic data-processing techniques have been used. Some of these files are used in real-time, since they seek to provide information in time to influence on-going events. The most familiar application is the dispatching of prowl cars to the scene of a crime as soon as it is reported. Most applications, however, are retrospective, in that they report and analyse what has happened. The analysis, which usually involves the search for a pattern that will improve future operations, relies on a good data base full of well-organized, accurate, and relevant facts which must be kept up to date.

In the remainder of this chapter we shall describe some of the computer systems used for police work today. There are three main classes of application we shall consider:

1. *The rapid transmission of police information to where it is needed.* Good police work relies on rapid and reliable communications and message switching.

2. *The rapid provision of information to police on their beats or in patrol cars.* Appropriate files must be searched in real-time to give a policeman almost immediate information about suspected criminals, suspected stolen vehicles, and so on.

3. *The discovery of patterns in criminal behaviour.* Crimes

have definite characteristics and reveal differences and similarities which depend on the criminal, his neighbourhood, the time of day, week, or year, and many other factors.

NEW YORK STATE POLICE COMMUNICATION

It often used to take the old manual system four hours to 'flash' a general alarm to all New York State Police. The messages themselves were as likely as the criminals to get away and disappear without trace, for there was neither control of the messages nor guarantee of delivery. The files were cluttered with incomplete and duplicate records – or worse still, nearly-but-not-quite-duplicate records.

TABLE 5.1

Police Data Files

File

Crime-Related Applications

Criminal-offense file (type of crime, where, when, offender, *modus operandi* factors, etc.)
Criminal-arrest file (who, where, when, disposition, physical characteristics, *modus operandi*)
Juvenile criminal activity
Warrant file
Stolen property
'Hot-desk' applications (wanted persons, stolen autos, circumstances, participants, etc.)
Traffic citations (offender, disposition)
Parking violations (offender, disposition)

Police Operations

Police-service analysis (location, type of run, car, time, etc.)
Beat boundaries and patrol distribution (offenses, locations, day and time variations, data, etc.)
Communication switching
Jail arrests (who, cell number, disposition)

Location file (street addresses and city and patrol boundaries)
Auto registration (license number vs. owner name and address)

Police Administration

Personnel data (name, serial number, training, experience, etc.)
Inventory control file (fixed assets, location, description)
Vehicle-fleet maintenance accounting
Financial accounting
Budget analysis and forecasting

Uses

Statistical reports and analyses
Follow-up control, analyses, and special searches
Statistical reports and special studies
Wanted lists and special searches
'Hot lists'
On-line queries for special and timely data
Statistical reports, traffic studies

Traffic-court docket preparation, citation accounting
Mailing and follow-up for delinquents, parking-ticket accounting
Operational reports, planning data

Establishing patrol patterns and boundaries

On-line message switching and control, communications with other computing systems

Prisoner control
Dispatching
Special searches

In February 1966 the New York State Police installed a UNIVAC 418 computer with 134 terminals throughout the state. Messages could be sent from any one of these to each or all the others in an average of three minutes. In 1969 and 1970, New York City installed its IBM 'Sprint' system to facilitate the fastest possible police action in the crime-plagued metropolis.

A call for information might start from a prowl car with a radioed request to divisional headquarters. Here the details would be typed in, starting with a 'routing' instructing the computer where to send the requested information. The computer adds date, time, and – most important – a serial number so that no transaction entering the network can be forgotten forever. It then stores a copy and causes the recipient's typewriters to print the message. It is quite possible that the information the patrolman needs can be supplied from the computer's files, which hold descriptions of stolen, hit-and-run, and hold-up vehicles; guns; missing persons; assaults and homicides; robberies; and data from other criminal investigations.

SIMPLE PATTERNS

It is very easy for the system to recognize some 'patterns': those that point uniquely to an individual or object. Automobiles and guns have serial numbers and criminals may well be known to the police by a unique number.

Like many American cities, New Orleans has been using a computer (borrowed from the city Finance Department), to keep accurate and current information on traffic tickets. The prompt and impartial collection has resulted in lower costs. In New York one of the first victims tracked down by the computer was fined $370 for ignoring 27 tickets in the previous two years (his profession: computer engineer!). In Paris the police use punched cards to record parking violations. After a card is placed under a windshield wiper, a duplicate recording time, date, and licence number is fed to the computer for filing. The offender returns the card he was given, with the fine, and the file is searched and the record deleted. Later, those records still on file after the fine is due are displayed in an 'exception' report. The registration number is found in the record automatically by the computer, which searches the file of car-owners and sends a peremptory request for immediate action and further payment. This technique is typical of the way in which attention can be devoted to the few exceptions, here non-payers, while the orderly majority are dealt with cheaply and easily.

Data preparation by punch operators is a very expensive way of feeding a computer. In London an optical reading device that can detect marks and read printed characters on documents at a rate of 300 documents a minute is, we believe, already in operation to cope with the 2 million yearly offences to which a fixed penalty applies. These include parking violations and failure to display excise licences. Necessary letters to vehicle-registration authorities, vehicle-owners, and local police are produced by the computer if payment is not made in time.

FILE-SEARCHING

The file-searching methods needed to follow up the exceptions are equally applicable to firearms or people or anything else with the unique identifier. The most straightforward method of filing is to associate with each serial number – of, say, a firearm – the owner's name, or a series of cross-reference numbers from, perhaps, the identifier of a firearm to the identifier of the owner, to the identifier of a crime involving the gun.

The computer program would read the serial number of the gun sent from an inquiry terminal, and print out on the inquiring teleprinter all the details without noticeable delay at the receiving end. With this speed of response it is realistic to station prowl cars at each end of Brooklyn Bridge, and radio to the computer the registration numbers of automobiles about to cross. The computer checks for traffic-infringement history, and flashes results to the other end of the bridge before the car gets there so that the driver (if he is a delinquent violator) can be apprehended.

COMPLEX-PATTERN RECOGNITION

A serial number is a simple pattern which can be positively recognized. Identifying fingerprints, the correct names of those who use aliases, and individuals from verbal description is much harder.

Consider the problem of locating in a telephone directory a man who pronounced his name 'zhān vēa'. This would be a Frenchman who spelled his name 'Jean Wieille'. A Brooklyn policeman might note the spelling and then enquire for 'Jen Welly'! With the great mixture of countries of origin in the American melting pot, the link between pronunciation and spelling of proper names is tenuous.

At the International Police Exhibit in Hanover, Germany (27 August–11 September 1966) IBM demonstrated the retrieval of a name using a phonetic-coding technique. A teleprocessing terminal transmits last name, first name, and date of birth. The last name is converted to a phonetic code, the file searched, and each phonetic duplicate checked for the identified first name and date of birth. If a match occurs, a record number is printed and this can be used to retrieve further details. For instance, in Austria all the following names are similar in pronunciation:

Maia, Maier, Mair, Maja, Majr, Maya, Mayer, Mayr, Meia, Meier, Meir, Meja, Mejer, Mejr, Meya, Meyer.

By coding *ai*, *ei*, *ay*, *ey*, *aj*, and *ej* into '*eo*', and coding *ar*, *er*, *a*, and *r* into '4', all the names become '*meo*4'. Using this system, the Austrian Ministry of the Interior has set up a file in which all names starting with the same sound are found in the same group; given the pronunciation they can rapidly find all they have about a named individual.

Many people are far better known by nicknames than by their given names, and criminals frequently resort to aliases. In Oakland, California, the police files include juvenile-nick-name cards with true names and physical characteristics. Similar files can be set up for aliases and individuals thereby identified for their records to be retrieved.

FINGERPRINTS

A spoken nickname overheard by a witness is rarer than a fingerprint left behind. Fingerprints were among the matters investigated by the 'Task Force' (the Science and Technology Task Force which reported in 1967 to the President's Commission on Law Enforcement and Administration of Justice).[1]

The problems of fingerprint identification were described as follows:

Effective police work uses fingerprint identifications both to apprehend those who leave 'latent' prints at the scene of a crime and to identify positively persons held in custody.

Positive identification of persons already held makes use of files set up according to a 10-print classification system, since all 10 prints can be obtained from such persons. Manual techniques of 10-print classification and search have been used for more than 50 years. The major limitation in their use is the time it takes to search the files due to the large number of prints in modern files – the FBI file now contains over 16 million sets of different criminal prints plus about 62 million sets of different civil service and military prints. The classification and search problem is compounded by the large volume of prints which must be processed – each day the FBI receives about 30,000 sets for processing, of which approximately 10,000 are based on arrest. Advances are needed to increase the workload capacity and to reduce the costs and time delays of the fingerprint classification and search processes.

In order to search the regular fingerprint files using the classification formula now in use, a full set of 10 fingerprints is needed. When a criminal inadvertently leaves fingerprints at a crime, only one or a few fingerprints are usually available to law-enforcement officers. Once a suspect has been taken into custody, his fingerprints can be compared with even a single print recovered from the scene of the crime. By the same token, a single print can be matched against complete prints of a short list of likely suspects. But the process is now entirely manual and so time-consuming that it cannot be used to check less than a full set of prints against a national file or even a substantial local file of previous offenders. As a result, single-print files tend to be very small, generally containing only a few thousand prints (compared to millions in the larger 10-print files), and are very infrequently used. Most large police departments maintain a specially organized file of single fingerprints of several thousand persistent criminals. Probably more than 100 different manual systems are in use today for searching files of single fingerprints of persons who have been judged likely to violate the law persistently.

It will be a great step forward when fingerprints are stored, retrieved, and matched by computers. The primary problem here is in classification, not in retrieving matching prints once they are classified. There is a lot of information in a fingerprint, and the essence of it has to be reduced to a string of numbers. The human eye is good at noticing interesting minutiae like islands and ridge endings and bifurcations, but not so good at measuring and recording. In a semi-automatic procedure, an operator might sit at a console on which the print to be classified would be displayed. Using a light pen, he would point to each interesting detail and identify it. The computer would record each such point position relative to a grid on the screen, an easy measurement for a machine. It would then use a mathematical analysis to determine the shape of the whorls and so on, and count the ridges to determine separation. Using these data, it would generate a model print for comparison with the original, displaying them both to the operator for examination using a suitable comparing viewer. If the computer cannot recreate the print, then it has not got enough information to identify it uniquely.

By 1975 most adult Americans' prints will be available to the FBI and ten-print screening of a suspect will be quick and efficient. Latent prints will start to contribute to the quicker apprehension of the incompetent burglar.

VOICE PRINTS

Recent studies of voice analysis and synthesis, originally motivated by problems of efficient telephone transmission, have led to the development of audio-frequency profile or 'voice print.' Each voice print may be sufficiently unique to permit developing a unique classification system and perhaps even to make positive identification of the source of a voice print. This method of identification, using an expert to identify the voice patterns, has been introduced in more than 40 cases by 15 different police departments. As with all identification systems that rely on experts to perform the identification, controlled laboratory tests are needed to establish with care the relative frequency of errors of omission and commission by the experts.[2]

At present, there is no conclusive evidence to support manufacturer's claims that the voice prints recorded by their equipment identify an individual almost as uniquely as do fingerprints.

The first reliable 'sound spectrographs' were developed in 1940 at the Bell Telephone Laboratories. In 1962 a member of these laboratories claimed that a person's voice is as unique as his fingerprints. However, it is uncertain whether this uniqueness survives the changes in voice that occur as a person grows older. It is possible for people to disguise their voices and mimic others. Furthermore, a spectrogram used in court evidence may have been recorded by telephone tapping, which is liable to distort the sound.

The US Department of Justice has had tests made at Michigan State University in the course of which 'investigators verified the voice print concept',[3] according to a report in a publication of the Association for Computing Machinery. A correspondent a few months later referred to studies by the Technical Committee on Speech Communication of the Acoustical Society of America which concluded 'that the available results are inadequate to establish the reliability of voice identification by spectrograms'.[4]

IDENTIFYING MODUS OPERANDI

The 'signature' of a criminal can be detected in the pattern of criminal activity. In Italy the General Management of Police catalogues crimes on a nationwide scale. The manner and time of the organization and execution of a series of offences can point to a particular person. *Modus operandi* searches are now automated and computer lists supplement intuition and human memory.

In Britain the techniques are becoming even more sophisticated. The following appeared in *The Times*:

A Home Office research group is working on a scheme to help crime-prevention squads to get on the track of the men behind highly organized criminal operations more quickly.

The scheme, known as Supertec, has become possible through the work of computer specialists at the National Physical Laboratory who are using machines for translating Russian into English.

The idea behind the automated detective is to catch the big criminal who is never on the scene of a crime and has no known record. Many men escape for that reason. Police estimate that more than half the 100 men who would rate as the underworld's 'tycoons' avoid detection because they have a clean record.

The machines will collect many items of seemingly trivial information until, through a process of logical deduction, a pattern of behaviour that narrows the field of investigation is found. Although such work is ideal for a computer in theory, the job of building a data bank in a machine and devising a scheme for retrieving the information is very complex.

Scientists at the National Physical Laboratory were up against this difficulty in their Russian–English research. Early developments in automatic translation started with the compilation of a dictionary for both languages which could be 'read' by machine. By direct substitution of words, this approach worked for languages with similar roots. But the method foundered when two very different languages were involved; translation could then be achieved only by detailed linguistic analysis.

Even by machine this called for a very advanced method of computer programming and an expensive computer able to store millions of items of information. The process of translating by detailed analysis involves logical deduction from thousands of items of descriptive detail. These are filed away in various parts of the electronic system.

The automated detective will be making logical deductions by piecing together scraps of information gathered over months of painstaking investigation. Such a fact as a suspect having lunch with a man with no criminal record may lead to the build-up of a picture that no detective could hope to piece together on his own.

This approach to crime prevention is now possible because of other developments in computer technology. A system such as the one the Home Office is examining is practicable only if it is available to many investigators. This means using the new ideas of a typewriter and a telephone line.

Another important facet is the means of keeping secret the information filed in a computer.[5]

The information retrieved from police files can never be better than that which is put in. If the files are centralized, the data they contain may be kept more accurately and continuously updated by *contact reporting*: every time an individual is in contact with the law enforcement agency, his particulars are gathered, reported, checked against any previous record, and retained. Studies are underway to allow policemen's log books to be indexed automatically. Robbery reports are converted word for word to magnetic tape, and stored with a concordance on an index tape. Investigators, *modus operandi* experts, or patrol supervisors can enter plain-English statements like 'liquor store, ask for pack of cigarettes'. The computer hunts through its index for all these words, and prints out automatically prepared abstracts of those in which several key words appear.[6]

At any time subsequent to 'processing' a juvenile, the police in Oakland can determine from their files – besides name, sex, and address – a suspect's nicknames, school, grade, age, height, weight, hair colour, complexion, race, tattoos, scars, place of birth, parentage, maternal and paternal figure, and associates along with the details of any offence and subsquent court action.

A hint to the use that can be made of this material is SIMBAD, (SIMulation as a Basis for Social Agents' Decision). At the University of South Carolina's Youth Studies Center, a file of delinquent records has been built up. It contains, along with the juveniles' characteristics, information on treatment and success in each case. From the file, a model is being constructed; when this is fed with material on a new case, it will show the historical probability of the success of various types of treatment, and also indicate the need for further research into the background of the individual if it suspects the absence of vital facts.

Similar techniques are applicable to adult offenders, motor accidents, and so on. All can be analysed by the computer in search of a pattern which will point to the causes of crimes or accidents and hence to their prevention and cure. Time of day, week, and year and the weather all affect criminal behaviour.

From gathered data, trends in types of crime associated with, though not necessarily *due* to, varying demographic, geographic, and temporal factors can be determined, and corrective action initiated early. Patrols can be strengthened and the type of personnel changed to increase efficiency.

Such systems naturally provide the police administrators with operations research data and comparative policeman efficiency as well. Vacation plans and patrols can be made to suit local 'events', such as parades, well in advance.

The raw data can be used with a simulation model to investigate the probable consequences of changes in patrol activity, beats, and types of vehicle. Besides avoiding the long time it would take to ascertain the effects of a real-life change in, say, patrol routes, such models avoid risks to the public had the alteration to the force itself been unsuccessful.

To make optimal use of a police force, men need to be assigned where they are most effective. Unfortunately, this is never easy to estimate.

The Task Force undertook a preliminary analysis based on limited data contained in the 'Statistical Digests' of the Los Angeles Police Department from 1955 to 1965. The statistical technique of *regression analysis* was used to relate the number of reported serious crimes in each of the department's divisions to the number of patrol officers assigned to the division.

The manpower allocation for 1963 in the department's four divisions was examined and a study made to determine if a different allocation might have resulted in an overall decrease in reported serious crimes. In 1963 there were 824 police officers assigned to the four patrol divisions. The approach was to determine a reallocation among the divisions which might reduce the predicted number of crimes. To avoid major perturbations, no reductions of more than 10 per cent in the number of officers per division were considered. The analysis suggested that a shift of officers might have led to a net decrease in reported crimes for the four divisions, primarily due to what appeared to be a particularly high officer effectiveness in one of the four divisions.

In St Louis four of the city's nineteen police districts have their patrols assigned by reference to a crime-density map prepared every eight hours from crime-location data fed straight to the computer centre. The set of maps shows contours of crime density or intensity and is supported by tables showing predicted and actual calls and response times. Manpower allocation is now a matter of hours of work, not months, and the cars can handle nearly twice the volume of work they used to.

THE DATA BANKS

Rapid communications and data-gathering come together at the 'inquiry file', or bank of immediately accessible data. In Chicago the police hold descriptions of 100,000 persons: wanted men and missing women, suspended drivers, and suspect criminals are 'retrievable' phonetically, by alias, by description, or simply by name. All day, every day, the computer can respond instantly, and there are nearly half a million inquiries a year.

Distance is no barrier to inquiry. In Oakland 300,000 names and 25,000 vehicles are recorded. Nine thousand miles away in IBM's London office, in a demonstration of the system, visitors inserted names and got back information in a few seconds. With 140 terminals in California, 93 police agencies put in thousands of inquiries each day.

California police also have access to Autostatis (AUTOmatic STATe-Wide Autothief Inquiry System), which records stolen cars, plates, parts, and vehicles involved in crimes. Many patrolmen owe their lives to it. In a 1965 incident an officer spotted a speeding automobile and radioed a description to his dispatcher, who typed the licence number and description into his communications terminal. The Highway Patrol's computer scanned thousands of 'marked cars' reported by local and neighbouring forces, and discovered that it had been used in a supermarket hold-up a few minutes earlier in another county. Within a minute, the highway patrolman was warned not to stop the getaway car, but to keep it in sight while roadblocks were set up. Two gunmen were crouching in the back with guns drawn when the car came to a halt, but they had no opportunity to use them against the prepared and superior force.

The Autostatis system was serving California, Arizona, and Nevada, handling 12,000 queries a day through 229 local terminals, as of the summer of 1968. As well as its inquiry facility, it also provides a 'car sheet' detailing the latest 270 car thefts reported by local officers. As a result, 350 local police departments have eliminated stolen-vehicle files, saving hundreds of thousands of dollars on clerical costs alone. The network is tied into California's Department of Justice network, other states' networks, the FBI, and the State's stolen-property records, by CLETS, the Californian Law Enforcement Telecommunications System. This high-speed message-switching network allows more than 900 teletypewriters to transmit and receive messages from each other, to broadcast to many stations at once, or inquire into the data banks.

Developments on the East Coast are no less ambitious. The State of New York has taken many effective measures in the war against crime – among the more recent the establishment of the New York State Identification and Intelligence System (NYSIIS). The concept of this system has its genesis in the investigation of the notorious 'Appalachian meeting' in November 1957 of more than 75 of the nation's top criminals. Even a summary account of this investigation dramatizes the nature of the information problem the system was conceived to solve.

Following the discovery that a convention of top criminals had taken place at Appalachia, the New York State Commission of Investigation was formed in 1958 to undertake an all-out inquiry into the meeting – to identify those attending and to determine its purpose. The meeting's participants, of course, remained silent. Thus the mechanical task of assembling information on these major criminals proved to be formidable; just one of them was the subject of as many as 200 separate official police files from a surrounding area several hundred miles wide. Two years later all of the State's files on these criminals still had not been accumulated or assimilated. Those files that were examined often were barren of original material – a relatively meaningless collection of newspaper clippings, copies of other files, and loose notes. The distillation of valuable content was minute.

As Chief Counsel of the Commission, Eliot H. Lumbard was ultimately concerned with that investigation and its frustrations. When he was appointed to Governor Rockefeller's staff in 1961 as Special Assistant Counsel for Law Enforcement, among the assignments he undertook was a search for possible remedies to the problems presented by the Appalachian investigation, as well as by the deluge of papers and files inundating operational agencies concerned with the administration of criminal justice. Out of his inquiries into the possible application of electronic data-processing techniques to criminal justice came the New York State Identification and Intelligence System project, which has as its long-range goal a computer-based central facility, located at the capital (Albany), that will store all the information pooled in the system. Qualified agencies concerned with the administration of criminal justice, located anywhere in the State, may contribute and have access to this information centre on a voluntary basis via a communication network.

The system's approach is based on the concept of information-sharing among all agencies involved in administrating criminal justice; these agencies fall into six broad functional categories – police, prosecutors, criminal courts, probation, correction, and parole. The respective agencies are necessarily independent and autonomous in their operations, yet none of these agencies provides a service that is a totality of and by itself: none functions alone. Each of these agencies has a part in a continuous operation: one subject and his accompanying file, for example, go successively through these respective agencies until the ultimate disposition of the particular case. With rare exceptions, none of these agencies, except the police, may take official action unless it concerns a matter in which another agency has previously acted; and then it is usually concerned with the preceding agency's 'product'. Each agency needs to have the same information, with some variation, about each subject, yet each is currently served by separate information sources.

In summary, the ultimate system as envisaged would provide four major new capabilities. First, suspects from files of known criminals can be identified to a degree and at a speed hitherto impossible. Second, outstanding open cases can be solved by automatic and comprehensive searches conducted on the basis of statewide information. Third, automatic abstracting, indexing, and retrieval of textual intelligence and *modus operandi* information can be used to conduct a comprehensive search of all documents stored in the system and a retrieval of only those documents that are relevant to the interests of the user, instead of the current time-consuming index-card files and bulky folder files.

Finally, the system offers the tremendous potential of being able to perform research: a pattern analysis of data to assist in evaluating new trends in crime, in testing new approaches to the administration of criminal justice, and in discovering patterns of structure and activity of criminal and special organizations in order to discover relationships in vast bodies of facts that would be almost impossible to develop through standard manual techniques of file searching.[7]

THE FBI NATIONAL CRIME-INFORMATION CENTER

Why stop at a statewide system? The International Association of Chiefs of Police saw no reason to do so, and so there was by 1971 in Washington a file of 25 million active records. Terminals in many states – and in some cases, local police computers – make tens of thousands of inquiries each day over as many miles of leased lines.

The system was planned in early 1966 and was implemented, with astonishing speed, by January 1967. It has required the standardization of records in many forces and much cooperation between large and small agencies. Every state, Canada, and a few major metropolises have direct connections to the centre, through which all questions must pass. The California

Highway Patrol's IBM 7740 was first to be coupled to the FBI's twin IBM System/360 Model 40s in April 1967 and provided 270 local terminals.

In the future the records can be expected to expand – more details of more criminals and suspects; links to other nations; higher-speed transmission lines to make the service available 'on the beat', not just through regional centres; graphic data and facsimile transmission, perhaps even into prowl cars. The data in the system all come from local agencies who are responsible for maintaining accuracy and purging irrelevant and out-of-date records.

The Department of Justice now has six other computer data banks besides NCIC. The FBI maintains a file of known and professional cheque-passers. There are Immigration and Naturalization files, Federal prison records, Bureau of Narcotics and Dangerous Drugs narcotics-users file, an organized-crime intelligence system, and a civil disturbance system. There is also a National Drivers Register listing 2·2 million people whose licences have been withdrawn in any State.

The Task Force Report analysed the cost of a centralized system for stolen vehicles and identifiable property, guns, and missing persons, and compared it with the total cost of local state systems linked together.[8] Automobiles constitute a somewhat different category in that 45 per cent are recovered within twenty-four hours, 70 per cent within the jurisdiction of the theft and nearly all the rest in neighbouring jurisdictions. Also, licence plates are specific to a state. The major costs are computation, computer storage, and communications.

Some three quarters of a million automobiles were stolen in the US in 1970 and 90 per cent of them were ultimately recovered. About half a million stolen cars were on file in 1970, with 2,000 being added and 1,800 removed daily. The national file was thus made up of 50 million characters – small by today's computer standards.

One hundred and twenty thousand wanted persons filled a similar file. Only one-third of identifiable stolen property is ever recovered, so the file would grow without end at a rate of two-thirds of 150,000 entries per annum unless purged of old records. Stolen guns presently number 100,000 annually. Another file of vehicles, property, guns, and persons would occupy half of a $5,000-per-month disk file and it would cost $8,000 a month for communication lines. The whole system would cost $1.5 million a year across the nation, most of it going into computers operating at local and central points. This cost is about the same as maintaining fifteen prowl cars and crews. A decentralized system would cost ten times as much. Nationwide the patrol function costs $800 million a year.

Police Computers in Britain

Britain is faced with an altogether different level of criminal activity from the United States. The two million inhabitants of Philadelphia, 'the city of brotherly love', commit more murders than the 52 million British. Most of these are committed with guns, including nearly all the hundreds of murders of policemen in the last decade. Lesser crimes are proportionately fewer in Britain too.

The British national police computer system is smaller than most American State systems. The policeman on the beat can radio in to his local control room to make enquiries and get answers in a minute or two, or sometimes a few seconds. A thousand teletypes and visual display units with trained operators are attached by data links to two Burroughs 6700 computers at Hendon. The files will eventually contain details of: stolen vehicles; fingerprints; names of criminals and wanted and missing persons; disqualified drivers; vehicle owners (provided by the motor vehicle licence system at Swansea); suspended sentences; stolen cheques; stolen property; cycle frame numbers; and modus operandi.

The criminal records office will be adjacent to the computer, but will be maintained manually with the computer indexing it through the CRO number in the criminal name file.

The fingerprint index, rather than the fingerprints themselves, is held in the computer. Research is going on leading to a print transmission service and into improved coding techniques. The old Henry system has been superseded by something less crude. Loading up past prints is expected to take several years.

The vehicle records can be searched on several factors to produce lists of suspect vehicle owners. The facility would have helped greatly in the search for the 'grey Austin Cambridge' in the Cannock Chase murder. The record also links chassis, engine and registration numbers to inhibit ringers who assemble 'new' cars from the component parts of stolen vehicles.

The system is the first of its kind to operate on a national basis anywhere in the world. Locally, the British police are experimenting with 'command and control' in Birmingham. A convicted criminal data bank containing details of habits, appearance, associates and so on is being tried in Slough, based on the 'Supertec' described above (p. 113).

LEGAL AIDS

When the police and their computers have brought home the villains, there remains the problem of convicting them. This can be a remarkably slow process. One reason for this in America is the complexity of the legal system.

A large number of civil suits reach the United States courts. The ease with which Americans resort to law is a source of continual astonishment to foreigners. One result of this indulgence is a flood of legal precedents which may or may not be significant to an individual case, and which make achieving a rapid and just settlement difficult.

The computer is an ideal tool for assisting lawyers in their search for precedents. The lawyer of the future may well use the terminal in his office to instruct distant computers to carry out a search for him and to display legal information. Setting up the necessary files for such a system, however, will take a vast amount of work and time.

It has been suggested that such procedures might be used by the court to insure that all relevant facts and precedents can be recalled by judge and jury. Although this could be no substitute where analysis, argument, and insight are required, it could assist in reaching wise and just decisions.

It may also be possible to analyse statutes for anomalies and inconsistencies, and to codify and consolidate wide areas in which legislation has marched ahead of rationalization.

Many attacks have been made on the problem of using computers in the search for precedents. In the US Air Force system, LITE (Legal Information Through Electronics), an abstract is automatically formed by recording the first sentence of a report, the last, and ten others. The intermediate sentences are chosen with respect to the presence of the largest number of highly significant words. Of these words, ten are chosen for the index by which the abstract, and hence the stored, text can be retrieved. If the lawyer finds, on retrieving the record, that the material is relevant, he is then directed to the original.

Law Research Service, Inc. has installed a central computer in New York and provides each subscriber with a teleprinter and coded citation directory based on important words and phrases. The researcher indicates the category of his case to the computer and in a few minutes the computer sends back the codes to look up in his directory. Preparing the directories for a million cases in New York alone occupied fifty lawyers for three years. To handle nationwide and ultimately international law by this means would require a massive force.

There are a variety of other such schemes. At present there are technical and ethical problems associated with them, but they indicate the direction of the future. Eventually the indexing and searching methods will become standardized and the law of a country will be stored in its machines.

In Britain the only significant legal information system is used to search and retrieve the full text of Acts of Parliament and Statutory Instruments relating to Atomic Energy. They total only 770 sections and 140,000 words, a drop in the ocean of 20 million words of English Statute Law. A search language has been designed for use by any intelligent person, and the system written in a high level language for easy transfer to machines other than the original KDF9 owned by the Atomic Energy Authority.

In the nearer future there is a danger that the small independent lawyer may not have the access to the computers that a large legal firm can afford. Like his counterparts in other professions, he will be at a loss if he does not have access to the new machines or the capability to understand them.

LEGAL DELAYS

Justice delayed is tantamount to injustice. A felony charge takes very little court time. The court time spent on a defendant who pleads guilty (approximately one-half of the felony defendants) probably totals less than one hour, yet in the District of Columbia the median time from initial appearance to disposition is four months.[9]

In January 1964, Pittsburgh had a backlog of 8,000 cases[10] which could be expected to take from two to five years before reaching trial. Data-processing was used to determine the causes of delay and in two and a half years the backlog was reduced to 5,000. A principal reason for delay was the shortage of trial lawyers and the clash of their cases, which kept causing postponements until all the protagonists were available. Now the trials are planned from punched cards, and the courts are run much more efficiently.

The Task Force built a computer model of the court system, which provides an interesting illustration of the use of simulation. Figure 5.2 illustrates the model. The computer imitates the real world by feeding hypothetical cases into this model, rather like a game of Monopoly. A suspect is arrested and

Fig. 5.2 A computer model of the US court system used for investigating the causes of delay. It shows the steps that defendants go through between arrest and sentencing. The model was manipulated in a computer to discover the best methods for streamlining the operation of the court system. *Reproduced from 'Task Force Report: Science and Technology', a report to the President's Commission on Law Enforcement and administration. Prepared by the Institute of Defense Analysis.*

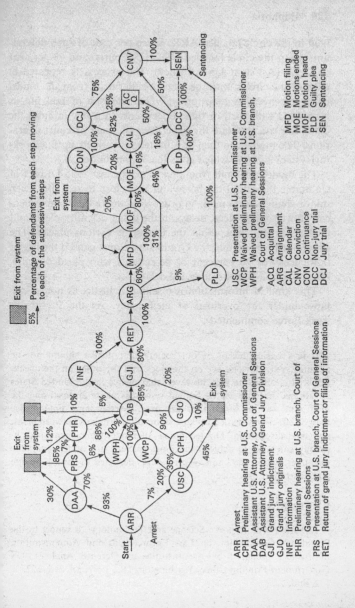

Exit from system

Percentage of defendants from each step moving to each of the successive steps

Exit from system

5%

ARR — Arrest
CPH — Preliminary hearing at U.S. Commissioner
DAA — Assistant U.S. Attorney, Court of General Sessions
DAB — Assistant U.S. Attorney, Grand Jury Division
GJI — Grand jury indictment
GJO — Grand jury originals
INF — Information
PHR — Preliminary hearing at U.S. branch, Court of General Sessions
PRS — Presentation at U.S. branch, Court of General Sessions
RET — Return of grand jury indictment or filing of information

USC — Presentation at U.S. Commissioner
WCP — Waived preliminary hearing at U.S. Commissioner
WPH — Waived preliminary hearing at U.S. branch, Court of General Sessions

ACQ — Acquittal
ARG — Arraignment
CAL — Calendar
CNV — Conviction
CON — Continuance
DCC — Non-jury trial
DCJ — Jury trial

MFD — Motion filing
MOE — Motions ended
MOF — Motion heard
PLD — Guilty plea
SEN — Sentencing

held in the circle labelled ARR. Seven per cent of such defendants go to the circle labelled USC (US Courts) and 93 per cent to DAA (Assistant US Attorney). The process is repeated at each of the circles until the suspect either exits from the model or is sentenced. A time interval, characteristic of real life, is associated with each of these operations. In a few minutes the computer can simulate the experience of years of court operations. The model can then be adjusted repetitively until a working system is found which gives the minimum delays.

The model was built from and tested against 1965 District of Columbia data until the behaviour was close to real life. Then minor changes were made in order to evaluate their consequences: fewer guilty pleas, delays in the entry of guilty pleas, and changes in the rules about the timing of various phases. The simulator revealed that the Grand Jury Court caused the major bottleneck for 70 per cent of defendants who pleaded guilty. Awaiting trial was the other big delay.

This kind of information is essential if justice is not to collapse under the overload of inefficiency. As the President's Task Force commented:

It is a basic precept of our society that justice should not be administered with one eye on the clock and the other on the checkbook. It is often the fact, however, that justice in the United States is rationed because of the limited resources at its disposal and the inefficient way in which they are used. At the same time justice may be effectively denied because of inordinate delays between arrest and final disposition. The techniques of modern management technology can help to achieve the most efficient use of the available resources, within the limits of procedures designed to ensure the due administration of justice.[11]

REFERENCES

1. 'Task Force Report: Science and Technology', a report to the President's Commission on Law Enforcement and Administration of Justice, prepared by the Institute of Defense Analysis, US Government Printing Office, Washington, DC, 1967, pp. 218, 222.
2. ibid.

3. *Communications of the Association for Computing Machinery*, 14, 6 June 1971.

4. ibid, 11 November 1971, quoting *Journal of Acoustical Society of America*, 47 (1970), pp. 597–612.

5. *The Times*, 20 May 1968.

6. *Proceedings of the American Society for Public Administration*, 1964.

7. 'The Computer and Invasion of Privacy.' Hearings before a subcommittee of the Committee on Government Operations. House of Representatives. 89th Congress, second session. 26, 27, and 28 July 1966. US Government Printing Office, Washington DC, 1966, pp. 159ff.

8. 'Task Force Report', loc. cit.

9. ibid.

10. Hon. Henry Ellenbogen, 'A Twentieth Century Approach to Judicial Administration,' *FBI Law Enforcement Bulletin*, May 1966, p. 7.

11. 'Task Force Report', loc. cit.

6

TEACHING WITH COMPUTERS

We have to bring the entire human race, without exception, up to the level of semiliteracy of the average college graduate. This represents what may be called the minimum survival level; only if we reach it will we have a sporting chance of seeing the year 2200. – Arthur C. Clarke

Today's education is an inadequate preparation for tomorrow's computerized society. It is no longer possible to wind the mainspring of the mind in childhood and allow it to run down for the next half century. The computer provides both the requirement and the means for continuing learning into old age.

Education will no longer end when one leaves school or takes one's last degree but will go on throughout life in order to enable the individual to keep in tune with his constantly changing environment. Workers will need to be retrained for jobs that change not once, but several times in a lifetime. Professionals will have new facts and techniques available to them, and will continually be under pressure to learn because of the rapid obsolescence of their existing knowledge. The 'man in the street' will find his world constantly slipping into new forms like sands as the tide rushes in; but he will also have more leisure time, and learning will be a way to absorb the newly gained hours. To a major extent, the key to success in this new world is improved and continued education.

The teaching profession, however – even with today's inadequate level of education – is short-staffed. In the US, it has been estimated that there is an annual shortage of 8,400 elementary- and secondary-school teachers and of 90,000 college teachers with doctors' degrees. The underdeveloped countries are much worse off. Part of this problem will be solved by the computers themselves being programmed to teach.

COMPUTER-ASSISTED INSTRUCTION

The letters CAI are now well known in the computer world and stand for 'computer-assisted instruction'. 'Computer-*assisted*' implies that the machine does not replace the human teacher but supplements his efforts. As in other areas, it is the combination of human and machine capabilities that provides the optimum rather than either of these alone.

Computer-assisted instruction made its debut in elementary schools in 1966. A hundred small children entering the first grade in Brentwood School in East Palo Alto, California, found themselves confronted with computer terminals with screens and light pens. The machines were used to assist in teaching reading and arithmetic. Brentwood was an interesting choice of school because it is in a slum area, had an 80 per cent Negro enrolment, and the average IQ of the children participating in the reading experiment was 89.[1] The technique was very effective. The children loved playing with the terminals and their teachers had to 'peel them off the machines' to get them back to their other lessons.

The pupil, when being taught by a computer, carries on a two-way exchange at one of its terminals. A wide variety of different terminals has been used for this purpose. With a typical terminal a pupil can receive instruction from two screens, one of which shows coloured pictures selected from a reel of film inside the device. The other screen is a standard computer screen displaying writing, numbers, and simple drawings. The pupil also has a pair of headphones by means of which the computer can 'speak' to him, if so programmed. The pupil must be able to answer back: two-way communication is the key to success in computer-assisted instruction. To do this he can use either the typewriter-like keyboard, or a light pen with which he points to boxes or items on the screen.

Computer-assisted instruction is proving its potential at various levels of education, ranging from very small children to graduate students and professional men. The key to its success, as in other uses of computers, lies in the skill with which programs are written. Whereas in many other areas moderate programming competence will suffice, here considerable talent in using the new medium is needed. A poorly written teaching program can be tedious and exasperating, whereas one produced with an understanding of the subtle psychology of the medium can convey information quickly and insure a high degree of retention.

SUBJECTS SUITED TO CAI

Some subjects lend themselves well to computerized teaching and others do not. Subjects involving large amounts of routine detail or facts are natural for computers, as are subjects with elaborate but standardized logical procedures. CAI is ideal for teaching spelling, simple mathematical techniques, the mechanics of foreign languages, statistics, computer programming, electronics, and so on. It would have more difficulty teaching philosophy, carpentry, basic principles of calculus, or appreciation of music, although even here it could assist a human teacher. In teaching a student to write good English, draw, or play golf, the computer is hardly worthy of consideration. We were tempted to write that the computer would be unlikely to be used to teach a pupil to sing, but then read of an IBM 1620 at Stanford University which is doing just that! This computer prints out a series of notes and the student sings these to the beat of a metronome. The computer is programmed to compare the student's pitch with the true pitch, and decide whether the student must repeat the exercise or go forward to other material. The *Journal of Research in Musical Education* reported very favourable results, as the machine has a very precise sense of pitch.

The point is that, because the computer is ideally suited to certain types of teaching but not to all teaching, it will never *replace* the human teacher; thus the teaching profession must recognize both the usefulness and the limitations of computers in education. No doubt as time and technology advance the versatility of computer-teaching methods will increase. Today, however, in schoolteaching, as in the work of other professionals, there are vast areas where the computer can help greatly.

Given a suitable subject and the requisite skill in the programming, computer teaching can have significant advantages over the conventional classroom. Teaching in a classroom must be instructor-centred. The students have to proceed at the speed and level of complexity of the instructor. The brighter students, finding the pace too slow, are bored. Those not-so-bright, on the other hand, often become lost or fail to understand part of what is said. Any student, bright or dull, can miss sections through lapses in attention. With CAI, the process is pupil-centred, not instructor-centred and the machine adapts its pace to that of the student. The dull student can ask for endless repetitions without embarrassment and the machine will retrace its steps with infinite patience. The quick student or the student who already partially knows the material can skip a segment – with the machine questioning him to check that he does, in fact, know it. A student feeling that he needs more practice in a certain area can obtain it.

With a well-produced CAI program, the student is unlikely to become bored or distracted. The machine will be programmed to elicit frequent responses from him, and grade the difficulty of its demands according to the ability he displays. If he is a long time replying to a question, the program will send messages to prod him. The student cannot race ahead without a true understanding of the material because the program constantly 'examines' him to check that he is understanding. What the machine lacks in human intuition it makes up for in its tireless ability to provide a near-optimum response to almost every situation in the learning process. These responses have been carefully planned, with special regard to the psychology of learning, at the time a particular program is written. Furthermore, they can be easily revised as experience reveals deficiencies.

THE PSYCHOLOGY OF LEARNING

When the student answers the teaching computer's question correctly, he receives immediate *reinforcement* of that correct answer, since the program advances to a new 'frame' as the student hoped, and frequently issues a congratulatory message. Some much-publicized work by B. F. Skinner[2] of Harvard demonstrated the importance of such fast reinforcement of correct responses. Skinner initially studied learning using pigeons as subjects. He found that the key to teaching pigeons was to reinforce correct responses *as quickly as possible*, in this case with food. The speed of the reinforcement was found to be very important. He was able to teach the pigeons quite complex sequences of actions – for example, to play a modified form of ping pong. (His methods have been adapted in the training of circus animals.) Skinner and his followers, believing that these principles could be applied to human learning also, built machines which lead students through the material in small easy steps. A sequence of questions is asked, each of which is a small step beyond the previous question. The student finds out immediately whether his answer is correct, and the very fast reinforcement is thought to assist the learning process greatly. Many of today's CAI programs adopt the same approach.

The danger of the Skinner approach is that more intelligent students may become bored or disgruntled if they think the questions are too simple. Another technique, used in early systems built by N. A. Crowder,[3] uses larger steps in the programmed material: if the student answers questions correctly, he is moved on to more advanced material; if not, he is 'branched' to a remedial section.

On some CAI programs both approaches are used: if the student assimilates the material successfully in the large steps, then he progresses quickly; if not, he is branched to a Skinner sequence – of simple steps and frequent questions, the majority of which he is likely to answer correctly.

An elaborate program may have many branches and alternate paths. The response to a question may elicit a variety of differing actions. If correct, the main teaching routine will proceed. If partially correct or almost correct, a remedial sequence will be followed. If wrong, the instruction leading up to that question may be repeated, or an alternative, more detailed sequence followed. If the answer reveals a lack of understanding of an earlier point, the program may backtrack some distance. If the wrong answer is the culmination of many such errors, a branch may be made to a different teaching approach.

PROGRAM PRODUCTION

Writing these elaborately structured teaching programs requires much time and considerable skill. Furthermore, the program is likely to be modified substantially in response to the reactions of its users. For this reason, CAI programs are often written to record in detail their users' answers. A printout will show the author the students' exact replies to each question and will indicate how long they delayed in answering. The author may then make many modifications to his program. In addition to this feedback, he will normally want to talk to the students and observe their reactions at the terminal. Being a program, the CAI routine is capable of endless modification and polishing. It can be changed, not only by the author but by the teachers who employ it. Furthermore, when a program acquires an outstanding reputation, as a few have, it can be duplicated, studied, and used in machines throughout the world.

An outstanding CAI program is a work of great art. A strong sense of style is needed. This is a new medium quite different from any that preceded it, and, just as writing needs rules concerning style, so does CAI. As yet, no acknowledged sense of style has developed for CAI. The medium is too new. One can read books on literary style but not yet on CAI. No doubt a style guide will come in time and will probably change with time just as style in other media have changed. CAI may be even more susceptible to change because of the intimate two-way interaction which no other medium has and because of the immense programming versatility that is possible.

In the meantime, however, some singularly unstylish CAI programs are being written. Programmers, hurriedly attempting to demonstrate the new machines and infatuated by the ease with which they can make their words appear on the screen, are producing programs as bad as the home movies of an amateur with his first 8mm. zoom-lens cine-camera. Time is short, and the great thoroughness in reacting to student responses, which the medium demands, is sometimes not being exercised. It is *much* more difficult to write a worthy CAI program on a subject than to write a textbook on it. Nevertheless, programmers who would not dream of writing a textbook are gaily wading into CAI programming. This is probably a temporary dilemma: teachers rather than programmers are taking over the programming work as the technology spreads.

It would be unfortunate if the public, or the teaching profession (most of which is avidly looking for reasons to hate CAI), formed its opinions on the basis of those bad programs that are now circulating. A reasonable judgment of the potential of the medium should rest on the best of today's programs. Already, there is a handful of *exceedingly* effective teaching programs to consider, but they are submerged in the welter of shoddy products of average programmers. Today's best will probably look as crude as pre-First World War movies look today, when eventually the production of CAI programs becomes as elaborate and as professional as movie production is today.

LEARNING RESEARCH

The psychology of learning through CAI is imperfectly understood but the computers present us with a superb research tool with which to study it. The result could be a revolution in educational research, which until now has been complicated by the classroom situations. The diverse conditions in learning experiments have tended to camouflage the effects sought, and where experimenters have been able to tightly control the conditions the experiments, for the main part, have been oversimplified. Using CAI, the experiments can be well controlled, realistically complex, and the student responses can be analyzed in detail. The results should lead to a body of knowledge which indicates how to write CAI programs and also gives an insight into learning in general.

Languages have been devised for teachers who wish to write CAI programs, but who know little about computers or programming. Using these languages, the instructor-writer specifies in a simple form the contents of 'frames' of information which appear on the student's screen or other device. He specifies what questions the machine asks the student, and the frame to which the program will branch for each reply. These languages are very easy to use, and the computer itself translates what is written into its own program steps. Using them, any schoolteacher, psychologist, technical writer, unemployed housewife, or any specialist on a given subject can produce CAI programs. Professional educators, unlike programmers, should raise the standard and improve the style of CAI, since they are more concerned with ends than means.

TEACHER ASSISTANCE

Many CAI programs are written for use in an environment in which the student receives external help if he needs it. Typically, a classroom may have fifteen to thirty student terminals and an additional 'proctor' terminal. The latter is operated by a supervisor who is in charge of the operation and who can answer student questions. Messages can be sent directly from the proctor terminal to the student terminal, or direct verbal assistance may be given.

Whereas help and supervision from a teacher is most desirable in a classroom, one of the great advantages of CAI is that it permits education to proceed *out* of the classroom. Whereever there is a terminal, a student can dial a computer to continue his lessons. This is particularly valuable for professional men. A doctor or a systems engineer with a spare half-hour after dinner could use a home computer terminal to continue the essential process of self-education. With no human tutorial help available the program must be self-sufficient. Perhaps, in the future, we shall find two types of CAI programs: one for the classroom with a teacher, and one for the isolated individual on the other end of a telephone line.

TYPES OF PROGRAMS

Computer programs that assist in teaching take a variety of forms corresponding to the differing needs of the educational process. We can distinguish between a number of types of uses, as follows:

1. *Tutorial Programs*

The machine programmed for tutorial use presents material to the student and then asks him questions about it in the manner discussed above. The student may have a printed text to supplement what the machine says, thereby saving part of the cost and giving the student something to study when he is away from the terminal.

Figure 6.1 (pp. 137–40) shows how four students fare at the same point in a program for teaching statistics. This course was written by a behavioural psychologist at IBM's Thomas J. Watson Research Center. Notice how the computer varies its reaction to the different responses. In this illustration, the students had a printed text for the course. The terminal was a typewriter-like device and could well have been a simple teletypewriter.

A teletypewriter, however, can be frustratingly slow for computer-assisted instruction. A fast one would take more than three minutes to print a page of this book, and a slow one more than seven minutes. A student can hardly be expected to wait this long for 'frames' of information. A screen unit can produce the same quantity of information in a few seconds, and this information could be transmitted over a telephone line.

The rate at which data can be transmitted over a telephone line is a good deal faster than man's reading speed and so such a link is well suited to machine teaching. CAI workers have commented that using teletypewriter speeds is like having a teacher with a speech impediment.

2. *Drill and Practice Programs*

Many skills can only be acquired by repetitive practice. Spelling, arithmetic, the vocabulary and grammar of foreign languages, and the names of chemical elements are among the chores of teaching which can easily be assigned to a machine. The computer will exercise the student and report on his progress. A report on the *entire* class will also be given to the teacher, indicating any deficiencies or problems the class has with the material. The teacher can use this information to modify her instruction.

In a group of elementary schools near Stanford University, teletype machines have been installed and connected to a Stanford computer by telephone line. Each student uses the terminal for a short period each day for drill in spelling, mathematics, and languages. He begins by typing his name. The computer then looks at its record of his past performance and selects a section of the drill program accordingly. A small child learning to spell would receive vocal instructions from the computer and would be told to spell a given word. If the word were spelled correctly, the machine would say so and proceed to the next word; if he spells it incorrectly, the machine would take corrective action. Alternatively, the computer could show the child a picture of an object whose name he is asked to spell, or it could be programmed to ask him to point to the object in question with a light pen and then point to the correct spelling. Computer drills in mathematics are somewhat more complex than this, but still constitute a routine process.

Student No. 1:

COMPUTER: 51. Therefore, the shape of the sampling distribution of t depends on —

STUDENT: area.

COMPUTER: Remember that we have just seen that different — samples result in different distribution curves.

STUDENT: size.

COMPUTER: 52. Notice that the sampling distributions of t in Figure 2 are all symmetrical, but not normal, with a mean of —.

STUDENT: X

COMPUTER: If we convert our scores to standard scores, what should the mean of the distribution be?

STUDENT: Zero.

COMPUTER: 53. The means of the sampling distributions in the previous question tend to be zero because all the raw scores have been converted to — scores.

STUDENT: z.

COMPUTER: This answer is partially correct but a more inclusive answer would be — scores.

STUDENT: t and z.

COMPUTER: What is another name for t and z-scores?

STUDENT: Deviation.

COMPUTER: Since t- and z-scores are standard scores which result in a mean of zero for the distributions, the answer is — scores.

STUDENT: Standard.

COMPUTER: 54. Now refer to the last page of this chapter in the text. There are two distributions with n's of 5 and 50. The curve based on an n of 5 is (Y/Z).

Student No. 2:

COMPUTER: 51. Therefore, the shape of the sampling distribution of t depends on —.

STUDENT: n.

COMPUTER: 52. Notice that the sampling distributions of t in Figure 2 are all symmetrical, but not normal, with a mean of —.

STUDENT: Zero.

COMPUTER: 53. The means of the sampling distributions in the previous question tend to be 0 because all the raw scores have been converted to — scores.

STUDENT: t.

COMPUTER: This answer is partially correct but a more inclusive answer would be — scores.

STUDENT: t and z.

COMPUTER: What is another name for t- and z-scores?

STUDENT: Standard.

COMPUTER: 54. Now refer to the last page of this chapter in the text. There are two distributions with n's of 5 and 50. The curve based on an n of 5 is (Y/Z).

Student No. 3:

COMPUTER: 51. Therefore, the shape of the sampling distribution of t depends on —.

STUDENT: Number.

COMPUTER: 52. Notice that the sampling distributions of t in Figure 2 are all symmetrical, but not normal, with a mean of —.

STUDENT: 0.

COMPUTER: 53. The means of the sampling distributions in the previous question tend to be zero because all the raw scores have been converted to — scores.

STUDENT: z.

COMPUTER: This answer is partially correct but a more inclusive answer would be — scores.

STUDENT: Standard.

COMPUTER: 54. Now refer to the last page of this chapter in the text. There are two distributions with n's of 5 and 50. The curve based on an n of 5 is (Y/Z).

Student No. 4:

COMPUTER: 51. Therefore, the shape of the sampling distribution of t depends on —.

STUDENT: Area.

COMPUTER: Remember that we have just seen that different size samples result in different distribution curves.

STUDENT: Tails.

COMPUTER: Since the shape of the sampling distribution is dependent on sample size, the best answer here is 'n'.

52. Notice that the sampling distributions of t in Figure 2 are all symmetrical, but not normal, with a mean of —.

STUDENT: n.

COMPUTER: If we convert our scores to standard scores, what should the mean of the distribution be?

STUDENT: x.

COMPUTER: Reflect back to the mean of the normal curve of z-scores for a possible solution.

STUDENT: 0.

COMPUTER: 53. The means of the sampling distributions in the previous question tend to be zero because all the raw scores have been converted to — scores.

STUDENT: Standard.

COMPUTER: 54. Now refer to the last page of this chapter in the text. There are two distributions with n's of 5 and 50. The curve based on an n of 5 is (Y/Z).

Fig. 6.1

3. *Problem-Solving*

A student given a problem for homework may discuss the results with a computer. This may be done with quite complex problems, such as those met in undergraduate physics or engineering courses. Students usually need individual attention if they get such problems wrong, but are rarely given it in a conventional classroom.

The dialogue might go something like this:

STUDENT: Systems design – problem 384.

COMPUTER: What did you obtain for the overall response time?

STUDENT: No answer.

COMPUTER: Let us see where you had difficulty. Did you obtain the communication-line loading?

STUDENT: Yes.

COMPUTER: What was your value for line utilization?

STUDENT: 0·372.

COMPUTER: That is the utilization for input traffic only. The line is a full duplex line and so you should have included the output traffic, also. Have you forgotten the output traffic?

STUDENT: No.

COMPUTER: Because it is a full duplex line, input and output traffic will travel over it at the same time and so both contribute to the congestion. The facility-utilization figure you calculate must include both. What would be the figure due to output traffic only?

STUDENT: 0·25.

COMPUTER: That is correct. This gives a total line utilization 0·372 + 0·250=0·622. Do you follow this?

STUDENT: Yes.

The student presses a key to terminate the conversation, tackles the problem again, and then approaches the computer with another – hopefully correct – solution.

In this case, the program writer must have understood the types of errors which students would be likely to make in tackling this problem. He thus programmed the machine to look for possible wrong approaches and to give such explanations as above when it found one. If the machine could not find out where the student had gone wrong, it would explain how the problem should be done, step by step, giving the student the option at each step of completing it by himself. As more students use the program and the writer reviews their responses, he will see how completely he has covered the possibilities. This 'class-testing' usually prompts him to modify his program until it covers the probable reactions of students very thoroughly. In effect, the program 'learns' by feedback and reinforcement, just as the students do. It is this capacity for self-improvement that makes CAI programs so much better than the 'programmed instruction textbooks' now being produced in increasing numbers.

4. Dialogue Systems

It is possible that CAI systems of the future will be programmed to carry on much more elaborate dialogues with the student than the ones above. Researchers are presently trying to make computers converse with their users on a limited subject area in more-or-less natural English. The main problem is programming the machine to analyze what is said to it. It can recognize words easily, but it must also determine meaning from the context of sentences, a process which may be beyond even complex logic. The English-language dialogue systems that exist as of this writing can only be classed as tentative and experimental. Nevertheless, they are impressive because the machine gives the illusion of behaving with near-human intelligence.

Figure 6.2 (pp. 143–5) gives an example programmed at Bolt, Beranek and Newman, Inc., who refer to this as a 'Socratic system', because the computer and student enter into a dialogue supposedly reminiscent of ancient Athens.[4] The computer poses a problem, and the computer and student together 'discuss' it. The student may ask for clarification of data he believes to be relevant, and as he works his way through the problem the computer can, if necessary, alter the course of the student's working or even admonish him. In responding, the student is restricted to words from a vocabulary which is, however, quite large.

In Fig. 6.2 the system is presenting a case study developed by Harvard Business School. The conversation takes place again on a remote teletypewriter. When students approach the problem in ways other than that illustrated, the computer sometimes gives them different data to work with, and different decisions can satisfy the 'tutor'.

While this example gives an idea of the eventual power of the medium, it is likely that most computerized teaching in the next decade or so will involve much less elaborate dialogues.

5. Simulation

There is no need to have a laboratory today in order to conduct many scientific experiments; they can be performed in a simulated fashion with computers, and students can observe the results of their actions on the screen of a terminal. In physics and chemistry, programs have been written to obviate the need for experiments with wires and test tubes. In the Bell Telephone Laboratories and many other research establishments, work that would earlier have needed physical equipment is now being done with computer models. As a teaching aid, simulation can be extremely valuable. In medicine, for example, the student can observe a diagram of a moving organ for the effects of different actions upon it.

Simulation adds new dimensions to experiments and to their use in teaching because it makes visible many very small, very slow, or very rapid changes which cannot be observed in actuality. The time scale of events can be expanded or contracted thousands of times. The medical student can manipulate physiological models in ways which might kill a real patient. The physics student can see 'waves' travelling along a transmission line, or can explode an atomic bomb in a simulated fashion while watching the nuclear reactions in ultraslow motion.

Dr Eugene Fubini described dramatically an impression that this form of teaching made on him:

This is the experiment: You take a computer and send a simulated stream of particles from above, into the same cathode-ray tube. All these particles are the same size. You then tell the computer to say that these particles attract each other with Newton's law, and the ordinary laws of mechanics apply. And then you let the experiment run. It's breathtaking to see what happens. I can't forget it. Right in front of you the universe is forming. You see spiral galaxies coming out; you see all the types of galaxies that are present in the world today – they are being created in front of you through this simple mechanism. I never really understood how a galaxy was formed until I saw that simulation. And it is with experiments of this kind that I think we can do a type of education that we have never been able to do before.[5]

6. *Use of a Large Data Base*

In some fields the use of computer-assisted instruction may be linked to a very large file of instructional information. Today the student cannot be expected to learn more than a fraction of the total information pertinent to his subject. Instead, he must 'browse' selectively through the information, a retrieval process with which the computer could greatly assist. Such a data base could be a file of medical or drug information, legal information, engineering data, data on computer programs, and so on. At present, we know of no such systems in operation, but undoubtedly a number will be set up sooner or later.

COMPUTER: You are a management consultant called in to advise a toy company on a proposal to change their schedule of production. In the past the company's production schedule has been seasonal, reflecting the highly seasonal pattern of sales.

The company is currently considering the advantages and risks involved in a possible shift to a level schedule of production over the year. Your judgment is being sought. Ask for the information you want by typing in items from the vocabulary list that you have been given ... Begin your investigation as soon as you are ready.

STUDENT: Machine maintenance costs?

COMPUTER: The estimated machine maintenance costs are $23,000 for seasonal production and $10,000 for level production ...

STUDENT: Pro forma income statement?

COMPUTER: You haven't got all of the data needed to make up the pro forma income statement. You don't know the cost of goods sold.

STUDENT: Cost of goods sold?

COMPUTER: The cost of goods is 60 per cent of last year's sales.

STUDENT: Production volume?

COMPUTER: The projection production volume? That's your problem. Tell me when you know.

STUDENT: Income taxes?

COMPUTER: Do you mean federal income taxes?

STUDENT: Yes.

COMPUTER: The federal income taxes are 50 per cent.

STUDENT: Projected sales?

COMPUTER: The projected total sales figure is $1 million. Would you like to see the projected monthly sales also? ... Now that you have obtained the pro formas, you can state your decision favoring either seasonally varying production or level production throughout the year – whenever you're ready. Meanwhile, type 'help' if you become troubled by any of your findings.

STUDENT: Direct labor costs?

COMPUTER: The estimated direct labor costs are $217,000 for seasonal production and $170,000 for level.

STUDENT: Help!

COMPUTER: What finding bothers you? Is it your pro forma information?

STUDENT: Yes.

COMPUTER: You can have revised pro formas if you think of something that would change them. What new factor is relevant?

STUDENT: Overtime costs.

COMPUTER: It's fortunate that you had second thoughts here. The overtime amounts to $60,000 for seasonal production and none for level. These new data should change your pro formas and have very significant effects on your decision. . . .

STUDENT: Seasonal production.

COMPUTER: Now that you have made a decision, let's consider it in the light of the relevant facts. . . . Consider these corrected pro formas carefully. Then tell me – are you still happy with your current decision?

STUDENT: Yes.

COMPUTER: Tell me again, just to be sure that I understand, *which* production schedule do you recommend?

STUDENT: Seasonal production.

COMPUTER: Let's consider a hypothetical question. If you were guaranteed that you would not be caught with a lot of unsaleable inventory, would you change your decision to one in favor of level?

STUDENT: No.

COMPUTER: Then let's consider whether or not we are agreed on amount of savings that would result. Assuming all the inventory were sold, how many thousands of dollars in savings would result from going to a level production schedule?

STUDENT: 110.

COMPUTER: That's about right. The actual value, determined from direct labor costs, overtime costs, warehousing costs, and machine maintenance costs, is $108,000. That's a lot of money. Doesn't this change your decision?

STUDENT: No.

Fig. 6.2 Illustration of a 'Socratic' dialogue with a computer. This conversation takes place at a remote teletypewriter. (Omissions are indicated by dots.) *Reproduced from 'Computer-Assisted Socratic Instruction' by Richard H. Bolt of Bolt, Beranek and Newman Inc., in Conversational Computers, William D. Orr (ed.), John Wiley & Sons Inc., New York, 1968.*

COST

The primary drawback to CAI, at least at present, is its cost. Even with many students being handled at once by one computer, costs of the order of $10 per student hour are common. This is too high for the average elementary school. It may not be too high, however, for professional men to whom rapid and efficient education is very important. The present use of CAI, therefore, is growing rapidly in such organizations as medical schools, computer centres, and large corporations. IBM has established a nationwide network of terminals in the United States for training its field engineers, but elementary and secondary schools are mostly waiting for a major drop in cost.

Costs will almost certainly be reduced, however. (Figures 1.2 and 1.5 showed how quickly computing costs are dropping.) A cheap mass-produced terminal will probably be available when the market becomes large enough (and this will be a gigantic market). Computers will probably be designed to handle thousands of terminals (like the airline reservation systems), and if, as in a city, such a computer were not far from the terminals, the communication costs need not be too high. The telephone lines connecting private homes, which lie idle 99 per cent of the time today, could be used for CAI.

If the cost drops from $10 to $1 per hour, CAI will no doubt become a highly popular leisure occupation. It will be cheaper than going to the movies and far more instructive.

THE WORK NEEDED

To make CAI really fruitful, a *tremendous* quantity of programming work is needed. At a New York City elementary school, thirty children spent five minutes a day over the 160-day school year learning mathematics at a terminal. They made two years progress in one year, but 96,000 exercises had to be prepared, a task that took 1600 man-hours. So more than 100 hours of programming time are required for each hour of student time at the terminal. We suspect that for many subjects the ratio for first-rate programs is higher than 500 to 1 (two or three times higher than the ratio for equivalent textbook writing). Writing all the programs for all the subjects amenable to CAI techniques will take millions of man- (and woman-) years.

Eventually, enormous data banks of educational programs

will be instituted and the public, whether in schools, colleges, corporations, private homes, or special cars on commuter trains, will be able to dial, and converse with, a first-rate psychologically planned computerized tutor.

Underdeveloped nations will be able to share first-hand the educational riches of the most developed countries. Perhaps, in a clearing in the jungles of the Amazon, a school hut will have cheap terminals connected by satellite or microwave to a CAI computer using programs written in the US or the USSR. We will postpone, until Section II, any speculation about the material that either country might make available for the teaching by CAI of history, economics, or political science.

A revolution in technique requires a revolution in the attitude not only of teachers, but also of parents. Today's conventional teaching is teacher-centred and teacher-paced; tomorrow's instruction will focus on the student and will progress at his pace. All children will learn better and faster, but the brightest will outstrip their more normal contemporaries. What the average pupil learns in twelve years, the very able might do in six. The egalitarianism that causes children to advance one grade each year and blurs the distinction between chronological and educational age will be obsolete. Children will demonstrate their inequality and parents, teachers, and society as a whole will have to cope. We shall return to this problem in Chapter 23.

REFERENCES

1. Figures taken from 'There's a Computer in Your Future', *American Education*, November 1967.

2. See *Harvard Education Review*, Vol. 24, No. 2, 1954.

3. See Crowder's article in *Teaching Machines and Programmed Learning*, A. A. Lumsdaine and Robert Glaser (eds.), National Education Association, 1960.

4. Richard H. Bolt (of Bolt, Beranek and Newman, Inc.), 'Computer-Assisted Socratic Instruction', in *Conversational Computers*, William D. Orr (ed.), John Wiley & Sons, New York, 1968.

5. Eugene G. Fubini, speech given to the IEEE 1969 International

Convention's Highlight Session, reprinted in *IEEE Spectrum*, July 1969.

OTHER READING

1. William D. Orr (ed.), *Conversational Computers*, John Wiley & Sons, New York, 1968. See Section III 'Instructional Modes.'

2. Don D. Bushnell and Dwight W. Allen (eds.), *The Computer in American Education*, John Wiley & Sons, New York, 1967.

3. R. W. Gerrard (ed.), *Computers and Education*, McGraw-Hill, New York, 1967.

7

HOME SWEET HOME

Imagine a large screen, 4 ft by 5 ft, on a living-room wall. The owner has a typewriter-like keyboard at the side of his favourite chair. He types a few characters, which appear on the screen, thereby inquiring of a computer what movies are available that evening. There is a fairly large selection and in particular a Godard festival is taking place, so he asks the remote computer for a list of the Godard films. He selects one he has not already seen – *Alphaville* – and requests a synopsis. The 12-line abstract which appears on the screen looks interesting, so he sets up the evening's entertainment. The advertisements are short and designed to appeal to the type of personality that would select *Alphaville*. He can stop the film for a while if he wants to make coffee, or abandon it if he does not like it. And his bank balance will, once again, be automatically debited.

When the film is over his wife wants to continue planning their vacation. She dials the local travel-agent computer and a message appears on the screen requesting politely that she type in her identity number. She does so and the screen displays the message:

```
GOOD EVENING MRS SMITH.
WOULD YOU LIKE A CONTINUED PRESENTATION
ON ONE OF YOUR PRIME SELECTIONS?
1: PERU?
2: BOLIVIA?
3: HAWAII?
4: EASTER ISLAND?
IF SO PLEASE TYPE ABOVE NUMBER.
```

She types '4' and the screen says:

```
EASTER ISLAND.
WHICH OF THE FOLLOWING WOULD YOU PREFER?
1: SCENERY?
2: ENTERTAINMENTS?
```

```
3: HOTEL INFORMATION?
4: SHOPPING?
5: TIME-TABLE?
6: TO MAKE RESERVATIONS?
PLEASE TYPE ABOVE NUMBER.
```

She types '1' and the screen says:

```
SCENERY.
PLEASE PRESS ENTER KEY TO CHANGE
SLIDES AND TO BEGIN PRESENTATION.
```

She presses the ENTER key and a magnificent view of Easter Island appears on the screen in colour. Husband and wife look at breath-taking pictures for the next half-hour and then she presses the END key. The screen asks if she wishes to make a booking. She indicates 'NO' and switches the unit off.

How long will it be before we have such facilities? With the exception of colour television the last decade has brought surprisingly few new types of inventions for the home. This deficiency is probably going to be made up in the next decade or so; all manner of devices based on computer logic circuits, telecommunications, memory devices, and videotape are presently envisaged. An executive of Texas Instruments estimates that the homes in the US may soon contain as much as $10,000 worth of electronic devices.[1] The 15 March 1967 issue of *Forbes* predicts that a $30-billion-a-year market for such equipment challenging the motor industry for the largest single share of consumers' income by 1980.[2]

INEXPENSIVE DEVICES

Not all home uses of terminals need be expensive. With a relatively straightforward device connected to the telephone line, many applications are possible.

Already by 1966, it was possible for a New York City schoolchild to do his homework with the help of an IBM computer 50 miles away. The device used is inexpensive and looks like the standard touch-tone telephone keyboard that has replaced the dial on many American telephones. To perform a calculation the user must first 'dial' the computer's number in the same way that he would dial a friend's telephone number. He then presses the appropriate touch-tone keys to give the machine details of the calculation. The computer does the calculation and replies over the telephone with a clear human voice. The system can do any calculation that can be done on a desk calculator, and can work out square roots with ease. A minor modification enables it to do all manner of things that a desk calculator could not. If the computer detects that the user has pressed the wrong keys, it would inform him, again with a spoken reply. The children involved learned how to use this scheme quickly and enjoyed playing with it very much.

The Bell System is now marketing a twelve-key keyboard like this on normal home telephones instead of their earlier ten-key keyboard. A set of plastic overlays could be used to label the newer keyboard in different ways, permitting an almost endless variety of applications in which the computer replies with the spoken voice. Thus each overlay would have a unique number, and the computer might then begin its operation by saying, 'Please key in the number of the keyboard overlay which you are using.'

Different overlays need not be limited to programs in just one computer, for if such a scheme becomes accepted it will no doubt be possible to dial a variety of talking computers. Each overlay would have written on it the number of the computer to dial along with the number of the overlay. One overlay might be used when dialling one's bank computer, another to obtain baseball scores, another to reserve theatre tickets, others for playing a variety of games with the machine, and so on. For this, the cost in home electronics is no more than that of a twelve-key touch-tone telephone, plus the sum charged for using the computers.

OTHER HOME TERMINALS

Computer voice-answerback is inexpensive because it can be used with a conventional telephone, but it restricts the dialogue that can take place. Commercial applications of real-time systems have shown that the dialogue must be of an elementary nature when voice-answerback is used, but that it can be much more elaborate when the machine gives a printed response, and that the best dialogue of all takes place when a screen is used.[3] The household television set could be converted to display verbal computer responses, coming over a telephone line, in addition to its normal function. The set would be designed so that it could store digital signals from the computer and display them on the screen. Although a touch-tone telephone could still be used to send messages to the computer an *alphabetic* keyboard would extend greatly the range of practical applications.

Conventional telephone lines will have to be used to transmit the data to and from the computer when such services first become available. These lines can carry about 170 letters or figures per second, many more than the human eye and brain could read if they were all significant. In practice, however, many of the characters transmitted are used merely to display the message neatly on the screen. A new 'frame', or page of twenty 50-character lines, takes an exasperating six seconds to appear. New equipment first installed in the late sixties has four times this transmission speed and editing facilities, and so can 'turn the page' in less than a second. Yet higher speed lines are planned, and these will be able to transmit pictures.[4]

The home terminal will not be for entertainment at first, but for business. We expect its introduction to be phased – starting with industrial users, then amateur enthusiasts, paving the way for the mass of ordinary people.

HOME TERMINALS USED BY INDUSTRY

Computer terminals in homes *today* are not usually there at the user's expense. These users are mostly university staff engaged in research, programmers employed by a large corporation or sometimes a 'software house', a handful of executives enthralled by the new technology and some travelling salesmen who use their terminals when they arrive home to transmit their orders to a computer.

The first widespread use of home terminals will probably be sponsored by employers who expect to benefit thereby. Certainly the authors could make good use of home terminals installed by their companies, and there are thousands of other systems engineers and technical staffs who could similarly benefit. What would they use them for? Writing programs, developing their own collection of programs for systems analysis or other work, obtaining technical documents, looking up facts, running teaching programs, and exercising newly learned skills. They may decide they need to learn a new computer language – for example, PL/1. They may spend an hour for several evenings with a manual and at a terminal using a teaching program. However, the only way *really* to learn a computer language is to program in it. So they may dial up a computer which enables them to program with a given language on-line and then spend a period each evening developing their ability to use their new tool.

Many technically minded people would find it pleasant to sit down for an hour or so after dinner and tinker with a program they are working on. Often, too, ideas come at strange times – when the mind is relaxed. One wakes up in the morning and suddenly sees how to do something that was puzzling him the night before, or realizes where the cause of an error lies. So one would wander up to the terminal in one's pyjamas and check the thought or modify the program then and there.

In writing this book, the authors could have typed the text directly into the same home terminal. It would have resided in the memory of a distant machine. We could then have modified it, edited it, restructured it, snipped bits out, corrected each other's work, added to each other's ideas, and instructed the terminal to type out clean copies when we were ready. A team of authors in Washington is writing a technical report in just this way. Possibly magazines will be edited with such aids in the future.

When the ability to use computers from the home becomes common, it seems logical that some employees should spend at least part of their time working at home rather than travelling to an office. The overhead costs of providing staff with offices and desks is very high, especially in big cities; some of this will be saved. A manager can see what remote staff members are doing by telephoning them, dialling up the computer they are using and examining their work, and sending them instructions on their home terminals. It is possible, indeed, that in the future some companies may have almost no offices. A software company for producing programs may cut its costs significantly if most of its personnel work at home. Mothers who participate in such a scheme may be relieved of much of the boredom they feel when they are unable to leave their children.

The growth of computerized teaching – from today's experiments to tomorrow's industry – will need, as we indicated in the last chapter, a tremendous amount of human thinking and development of programs. Much of this may also take place in the home. Step by step, a teacher can build up his lessons on his home terminal. Occasionally, he may dial a colleague to ask him to try out what he has produced on *his* terminal. When it is nearing completion he may try it out in a classroom, study the reactions of students, and return home to build in appropriate modifications.

We see here a beginning of a return to 'cottage industry' – a trend that will probably increase greatly during the next few decades. Nevertheless, at the moment a few factors are counteracting this trend. The first is the reluctance of some companies to give their systems analysts or other employees home terminals because such a step seems an unprecedented and potentially unpopular encroachment on leisure time. The second is a feeling that, regardless of what they are doing, employees ought to be at their desks from 9 a.m. to 5 p.m. (At home – who knows – they might be watching television!) The third is a feeling that people cannot work at home because the environment is not suitable. Perhaps some homes are unsuitable, but the authors of this book had no difficulty in writing it at home with a rate of productivity that would be regarded in industry as high. We believe that these three views are not generally held by intelligent and enthusiastic members of the community, and that this is an era when we must throw off the mores of tradition and rethink what is applicable to the age of teleprocessing.

Certainly if such 'cottage industry' spreads extensively in our society, it will drastically change social patterns.

COMPUTER AMATEURS

Perhaps the next group to help introduce terminals into the home on a large scale will be the computer hobbyists. The equipment that ham-radio enthusiasts and other such groups now have at home is more expensive than basic computer terminals such as teleprinters or alphabetic screen units. Furthermore, the interest and excitement that is likely to be stirred up by being able to dial up and work with an ever-growing number of computers is likely to be far greater, and far more time-absorbing than the attraction of hobbies such as ham radio and videotape. It is probable that as the available computers increase, as education about computers spreads, and as leisure time increases as a result of automation, the computer amateurs will become a growing body. Magazines will be produced for them. Industry will encourage them and enthusiastically sell to them.

Computer hobbyists may fall into a number of different groups. There will be some who hope to produce and sell their own programs or make them available to other amateurs. There will be some who are less creative and mainly interested in education; they will dial various instructional and information-retrieval programs. Others will be interested in playing games, doing puzzles, or indulging in mathematical recreations. But perhaps the majority will fall under the narcotic spell of programming. Working on ingenious programs (rather than the routine of commercial programming) is, to a certain type of mind, endlessly captivating.

The computer amateur will have significant contributions to make to the development of this technology. Most technical fields are too complicated or specialized for the amateur to make his name, but in programming there is endless scope. In every direction, new territories await the ingenuity and care of a dedicated amateur.

As we have commented before, *enormous* quantities of ingenuity and programming are essential to this new era – not the work of geniuses, but ordinary step-by-step construction and testing. The work requires a high order of craftsmanship. In general it is creative, enjoyable work; work that wives and children can do; work that disabled and in some cases, blind people are doing. It is work to which the hobbyist, or the enthusiast making money in his spare time, will contribute enormously.

DOMESTIC MASS MARKET

The time will come when the computer terminal is a natural adjunct to daily living. Sooner or later computing is going to become a mass domestic market and the computer manufacturers' revenue is going to soar. The airline industry, the automobile industry, telecommunications, and other complex technical industries all spent two decades or so of limited growth but then expanded rapidly when the general public accepted and used their product. Data transmission is going to make this happen in the computer industry also.

The eventual uses for terminals in the home are endless, and numerous computer systems will be set up for domestic and entertainment purposes. Whether this will come about in the next ten or fifteen years is difficult to tell, although we tend to doubt it. Let us discuss the possibilities, anyway – after all, the computer industry moves with surprising speed.

The home user will eventually have access to a wide variety of data banks and programs in different machines. He will be able to store financial details for his tax return, learn French, scan the local lending-library files, or play games with a computer. If he plays chess with it he will be able to adjust its level of skill to his own. It has been suggested that news will be presented in this way in the future; the user will skip quickly through pages or indexes for what he wants to read on his screen. As the machine's files are very large, newspapers from other countries transmitted by satellite could also reside in his local machine. He may have a machine to *print* newspapers in the home, although use of the screen might be preferable. If he wants a back number, he will be able to call for it, using a computerized index to past information.

A man interested in the stock market could dial up a computer holding a file of all stock prices, trading volumes, and relevant ratios for the last 20 years.* Possibly stockbrokers would make the information available free and would provide analytical routines to their clients. When a client bought or sold stock he could give the appropriate orders directly to the stockbroker's computer with his terminal.

On the other hand, someone without the money to buy stock could *play* at buying it. The computer would calculate the effect of his buy and sell orders, permit him to trade on margin, pretend to give him loans, but no actual cash transfer would take place (apart from the cost of using the machine). When friends came round to visit he would dial up his records to show them that he started with $100,000 and in six months had made $78,429! Perchance to dream in glorious detail. At least he will have had lots of practice ready for when he does eventually become rich – good training for one's children.

SHOPPING

Whereas a man might use a home computer terminal for scanning newspapers or for stockmarket studies, his wife might use it for shopping. In some countries punched-card supermarkets have come into operation. The housewife at the supermarket picks up a card for each item she wants to buy and these are then fed through a tabulator. She pays the bill and the goods are delivered from the stock room. The advantages of the system for the supermarket are that it needs less space and less capital outlay, and that there is less pilferage. Why

* See Chapter 4, p. 92.

must the housewife come into the store at all? She could scan a list of the available goods and their prices at several different shops on the home terminal, and then use the terminal to place her order. An organization selling in this way could cut overheads to a minimum by eliminating stores.

The customer who buys through such an automated catalogue avoids the exasperation of fruitless searches for hard-to-find merchandise. Besides replacement bulbs for projectors condemned by planned obsolescence, spare parts for automobiles or rare gramophone records – all items likely to be found, if at all, in only one store in the city – there are tedious searches to be made for unique items like summer houses for rent, theatre tickets for a particular show, or boats with a particular specification. Presently, these are found through 'classified ads' in newspapers or through agents.

Much of the work of such agents, and of the advertising columns, could be more cheaply and more efficiently done by a data bank accessible from terminals. Until home terminals are common, it is unlikely that those who now make their livings from putting buyers in touch with sellers will cooperate with each other to set up systems that will make themselves redundant. A British service to home-seekers foundered when most real-estate agents refused to place details of properties on their books in the computer system as well, even though the purchaser would have to use the agent to reach the seller. Theatre ticket systems are doing better.

Many newspapers and magazines, already having to compete with television for advertising revenue, may go out of business. Printing unions are well organized to prevent the automation of printing itself, so many newspaper publishers may be unable to avoid closing down when the 'classified ads' emigrate to a computer.

The above services may be provided free by the advertisers; many others will be charged for, almost certainly with the book-keeping being done on a central computer linked to the message switching system which connects the terminal to the appropriate data bank. An automated diary might cost one cent per entry – a small fee to insure that no appointments are

missed and no birthdays forgotten. Hunting through diction-
aries, almanacs, abstracts of literature, etc. might cost about
one dollar an hour, much less than a trip to a good library, and
much more fruitful.

As more such services become available the economic justification for terminals in the home will be strengthened. Presently,
only those to whom economics are irrelevant can own them.
Nieman Marcus, the world famous department store in Dallas
which caters for such people, featured the Honeywell H316 in
its 1969 Christmas catalogue. It is a small and elegant computer (not a terminal since there is as yet a dearth of data
banks), and it comes ready programmed with menus, diets, and
routines to handle domestic accounts. It costs $10,600, so
those who can afford it surely have no need of domestic
accounts!

SPORTS

Sport, too, will be aided by computer, and the sports fan will
be able to obtain all manner of information on his own home
terminal. In top golf tournaments today computers are used to
keep track of everything that is happening. Observers stationed
around the course report information to a computer station by
means of walkie-talkies. The machine digests all the information, operates a scoreboard for the clubhouse gallery, press,
and television; and displays on a screen hole-by-hole scores and
such information as greens reached in par, numbers of putts
on each hole, and lengths of drives on selected holes. Instantaneous comparisons between players can be produced along
with all manner of asides, such as remarkable runs of
birdies.

The terminal-owner of the future will presumably be able to
dial machines giving up-to-the-minute detailed information on
any sport and will not be restricted to the one or two items fed
to them on the television channels. One imagines a Sunday
afternoon of the future watching pro football television much
as today, but with the teleprinter chattering away at one's side

printing the commentaries it has been instructed to give on other games.

Horse Racing

In the winter of 1967–8 England was swept by a disastrous plague of foot-and-mouth disease which killed many millions of pounds (Sterling) worth of livestock. Most horse racing was stopped because of the epidemic, and working men who obtain much enjoyment from placing bets with local bookies were highly despondent. Fortunately, the *Evening Standard* realized that to bet on a horse race does not necessitate real horses galloping around a track – indeed, in the age of computers it is inefficient and a waste of manpower (to say nothing of 'horsepower'). Consequently, the paper made the following announcement: 'The *Evening Standard* today proudly announces that it has devised, and will stage, the World's First Electronic Horse Race.'[5]

It went on to say that the Massey-Ferguson Gold Cup, cancelled because of foot-and-mouth disease, would be run on the London University Atlas computer. A mathematical model of the horse race was programmed with the help of racing experts who provided details of the horses and their form over the previous two years, jockeys, distance between fences, and so on.

The computerized horse race met with the full approval of the National Hunt Commitee and the Cheltenham Racecourse Executive. The BBC Grandstand programme broadcast a full commentary on the race with commentators Clive Graham and Peter O'Sullevan sounding no less excited because the horses were not real. The commemorative Gold Cup was awarded to the winner, the jockeys received their normal fee.

Clearly this concept can be extended. With terminals in the home, the racing enthusiast can have a race any time he feels like it. It would probably be a great after-dinner entertainment. He and his guests could use the terminal to ask questions about the various horses' form, and the jockeys' past performance. Bets could be placed at the terminal and the computer might ask whether these are simulated bets or whether an actual cash transfer will take place. The race should be no less exciting than an actual race. There is no need to have the monotonous voice we sometimes hear on these machines. Even today's equipment can reproduce the voices and even the intonations of commentators like Graham and O'Sullevan.

Many other such games will be played with the terminals. Who knows what forms gambling might take in the computerized society, with the home-gambler's bank balance being automatically incremented or depleted. Perhaps one will be able to telephone the local Mafia computer.

HOUSEHOLD APPLIANCES

The telephone line, in addition to its normal use, will probably also be used for activating household appliances. A family driving home after a few days away will telephone their home and then key some digits on the touch-tone phone which switch on the heating or air-conditioning unit. A woman before leaving home for work will pre-program her kitchen equipment to cook a meal. She will then phone at the appropriate time and have the meal prepared. Or possibly a computer might telephone the equipment and instruct it step by step to perform a sequence given to the computer the night before – for example to switch on the oven at 3 p.m. to cook the roast that had been placed in it, and to move some vegetables out of the freezer compartment, leaving them to thaw. At the appropriate time the vegetables would be heated in their aluminium foil, and the dish-warmer would be switched on.

The lawn-tending machine might similarly work without human attention although this is a less probable application as people living in the world we describe will doubtlessly need the therapy of lawn-mowing. Operated by numerical control signals, the machine could roll out of its hut and mow its way around grassy paths and a lawn of complex shape, avoiding the crocuses in spring and removing the leaves in autumn. It would sense the edge of the lawn and trim it neatly. Before the machine can do this it has to be 'taught' the shape of the lawn, by being taken once around the circuit it is to follow. It 'memorizes' this, possibly by means of a magnetic tape unit like a domestic tape recorder, or possibly by storing its 'lesson' in a remote computer – being in contact with the household terminal via a short radio link.

TELEPHONE LINES

In the above discussion we have assumed that computer applications will require extensive use of telephone lines to the home. How expensive will this be? In terms of the physical facilities needed, the lines that today connect homes with local telephone exchanges would suffice. In most telephone systems two wires run from the home of each subscriber to the local exchange (central office). With the exception of party lines, these wires are not shared. It is staggering to reflect that in Manhattan, for example, there are two wires under the street for almost every private subscriber. Ninety-nine per cent of the time the authors' telephone wires are idle, as are most people's, a remarkable waste of an expensive facility. Furthermore, when they are in use these two wires carry less than one twentieth of the information they could carry.

Without any change in telephone lines, therefore, we could carry out all of the above functions if the appropriate computer's telephone lines were attached to the same local exchange (central office) as ours. There probably would have to be an expansion in the central-office equipment. The lines would be in use for a longer period of time, and so the number of simultaneous paths through the exchange would have to be increased somewhat. In general, though, there is no technical reason why our telephone bill for *local* computer calls should be much higher than it is today.

Long-distance calls to a computer are a different matter, for they do not involve a permanently connected pair of wires; so we can expect an increase in cost proportional to our increase in line usage. All this means that in such densely populated areas as New York or London the above schemes will not involve a high line cost because of the proximity of the relevant computer. For areas far from the population centres, line-cost will be high.

It appears that, initially, there will be a marked line-cost advantage in living in a metropolis.

VIDEO CHANNELS

If telecommunications line capacity continues to increase as projected in Fig. 1.6, our homes will have links of much higher capacity than a telephone line. The coaxial 'broadband' cables that bring closed-circuit and broadcast television to some homes today can transmit several thousand times more information than telephone lines. Unlike telephone lines, however, they cannot be switched so that you can dial other users. Broadband switching centres *are* presently being introduced to handle the 'picturephones' now being sold in certain cities by AT&T. When, as AT&T intends, picturephones are installed in American homes, each will necessarily be served by a switchable broad-band cable which could be connected to a computer as easily as to another picturephone. AT&T forecasts that by 1980, 1 per cent of domestic telephones and 3 per cent of business telephones will be 'picturephones.' The coaxial cables for television now being laid into many American homes have sufficient capacity to permit larger-screen TV. Eventually, the screens in our homes may occupy a major part of a wall. The distant computer and the user, comfortably ensconced in an armchair with a keyboard on his lap, will conduct their dialogue by writing on the wall.

This is only the beginning. In the advertisers' paradise of America, all manner of highly coloured catalogues will become available the moment the new medium arrives, and there will be varied enticements for exploring them. Very elaborate presentations of products will become possible. The Sears-Roebuck catalogue could now include film sequences, although the user would still be free to 'turn the pages', to use the index, to select and reject. As with American television, advertising would help to pay for the new medium. Perhaps critical consumer guides will also become automated to aid product exploration. Having scanned the relevant catalogues and inspected pictures of the goods in detail, the shopper could then use the same terminal to order items, the money being automatically deducted from his bank account.

Selling abroad would be facilitated. Some large stores and firms would offer automated presentations of their products in most of the cities of the world. The consumer relaxing in his New York apartment could go shopping in Paris or London or Tokyo. Planning exotic vacations will probably be a popular pastime whether the planner can afford to take them or not. Travel agents or foreign governments will make exciting catalogues available to the home computer user, possibly with film sequences.

PICTUREPHONES

We shall not see these schemes in the home for some time. However, today's cable TV and picturephone bring their day much closer. In ten or twenty years, most office buildings will include a picturephone room, and executives will have such machines on their desks. Much of the need to travel will be eliminated. Engineers and systems analysts will be able to discuss their block diagrams or drawings on the screen with distant colleagues. By aiming the picturephone lens at their pads of paper they will be able to sketch out their ideas and develop arguments which today need face-to-face conversation. In selling, too, the screen will be used. A picturephone conference room for group conversations will become a necessity. Boardrooms will have sets of screens with picturephone and computer links.

As with other terminals, the first domestic users will be businessmen who need office facilities at home. Like other technologies, this will improve, and before long will grow to the critical point when it achieves mass acceptance. Mass-production lines will be set up, prices will come down, and the home picture wall will supersede the television and telephone as the principle medium of communication with the outside world of people and computers.

REFERENCES

1. *Forbes*, 15 March 1967.
2. *Forbes*, 15 March 1967.
3. James Martin, *Design of Real-Time Computer Systems*, Prentice-Hall, Englewood Cliffs, New Jersey, 1967 (see Chapter 8).
4. James Martin, *Telecommunications and the Computer*, Prentice-Hall, Englewood Cliffs, New Jersey, 1968.
5. *Evening Standard*, London, 9 December, 1967.

8

THE USE OF COMPUTERS FOR CONTROL

FEEDBACK-CONTROL

Imagine a boy riding a bicycle along a winding country lane. There is a certain path, perhaps a white line in the centre of the lane, which he wishes to follow. If he is alert, he may deviate only slightly from this intended course. Next imagine that he is blindfolded by a friend and that they play a game in which he must steer his bicycle on the basis of instructions yelled to him from somebody following.

There will now be a *time lag* between his deviating from the correct path and his knowing that he should correct this error. Because of this time lag he will swing far away from the optimum course. If the instructions shouted at him are fast and precise, however, he can still manage to ride his bicycle roughly down the lane. If the instructions are not fast, or not precise, then his oscillations about the intended path will become greater. He may swing dangerously close to the edge of the road, in which case his guide will yell in a more alarmed tone of voice. Quite probably he will then *overcompensate*, swinging back across the lane too sharply. His guide will shout at him to rectify the overswing, but his course will now be a succession of large oscillations.

Oscillations also occur in the economy of a country, in the profit or loss of a corporation, in the amount of stock held by a distributor, or, on a much smaller scale, in the voltage of an improperly stabilized electronic circuit. All of these oscillations can be controlled to some extent if suitable machinery for control is devised. In order to control them it is necessary to *feed back* to the controlling mechanisms information that describes how the process is deviating from its intended goal, much as the guide shouts information to the boy on the bicycle. The time scales of the above processes vary widely, but in all the longer the time taken to feed back the information about the deviation, the greater will be the oscillations. Also, if the information about the deviation is *inaccurate*, the oscillations will be relatively large.

A wide variety of control mechanisms is used for governing processes and devices. Some of these are very simple, like the governor on a steam engine; others are very complex and require a computer. An important application of computers in the years ahead will be the control of complex processes – administrative as well as mechanical or chemical. Many of today's operations will be controlled more efficiently and new types of operations that would have been too complex to control before the computer era will become possible. To achieve control:

1. Data must be collected, or measurements made, of the performance of the entity being controlled.

2. The data must be compared with pre-established objectives.

3. Action must be taken to correct any deviation from the objectives and to steer or schedule future operations in such a way that stable optimal performance is attained.

One such mechanism that does this, using the feedback of performance information, controls the flow of electrons in electronic circuits. Others govern the functioning of most living organisms. Feedback-control provides the key to automation by enabling machines or processes to be governed without human intervention. A man driving his car is exercising feedback-control over it. A missile heading toward Mars is also being steered using feedback, but operated by an electronic rather than human link: electronic sensors tell it if it is deviating from its course, like the guide shouting at the boy on the bicycle but much faster, and it then compensates for the deviation. In controlling the flow of work through the shop floor of a factory, information about performance is fed back to those responsible for dispatching further work, and they control the flow of operations accordingly. In this case computers can collect data more quickly and accurately from the factory floor and provide them with a form of *management information system* enabling them to control the process more efficiently. Eventually the computer itself will exercise much of the control.

Feedback-control theory is of prime importance to engineers in the design of control mechanisms for electronic or mechanical devices, for chemical plants, for reactors, for the stabilization of guns on moving tanks, and so on. The concept is also of value to psychologists studying the mechanisms of the human body. Its applications to industrial and administrative organizations are perhaps less obvious, but here again feedback-control is needed to monitor and stabilize what is happening.

To achieve control of a situation, a *goal* or *desired state* must be specified. The devices used for control have the function of trying to achieve this objective.

Figure 8.1 shows a process being controlled. The input to the process leads to an output which is compared with the standard objectives. Because certain uncontrollable conditions affect the process, there may be considerable deviation from the intended output. The variation and rate of change of variation measured by the comparator device are fed back to an adjusting mechanism (which could be an administrative process that involves human decisions as well as mechanized ones). The adjusting mechanism varies the input and controls the way the process operates in order to achieve the desired output.

The specified objective may be to keep a temperature constant, to maximise the output of a chemical plant, to maximize the profit of an undertaking, or to realize a particular government objective. The goal or desired state may be fixed or varying. If varying, it must continually be re-established – either by a human agency or by part of the mechanism itself – to meet changing conditions. The boy riding down a winding lane represents a control system seeking a changing goal; a factory responding to varying customer demands behaves analogously.

The state of the actual system is normally not the same as the desired state. The difference sometimes occurs because the goal is changing or because there are disturbing influences on

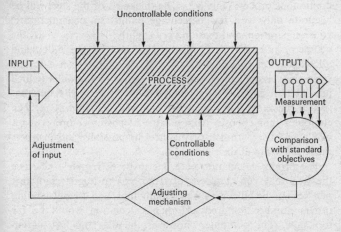

Fig. 8.1

the parameters affecting the system. *Feedback-control is designed to reduce the difference, or to bring the actual state of a system as close as possible to the desired state.* In order to do this, the difference must be measured so that it can be used to trigger an appropriate corrective action. The measurement which gives rise to the controlling action may be the *magnitude of the difference* or the *rate of change of the difference* or both. The action taken may be simple, as in the cases above, or it may require complex computation, particularly when it must measure and alter not one variable but many.

An interesting example of a complex calculation is the control of a flying shear in a steel-rolling mill. The billet of red-hot steel appearing from a final rolling process has to be cut into lengths to meet customer orders. The goal of this process is to have the minimum wastage of steel after the orders have been cut. In a simple case, if the length of the billet were 99 feet, cutting nine 11-foot lengths for customer B is better than than cutting nine 10-foot lengths for customer A and throwing away the odd 9-foot length.

Unfortunately, the length of the billet has to be calculated as it appears from the rollers, since cutting is an integral part of a continuous process (Fig. 8.2). The thickness of the steel will be accurate only to a few percentage points, since temperature, roller separation, and tension very slightly. However, given the geometry of the rollers, the original size of the billet, and allowances for edge and end losses, a computer can calculate from the speed of the front and rear ends of the billet (which are different since the same volume has to go through a different gap in the same time) what the final slab length will be. The computer thus chooses from a table of customer orders a set which will use as much of the steel as possible, and instructs the shear to cut at the right times.

The proper calculation here has to be extremely fast, since the steel may run at speeds of from tens to hundreds of feet per second, the billet may weigh many tons, and the outstanding orders number hundreds. Human calculation would be far too slow. The computer, however, handles this calculation along with many others designed to control roller separation in the reversing mill, slab temperature, and surface quality. In every case, it takes measurements, compares what is with what ought to be, and acts to improve performance in the split-second when the red-hot billet is hurtling down the steel mill.

TIME LAGS

Feedback-control is further complicated by *time lags* – one time lag between the sensing of an error, or a change in error, and the corresponding action being taken, and a second lag between initiation of the corrective action and the response of the system under control. In some systems the time lags are small compared with the rate of change of the variables being controlled, and so they have little effect on the operation of the system. In other systems, however, inertia is greater or the events affecting the system status may be slow. In these cases time lags significantly affect the performance or stability of the system.

The time scale we are talking about varies enormously, de-

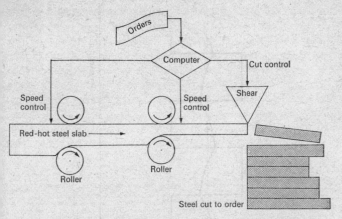

Fig. 8.2 Fast computation and control are needed to cut up a red-hot billet of steel, hurtling at tens to hundreds of feet per second, into the varying lengths that customers have ordered. Certain other computer systems are analogous to this.

pending upon the nature of the mechanism being controlled. In an electronic device the time intervals are measured in milliseconds or even smaller units. For a man driving a car, the responses may take a second or so; for a production control system, perhaps a day or so; for many administrative processes, weeks; and for governmental processes, months or even years. In spite of this vast variation in scale, the processes can be described in similar terms. A similar form of mathematical equations for feedback, stability, and so on, would be used even though the time parameters have widely different values.

OSCILLATIONS

Figures 8.3 and 8.4 illustrate the difference between a quick acting control system and one with a longer response time. The thermostat maintains the room at almost constant temperature, whereas the man permits large swings in temperature to occur because he does not act quickly or often enough. He

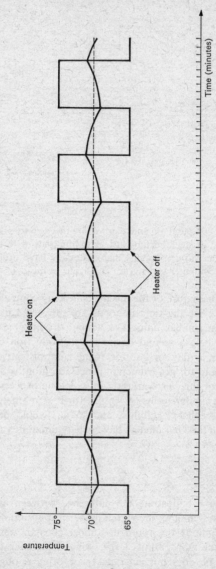

Fig. 8.3 Room temperature controlled by thermostat.

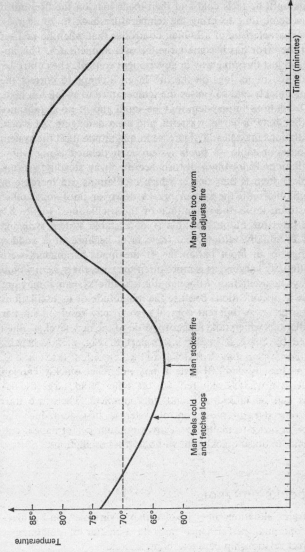

Fig. 8.4 Room temperature controlled by man stoking log fire.

waits until he feels cold and then loads logs on the fire, but it takes some time to bring the temperature back to 70 degrees. It is characteristic of a human controller that when he realizes that the error has become large, he overcompensates. This applies in just the same way in government control. The controller loads too many logs on the fire in an attempt to correct the error quickly and so causes the temperature to swing too high.

Oscillations in a system may be small and of no significance or they may be large, wasteful, and even damaging. At worst, oscillations may steadily increase in magnitude until the system becomes unstable – a *stable* system being defined as one whose oscillations following a disturbance die away steadily. An unstable system is thus one in which oscillations can increase in magnitude until the system breaks down or until some other constraint limits the magnitude of the oscillations. Figure 8.5 illustrates the output variable *A* of a stable system swinging and eventually settling to a new level because of a sudden change in an input parameter. If the input parameters were continually varying, as is more often the case, the system would always be oscillating. Although it is a 'stable' system, it may not meet its specifications because the magnitude of its oscillations is too great. A different control system may result in smaller oscillations which take longer to settle to the new level, as illustrated by curve *B* in the lower part of Fig. 8.5. A heavily *damped* system may drift slowly to the new value, as in curve C.

The heavily damped system may not react quickly enough to achieve what is required. On the other hand, large oscillations can be undesirable and uneconomical. There are thus two objectives of most control systems: first, to reduce the error in order to maintain steady, optimum performance; and second, to do so quickly but without gross oscillations.

PROCESS-CONTROL

Complex situations may involve many variables, and achieving optimum performance may be a matter of obtaining a correct relationship between many factors.

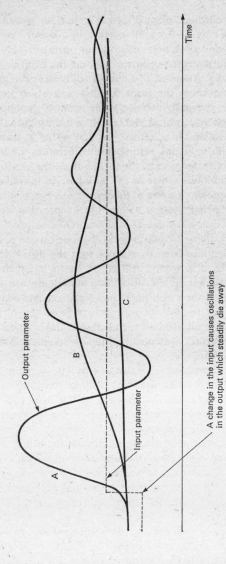

Output parameter

A

B

C

Input parameter

A change in the input causes oscillations
in the output which steadily die away

Time

Fig. 8.5

In a large chemical plant or power station an array of dials and controls may be brought together in a control room for the scrutiny and operation of a human operator. The operator makes adjustments to controls running the plant according to present rules. Although this human adjustment produces an operation that keeps the plant running and which has been acceptable in the past, because tighter control has been too difficult to achieve, in most plants it does not produce the best possible operation. It is rather like comparing a thermostat operation with a human setting the temperature, as in Figs. 8.3 and 8.4. Greater profitability arises from the capability to operate continuously closer to the optimum. Swings like those in Fig. 8.5 represent a waste of money.

A computerized system designed for operation at near-optimum efficiency is in use at American Oil Company's Whiting refinery,[1] where 240,000 barrels of crude oil are processed each day. Pipelines from nine states feed the distillation unit and 2,000 products from fertilizers to motor spirit are produced. In addition to the distillation unit there are a fluid catalytic cracker, in which finely powdered catalyst constantly circulates in the reaction chamber where molecules of gas oil are split, and an ultraformer for upgrading the octane number in gasoline, a process patented by Indiana Standard, American Oil's parent company.

Each of the processes is monitored by 200 or more instruments. In the distillation unit, these measure furnace efficiencies, tower loads, and product quality. The readings are sent via terminals along telephone lines to the computer. Each terminal has two typewriters, one routinely logging each measurement, the other giving alarms or indicating unusual conditions and permitting operator communication with the computer. The computer compares readings with values calculated to give the most profitable yield, and automatically adjusts the process variables, unless overruled by the operator from his terminal.

The control of continuous refining processes is easier than the control of batch processes such as smelters, since most of the task involves the *maintenance* of desired conditions, not achieving them to start with. In the first automated sinterplant in the non-ferrous metals industry, opened in May 1968 at the Avonmouth plant of the Imperial Smelting Corporation, an Elliott ARCH 2000 computer reads 200 instruments and manipulates 80 operation variables to control the blast furnaces. The computer starts with the composition of the charge, taking note of the specifications of the process-operators, of the availability of preheaters and of the sinter and lime in the bunkers. The transfer car is moved from filling station to filling station, picking up a load weighed out as required and then discharged into a hoist to charge the furnace, all under ARCH's supervision. Air distribution at the bottom of the furnace and gas temperature at the top are measured to maximize profit. Later, ISC expect to use X-ray fluorescent analysis to determine, with the aid of the computer, sinter quality, and thence to calculate changes in the weight of the charge components in order to get the best results.

In each of these cases the changes in efficiency are proportionately quite small, but the processes are so large and the total output so expensive that a 1 per cent improvement in yield easily pays the computer bill. Nowhere is this more obvious than in paper mills. At the Cellulose du Pin Kraft Paper plant in Facteur, France, electronic control of moisture, using a computer and sixty instruments in the block-long drying room, maintains constant paper thickness. At the Wolvercote Paper Mill near Oxford, England, paper weight is held to within 1·5 per cent of the required value, as against 2·5 per cent in the past, and changes of grade take five minutes instead of twenty. This precise control cuts waste and substandard output dramatically.

Computer control of start-up – or countdown, as it is called at Cape Kennedy – is becoming common, mainly because men are unreliable during lengthy processes and might forget something. Starting steam turbines can be almost as critical as starting rockets or nuclear reactors. Speed of rotation, temperature of steam, and turbine initial temperature have to remain compatible from the stationary position to full speed.

Sometimes processes have to be controlled in hostile environments where electronics would fail due to radiation or heat. Here, fluidic computers ('calculating bagpipes') have been used. In these the basic on-off switches are miniature pneumatic valves instead of electronic circuits. Honeywell has applied these to the maintenance of flying-suit temperature for pilots of aircraft and spaceships. Where air-conditioning cannot meet the computers' environmental demands, as in some machine shops, lathes can be controlled numerically in this way. Since they often use pneumatic controls and sensing devices anyway, the integration of the calculating logic is eased.

CONTROL IN ADMINISTRATIVE PROCESSES

Feedback-control in an administrative process attempts, as in a chemical plant, to maintain a defined objective or set of objectives. The oscillations that take place in administrative processes are considerable. Often they are very wasteful of resources and insufficiently understood by management. The discipline 'industrial dynamics' refers to the study of oscillations.[2]

The nature of oscillations that occur can be demonstrated simply, with pencil and paper. (See Chapter 14 of James Martin, *Design of Real-Time Computer Systems*, Prentice-Hall, Englewood Cliffs, NJ, 1967.) Each executive or decision-making person in an organization reacts to the information in his possession. He uses a certain set of decision-making rules, formal or otherwise, to order stock, to initiate work on certain items, to take financial action, and so on according to the nature of his job. Unfortunately, a given individual's information may not be up to date. There may be a time lag in implementing the decision he makes, or the effect of his decision may not be felt until some time later – a sharp movement of a yacht's rudder may not cause the yacht to turn as required for half a minute or so, and by then the wind may have changed again. Again, his decision-making rules are frequently not ideal. Often, like the boy on the bicycle, he overreacts.

In an organization there are several such decision-makers

operating nearly independently. Their behavioural characteristics or decision-making rules form a mechanism which is going to give rise to some form of oscillation, just as the man stoking the fire in Fig. 8.4 causes oscillations in temperature. Such mechanisms, however, are much more complex than that in Fig. 8.4, and the cause of the oscillations is going to be less obvious.

Jay W. Forrester and his co-workers on 'industrial dynamics' studied the events that occur in actual industrial situations and produced numerous curves such as those in Fig. 8.6. This figure illustrates a situation in which the demand for a product fluctuates in a random fashion. A retailer–distributor–factory system reacts in such a way that, although everybody appears – at least to himself – to be making rational decisions, the result is a severely fluctuating load on the factory. The retailer responds to his customer's demand by placing whatever order he feels best with the distributor. The distributor has rules for controlling *his* stock based upon how much the retailers order. Periodically, perhaps once a week, distributors place orders with the manufacturer. The factory fills these orders where possible from within its current production plans which are then adjusted on the basis of what is ordered. The factory withdraws raw materials and components from its stores as required, and these must be replenished regularly.

A model can be built of this retailer–distributor–factory system in which a computer copies (simulates) all of the decision-making processes. The model can be made to behave in the same way as the actual system, but at a much greater speed. Using such a model we can investigate how the system reacts under various circumstances.

A number of conclusions emerge from such a study. First, in a chain of events, oscillations in an early link in the chain – retailers' orders, for example – often give rise to swings of greater magnitude further down the chain. Second, lessening the time delays generally reduces the size of the oscillations. Many time intervals are unavoidable – production time, delivery times, and so on. However, some could be lessened – for example, those associated with decision-making and the

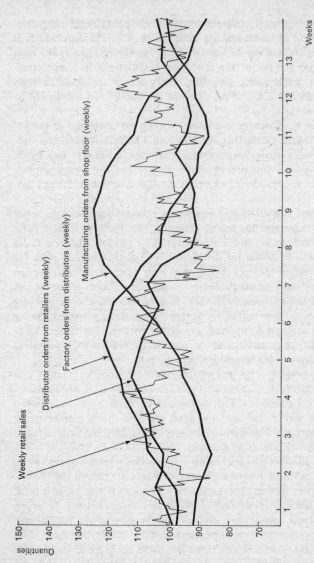

Weekly retail sales

Distributor orders from retailers (weekly)

Factory orders from distributors (weekly)

Manufacturing orders from shop floor (weekly)

Quantities

Weeks

Fig. 8.6 Response of systems to random variations in sales.

processing of information. In this example the greater the delays, and in particular those associated with decision-making, planning, and data processing, the greater the fluctuations that are likely to build up in such factors as stock levels, production, work in progress, and manpower requirements.

If sales are held constant in the model for a long period of time, the oscillations die away. If a sudden change in sales volume then occurs, a wave of changes passes through the system. Typically, if a 10 per cent rise in sales occurs, the retailer looks at his stock and finds it more depleted than usual, and furthermore it is depleted at a time when sales are up by 10 per cent. To make up the deficiency he therefore increases his weekly order by *more* than 10 per cent. The distributor sees the orders up by, perhaps, 20 per cent. He assumes that this might continue and he wants to hold enough stock to be able to fill orders several weeks ahead. He therefore places an order *larger* than 20 per cent with the factory in order to bring his stock up to a suitable level. The factory similarly overcompensates. Then the retailer finds that he has too much stock and drops his order for one week. The distributor now has too much stock and so overcompensates, and so on.

Once the model is built, a variety of decision-making rules can be used with it to see which is best. Most simple decision-making rules, however, give rise to some form of oscillations, because of the delays inherent in the process.

It is interesting to note that the *frequency* of the oscillations is a characteristic of the *system* and its delays, not of the sales pattern. Most organizational mechanisms, as with the electronic and mechanical systems discussed earlier in the chapter, have a tendency to oscillate with a frequency that is characteristic of the organization. The cycle time of the oscillations may be weeks, months, or even years, depending upon the time that elapses between an action and the results of that action. Forrester points out cases where swings caused by the organizational mechanism are incorrectly interpreted as arising from seasonal fluctuations in sales or from characteristics of consumer behaviour.

The oscillations we describe represent a substantial loss when viewed in terms of waste of productive capacity and of profitability. They cause an excess of capital to be tied up in stock; they give rise to bad customer service; they cause periods of underutilization of men and machines, and at worst could cause unnecessary unemployment and lay-offs. Tight mechanisms of control – as in the cases of the steel mill or petroleum plant – can save much money, particularly if product demand fluctuates violently and unpredictably, as with fashion goods or pop-music records. In such cases quick and sensitive reactions to changes are needed, coupled to a well-designed computation of the best course of action. The latter may involve computer model-building and simulation.

OSCILLATIONS IN GOVERNMENT

Oscillations of long duration occur in the economy of a country. Figure 8.7, indicating the misfortunes of the British economy, bears some resemblance to Fig. 8.6. The oscillations are caused in a somewhat similar manner, too, although on a much grander scale and with more complex mechanisms. The major effects on a country's economy of a credit squeeze or other government action are not usually felt until a year or two afterward. A trade cycle develops with intervals of some years between booms and depressions.

It is interesting to speculate to what extent the swings in the running of a country could be lessened with better computer use. The political press bemoans these oscillations almost constantly. The following leading article from the *Observer* is representative of the staple weekend diet of the British reader; one can find similar passages in the press of other countries about their governments:

We are once more enduring the spectacle of a Government wrestling with an entirely predictable crisis – with the Prime Minister on Tuesday delivering his packet of emergency cuts into the lap of the House of Commons with the air of a man dumping an unwanted baby on somebody else's doorstep. Why? Both parties have, over the past 20 years of recurrent crisis, often taken tough, unpopular decisions – the trouble is that too many of these have, like the Conservatives' Rent Act or Labour's wage freeze, failed in their purpose. Senior civil servants are not a bunch of classicists ignorant of modern methods of cost benefit analysis – the trouble is that the whole machinery of Whitehall is not designed to employ these sophisticated methods.

What then are the weaknesses in our system of decision-making which explain the fiascos not just of the past few weeks but of the past couple of decades? One is the permanent capacity of Governments to let events take them by surprise. The second is that having been taken unaware, Governments then react by presenting policies with an air of omniscient infallibility as though they had actually intended doing what they were forced by circumstances to do ...

What's the cure? The most urgent need, as recent events have demonstrated, is for Governments to be prepared for the unexpected. For the most depressing revelation of the present crisis is that Governments often are not even prepared for the expected: after all, devaluation had been widely discussed long before it happened – yet when it came, Ministers were caught with their policy trousers down.

Partly, this may reflect the congenital optimism of politicians in general and of Mr Wilson in particular. But largely it reflects a lack on the part of the Civil Service, not of ability but of the capacity to think through its main functions in today's circumstances ...

As in the brontosaurus and other now extinct monsters, its brain is too small for the still-growing body of minor functionaries.[3]

The time scale on such a diagram as Fig. 8.4, if it referred to the effect of government actions, would be in months rather than minutes. The effect of a budget change may sometimes not be felt for years afterward. When the government is 'caught with its policy pants down' and takes one or other emergency action, it reacts like our boy on a bicycle when he suddenly realizes that he is heading for the ditch and so overcompensates.

Two things are needed to correct this situation. The first is detailed, accurate, up to date information: the information currently available for decision-making is inadequate and out-of-date in many areas of government. The second is the ability to test the effects of different actions against such information. Without any doubt, both of these requirements would benefit greatly from the use of computers, as both involve highly complex processes.

In several countries computerized mathematical and simulation techniques are being used to study the behaviour of the economy. Step by step, the forces which change economic and social patterns are becoming better understood. However, much data of the type that would be collected by a government computer network is needed to test the economic and social models that are being built. (We shall be discussing 'National Data Banks' in Chapter 12.)

The second step to better government control, after setting up mechanisms to collect the requisite data, is to use such models. With experience and experimentation, their accuracy or validity will improve. Computers will thus be used to simulate the likely effects of various government actions. Thus, although all manner of eventualities may arise in the near future to upset a government's plans, most of these possibilities could be foreseen. Computer models will be used to investigate possible alternative futures that can be anticipated. Probability figures will be associated with these alternatives and such estimates will become the basis of government policy.

Computer control of less complicated processes may pass through three phases, a different system operating for each phase. The first phase involves setting up a system for collecting and analysing relevant data. In the second phase of 'open-loop' control (Fig. 8.8), in which the computer instructs human operators how to control the process, a set of decision rules and tables will have been established as a result of the extensive data analysis. The third phase – if it is ever achieved – is 'closed-loop' control, in which the computer, rather than human operators, takes action to control processes. This third phase will never be achieved in the government of a country, and the final decision about governmental actions will remain in the hands of the politicians. However, they will then be making decisions from a position of strength. The computers will be providing them with the data they need, an early warning of adverse trends, and a summary of the probable effects of different courses of action. We shall return to this theme in

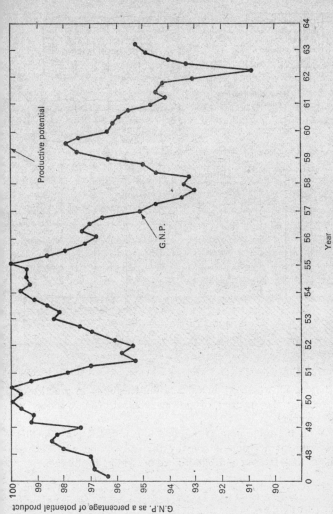

Fig. 8.7 Oscillations in a national economy. Britain's gross national produce as a percentage of potential product.

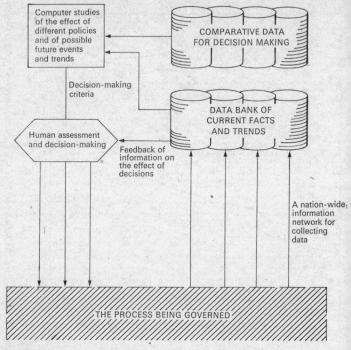

Fig. 8.8

Chapter 13, where we investigate the feasibility in practice of extending the theories of process control from business to government.

With the tools now at our disposal, it seems possible that government decision-making by the end of the next decade might look less like our blindfolded boy riding his bicycle.

REFERENCES

1. *Petroleum Times*, 10 May, 1968. *European Chemical News*, 10 May, 1968.

2. See Jay W. Forrester, *Industrial Dynamics*, MIT Press, Cam-

bridge, Mass., and John Wiley & Sons, New York, 1961. Professor Forrester discusses in depth the oscillations that occur in industrial organizations. He used a computer to simulate these and to evaluate methods of feedback-control.

3. Rudolf Klein, *Observer*, London, 21 January 1968.

9

THE COMPUTERIZED CORPORATION

Fortunes told to 17 decimal places of accuracy.
– *Gypsy in* Datamation *cartoon.*

Management-information systems are currently high fashion.
Hundreds of corporations have therefore promoted their present computer systems with a deft change of title, and continued to manage in their old, well-established – and usually successful – ways.

The computer literature is rich with potential and planned schemes to improve managerial performance. Nevertheless, few such schemes today have succeeded in giving management much of the information they need to function efficiently. Many are in the early stages of what might prove to be a lengthy evolution.

There are many other aspects of running a corporation in which computers play a part. To indicate where this technology is heading, let us invent a corporation and describe the part that computers could play in its functioning in the early 1980s.

Widget Spleeters Incorporated* is a corporation typical of the new growth industries of the 1970s and 1980s. With the development of new organic compounds grown in the hydroponic 'gardens' possible in the ultrahigh vacuum of low earth orbit, there came a great demand for fully spleeted widgets. To make these, a variety of naturally occurring raw materials have to undergo a closely controlled semicontinuous process in an inhospitable environment. The resultant product must be combined in batches with numerous different parts from many suppliers. These are then sold vigorously to other corporations in the widget industry.† Annual model changes keep the design

---

* The Random House Dictionary defines a *widget* as 'something considered typical or representative, as of a manufacturer's products'.

† Federal law prevents vertical integration of the widget business.

force busy and the tied credit company involves the corporate finance function in finding and allocating funds wisely. Faced with enormous expansion, internal management development and external recruiting are vigorous.

Widget Spleeters Incorporated makes massive use of computers in every area where is it economically feasible. Often WSI used computers in the 1970s in areas in which their cost could not at that time be tangibly justified. It did so in the belief that the hardware costs would drop in the next few years to a level at which the application would be cost-justified. This policy recognized that the lead time in program and data-base development was long and that it was necessary to build up a data-processing team with the relevant experience as early as possible – human talent was the bottleneck that was preventing the most profitable use of computers. This policy eventually paid off and to a large extent is responsible for WSI having become the leader of its industry.

Like many firms who use automation effectively in the early 1980s, WSI demands a high level of intelligence in its staff. One half of all its employees are college graduates, many culled from the less developed countries where the raw-material process is conducted. Of the 70 per cent of the population with an intelligence quota below 110, only a handful find employment at WSI.

Most of the laboratories were established overseas because of the high productivity and low pay rates (compared with the domestic sellers' market for researchers). The largest is in Japan. Several plants were also placed overseas near foreign markets because of tax concessions and foreign government restrictions.

An entire division of WSI is devoted to its data-processing; the head of this division has a very close working relationship with the president. Eight of the twenty-five vice-presidents are from the Data Processing Division.

THE INFORMATION-CONTROL ROOM

The showpiece of the WSI Data Processing Division is a control room at their Head Office that looks somewhat like the Houston control centre for the Project Apollo shots in the late 1960s. It is referred to as the 'War Room', and is staffed by about eighty men each sitting at a desk equipped with a keyboard, television-like screen, and small printer. Around the high walls are large computer-operated projection displays and a glass-enclosed visitors' gallery. This room is the nerve centre for data-processing operations throughout the corporation. It is connected to the Head Office computer complex, which maintains the massive control data base, and also, via a switching computer, to the data-processing centre in each of WSI's plants. The computer in each plant has its own data base and its own (much smaller) control room.

The War Room has a variety of functions:

1. The staff there are responsible for the corporate data base from which management obtain much of their decision-making information. Errors and wrong information inevitably find their way into the data base. There are many checks for detecting these, and when errors are found the War Room staff are responsible for correcting them.

2. Some management questions are too complex to be answered easily from terminals in remote locations; these are switched to the expert staff in the War Room. The War Room staff can cause any display to appear on the screen of a manager's office, and can monitor it on their own screens.

3. Emergency situations occurring anywhere in the corporation may be referred to the War Room, from which centralized expediting can be directed and the situation monitored.

4. When local management wishes to override a decision of their local computer, they contact the War Room for arbitration – as, for example, when a district sales office's computer says that a certain order that management considers important cannot be taken or met.

5. It has long been recognized that computers should not be permitted to make *all* decisions involving the control of a set of operations. The machine must be programmed to recognize when the intervention of a man *with experience* is necessary. The expertise for human intervention was collected together and made accessible in the War Room.

6. The Board Room, also equipped with terminals and screens, overlooks the War Room. This makes the highly specialized staff in the War Room accessible for answering questions that arise during board meetings and for generating appropriate charts on Board Room screens.

MANAGEMENT INFORMATION

As in other corporations, management at WSI yielded more slowly to colonization by the computer than any other territory in the business. When mere operating personnel pleaded that their functions required art and experience and could not be reduced to a mechanical, if high-speed, routine management used authority to overrule them. Computers were installed, albeit with mixed results. The managers were, however, much better placed to defend their jobs, because remarkably little was known about how they performed them – or even how their time was spent.

Managers make decisions – strategic, tactical, or operational – according to their place in the hierarchy. They must also train, evaluate, motivate, and control their subordinates, which is the real management art.

But there is also management science. By means of his decisions, a manager translates information into action. Immediate decisions are often made necessary by circumstances outside his control. Frequently, however, there is only one correct course of action, and this can be deduced from the information available at the time. In the jargon of management science, such problems are 'well-structured', and require only the application of a set of rules to calculate the answer. A typical example is determining the quantity that must be ordered to replenish stock and the level to which the stock can fall before signalling the need for the next order.

With a non-automated information system, management operates with a wealth of comprehensive reports about what happened days or weeks before. They hear about crises from colleagues or subordinates long after avoiding action should have been taken. A manufacturer needs to know when he will run out of parts in time to order and receive more, not when the storeman goes to an empty bin. Much managerial time is spent resolving crises that adequate forewarning could have averted.

At WSI, the management-information system is designed to give managers predigested information, not raw data. Instead of inch-thick daily reports, pertinent abstracts are provided *when the need arises*. Many of the reports reflect the fact that an approximation today may be of more use than precision tomorrow.

The system handles two classes of decisions automatically:

1. those for which the rules are known, such as payroll calculations;
2. those for which a policy has been developed, such as dispatching spare parts to customers at any cost.

The third major class of decisions – those for which experience, judgment, and intuition are essential, such as authorizing the investigation of new products – remains beyond the scope of the automatic data-processing system.

In many ways it is not so much an information system as a 'control' system, taking data from on-going processes and evaluating them to determine what actions to take to adjust the processes to their goals. Data-collection goes on in 'real-time' – that is, at the same time as the events which give rise to the data take place. Corrective action takes place only in 'relevant' time – that is, soon enough to be useful but not instantly (since this is usually very expensive). Relevant time varies from decision to decision. When parts run out on the production line or a machine breaks down, the foreman knows within seconds. When a delivery fails to arrive, the purchasing department knows by the next day. When a budget overruns, the manager knows by the end of the week.

Many events are recorded by the system, but few trigger action. Those that do are either events that should not occur, such as non-payment of an account, or 'last straws' that break some camel's back. The most common such event is when the stock of a particular part falls too low – when this happens action is essential. The information system is full of trigger levels of this kind. The computer merely logs most events, but acts when a trigger level is reached.

Management has little time to be interested in that which is going according to plan. Nevertheless, a few reports appear periodically saying 'all's well' merely for encouragement. But what matters are the exceptions.

DATA SECRETARIES

Much of the information that management need, they have to ask for. They must interrogate the computer systems and their data bases. For this purpose some managers have terminals in their offices. Others have television links to the nearest information-control room where the questions can be dealt with.

At WSI it was appreciated early that most managers had difficulty making good use of their terminals. Most learn to carry out simple operations on them, but a few did not even learn that. The functions that the terminal was capable of, however, became steadily more complex, and the level of expertise needed to take full advantage of it steadily rose. Precise language was needed in formulating questions, and a detailed knowledge of the available files and programs was required.

Because of this, a corps of specially trained personnel was set up. Referred to as 'data secretaries', their function is to inform and assist management in the use of the computers. Some managers have their own personal data secretary, others summon one from a pool when needed. The data secretaries train the managers to carry out simple functions at their terminals and perform the more complex operations for them. They explain to management the operations possible with the system and encourage and guide them in its use. Filing cabinets have largely disappeared and, instead, documents and data no longer on documents are stored in the computer files. The data secretary does a manager's filing and retrieval as the clerical secretary used to do.

The pliable human link has greatly enhanced the use of computers at WSI. At first most managers tended naturally to be somewhat frightened of the computers, and the data secretaries have helped bridge the gap.

Not all the effects of computing have been good at WSI, as we shall observe later in the chapter. First, however, let us describe some of the specific uses of the machines.

RESEARCH AND DEVELOPMENT

At WSI a large group of scientists works on fundamental widget theory. Abstracts of their papers are accessible at the retrieval terminals in all major scientific libraries, and originals are held on microfilm for facsimile reproduction. In their laboratories the scientists have a few small computers capable of rapid input and output as well as plotting, visual display, and keyboard entry. These machines will call in the company's master network for major 'number-crunching' operations or for large files of storage. The rest of the data are kept in self-sealing containers, in desk drawers away from the snow of tobacco ash. The scientists load these disks and tapes when they need them, since they do not know which major machine will back them anyway. Any of the system's general-purpose programs is available, however, and these are frequently cannibalized, copied, and modified for incorporation in local programs. Most of the users were taught by the machine itself how to write in simple computer languages such as APL and PL/2 as well as a variety of more specialized languages, and have long since passed the stage when writing a program took longer than solving the problem by hand.

Despite the importance of the basic work, it is the applied and developmental research that gives WSI immediate competitive advantages. Here the computer is used in several ways:

1. as a record-keeper for a widely dispersed team;
2. as a rapid means of processing experimental data and doing calculations;
3. for 'experiments' – in a simulated environment – that in the 1960s would have required a physical (and inflexible) laboratory environment;
4. as an extension of the designers' senses.

Like many widely spread international organizations, WSI links its laboratories by a teleprocessing net using satellite transmission. In one location an engineer may be producing a much improved version of a component widely used in many models of the company's line. In another location an engineer may be designing a new model to use the new part. Only one record of data on the old part is kept by the organization. Thus when a change in the part is contemplated – as in this case – the designer asks the computer where it is used in established and development equipment. The designers of both the component and the unit are notified by the computer, and each assesses the value of the improvement to determine whether it is an adequate model – its specifications, method of production, parts used, release dates, designers' names, numerical control tapes, packaging, and much more – is maintained, updated, for delivery to any plant on the day preparation for production begins. Some of this material will appear as routine tickets for the individual batches as they are made on a succession of machines. The rest seldom appear as 'hard copy', since the shop floor has display units for reading such information straight from the computer.

DESIGN WORK

Routine design used to absorb a lot of middle-grade engineering talent at WSI. Then an extensive data-retrieval system was instituted so that not too much good work would be wasted and so that most designs could be reused. They discovered, however, that making alterations to accommodate slight specification changes was inefficient, either because of relatively small improvement in product quality or because extensive alteration was required. So they decided instead to put the basic logic of the design process into the computer and program it to make the same fundamental decisions as would an engineer working on a well-defined problem.

Draftsmen and designers now have standard design displays on which they specify the basic performance and physical characteristics of each of many standard components. At the terminal they answer a series of questions; then the computer checks each response for validity, and bases its next question on what it is told. When all the specifications are complete, the computer checks for standard parts to do the job. If its search is successful, it reports at once; if unsuccessful, it signs off with a promise to provide an answer in one hour. It then designs a non-standard item and the material from which to generate drawings. When the engineer calls back, he is shown the drawings. If he likes what he sees, the drawing is microfilmed, a parts list is printed, and the basic data for the engineering paperwork are stored ready for recall as necessary.

At each stage, it should be noted, the machine waits for confirmation from the engineer before taking the potentially costly next step. Designers change their minds frequently during development, even in the two hours it now takes to go from specifications to working drawings and manufacturing instructions.

Much of the design output is produced in graphical form using techniques established by aircraft and automobile companies and shipbuilders in the 1960s. Plotters produce accurate drawings; cathode-ray tubes produce images for viewing and photographic recording; electronic scanners can record graphic input, and under program control analyze and selectively copy it.

Besides their *passive* computer graphics system, which

accepts input and provides output in visual form in batch mode, WSI also has an *active* system operating in real-time through interaction with the designer.* The close cooperation between man and machine enables the designer to describe and analyse his design more efficiently than ever.

For sketching, the natural language of design, the engineer uses a light pen. The majority of numerical data he keys in by typewriter, but he can alter a value that he previously inserted by pointing with his pen and pressing 'function keys' to raise or lower the value continually until he releases the pressure. Spelling has always been a weak point with many designers and miswriting of part numbers is endemic. To avoid trouble, the computer frequently provides lists and tables of acceptable entries which move across the screen till the designer sees the one he wants and brings it to a halt.

Geometry is the basis of much of engineering design. Aesthetics enter through the acceptance of good shapes and the rejection of bad ones. Although two-dimensional sketches are better than one-dimensional strings of data entered from a typewriter, tape, or cards, in structural work three-dimensional views are needed. In a typical operation, the draftsman draws a curve on his 'Sketchpad',† the computer fits a curve to it, the draftsman asks the computer to smooth it, move it slightly, increase its curvature, and fix both ends on prescribed coordinates. He then calls up some prestored shapes and places them where he wants them, rotates the design to see the other side, expands the picture and inserts a tiny component called from store, files the design away where he can recall it in the morning, . . . and goes home.

A major advantage of this system is its ability to insist on the complete details of a design despite the draughtsman's lapse of memory. Once it has what it needs, the system can produce control tapes for a profile-cutting numerically controlled

---

*The terminology here is that of R. A. Siders et al., *Computer Graphics – A Revolution in Design.* (American Management Association, 1966).

†This is the name Ivan E. Sutherland gave to the system – employing a cathode-ray tube, light pen, and TX-2 computer – which he developed at the MIT's Lincoln Laboratory in 1962.

machine tool or a wiring scheme for an electronic circuit while the engineer returns to creative work.

PLANT DESIGN

In the past decade WSI has expanded rapidly and now designs an extension to an old processing plant or a completely new one at frequent intervals. These include mazes of pipes whose courses are plotted by the computer so that they clear each other by distances great enough to avoid even thermal interactions. After one local government disapproved of a plant they had built, they initiated a program for taking orthogonal projections and generating perspective views for photographing and approving. Plant design, as opposed to plant-item design, is performed by a generalized flow-chart program, which can call upon all the hundreds of special-purpose item-design routines as required.

MARKETING

Widget Spleeters Incorporated sells in both the industrial and the consumer markets. Like many large corporations who supply the basic ingredients for their small customers' consumer products, they find that it pays to advertise directly to the final consumer who asks for WSI spleeters by name. Nevertheless, they use representatives to call, place trade-press advertisements, and keep the automated directory up to date with their product specifications.

Like marketing men everywhere, WSI's Sales Vice-President reckons that much of his advertising is wasted, but he is confident that this is never more than one half.

Operations researchers on Madison Avenue back in the late 1960s tackled sales forecasting and media selection, and estimated and measured advertising effectiveness. Much of WSI's market analysis avoids costly one-time surveys by buying data on tape from specialist market-research-and-measurement companies. These data are fed straight into the computer to describe that market segment which can be reached by a particular campaign. Point-of-sale advertising in an unfrequented warehouse will be cheap and little noticed; a television spot during a Western reaches kids of all ages; and a trade-press back cover is very expensive. But WSI gets a statistical picture of who would see its promotions in each case. They also have a similar picture of what kind of people buy widgets.

When a new campaign is to be launched, the computer is fed with the customer profiles, the media-exposure details and their costs, and the effects of saturation – when any more advertisements cease to help sales. The marketing team then specifies the characteristics of the people they wish to reach for the product line to be shown, and the computer prepares a schedule showing what space to purchase to reach the largest number of potential buyers and the probable success rate for each extra dollar spent. When the probable returns in extra sales match the extra cost of more advertising, the schedule is complete.

The same customer profiles are used for direct-mail advertising – WSI patronizes several agencies whose classified lists of names and addresses can be searched automatically by computer for such characteristics as age, sex, education, income, neighbourhood, and family. On receipt of a set of profiles detailing desired attributes and the text of circular letters to be mailed, the computer searches the files, matches the details, writes address labels, and then writes a letter in a personalized manner. It starts 'Dear Sir', or 'Dear Madam', or whatever is appropriate, and tailors each paragraph according to who will be reading it, inserting more or less technical data on the types of widget and so on.

After each campaign, considerable efforts are made to evaluate its results and to improve the statistical profiles. Although even the 'creative' men now think in quantitative terms, differing interpretations of the statistics lead to furious arguments in conference. These are always worst when WSI buys a package of television time and has to split it between its various lines. Each of the product managers argues for changes in his customer profile so as to get the largest amount of prime time, even though each line's profit margin, or return per sale, is a known quantity.

Quite a few product managers, newly converted to the merits of mathematical marketing, had their faith shaken by a follow-up exercise that revealed amazing results: one of WSI's competitors bought a large number of widgets during a test marketing campaign and, before returning the punched-card warranty forms, added a few strategically placed holes to each. A large proportion of buyers were revealed to be nonagenarians with two children of pre-school age!

Market research is thoroughly conducted before introducing a new product. Preliminary design does not start at WSI until the analysis of current uses is complete. Using the complex, complete model of its industries and customers which is maintained permanently, the consequences of a new product are simulated.

CUSTOMIZING OF PRODUCTS

Customizing its products is natural to WSI. It could profitably produce widget spleeters with minor difference of specification for a year without manufacturing a single pair of identical models. Customers who come to a sales office are able to design their own widget spleeter by selecting items from a list of options displayed on a screen coupled to the computer. The system insures that all the options chosen are compatible, and displays the price at all times, so that the customer can get value for his money by 'testing' the options he wants. Finally a delivery term is quoted.

Borrowing a technique used by Boeing's commercial airplane division, WSI makes it possible for big customers to select a range of equipment that suits their production needs and operating pattern by simulating their requirements and determining the lowest cost combination of spleeters. Such a service is expensive, and only large users warrant the cost. Some cheaper, similar service is provided for the smaller user.

Back in the late 1960s, life-insurance brokers first started comparing policies by computer in their clients' interests. All the rates and records of numerous underwriting companies were tabulated and a search of the tables for the best policy for an individual wanting a particular type of coverage, span of years, and share of profits at his age was easily made. Brokers in other lines of insurance followed and by the early 70s a large service bureau offered clients access, through local terminals, to the rate books of many contributing underwriters.

In underwriting, the conventions of the world's great insurance markets had standardized the style and terminology of the business. Attempts by bureaux to offer the same service for a wide range of industrial components and raw materials foundered on the absence of an agreed classification system. Initial enthusiasm had been widespread among producers. The system, for which they paid through a listing fee, was set up to show a purchasing officer in a consumer company the whole range of products suited to his needs. If the purchasing officer found what he wanted, the machine deduced his address from the code number he put in on calling up the system, sent him a one-page facsimile of the specification it held, and sent the producer a card with the potential purchaser's address and interest.

The scheme collapsed when one producer who was also a consumer used his purchasing-department terminal as a simulator to determine the search technique by which the system found items to display to inquirers. It was then able to classify its own products in such a way that each would have a high probability of being the first to appear in its class. The bureau took a while to recover from this scandal. Eventually, it developed blocking facilities which would prevent company B from inquiring about company A's products at A's request. It then randomized the presentation sequence of items conforming to specifications. Finally, it led the procurement officers through a set of questions and used the answers to deduce the classification and specification instead of requiring them to look up a code to type.

AUTOMATED CATALOGUES

Confident of the superior performance and price of its product, WSI now does a lot of selling through such automated catalogues as described above, available to users on their terminals.

Its regular catalogues are thoroughly automated too. When the designers have finished with a product, all salient characteristics are fed to the design automation system for storage. As well as going to production engineering and cost accounting, these data are delivered to the technical writers in the marketing group. Changes affecting price or performance are also signalled automatically so that sales literature cannot be printed if it is out of date. The writer of descriptive material types his copy straight into the computer, selecting typefaces and sizes, justifying lines, inserting illustrations, and making alterations as required. Preliminary page proofs are provided from visual displays. If these are satisfactory, copy of photographically-reproducible quality is prepared off-line and up-to-date literature is readied for printing as required within hours of a minor design change.

The short lead time from product conception to first sale is characteristic of marketing at this time. In the 60s it was usually the flow of information which lay along the 'critical path' to the introduction of a new product. Tooling and purchasing materials are easy compared with preparing working drawings and routings, locating and then organizing spare capacity, and handling the orders flowing to a new advance in design.

WAREHOUSE SITING

The customers of WSI want goods delivered faster than they can be produced following receipt of an order. Warehouses with ready stock are therefore sited strategically – by the computer, naturally. Every delivery made from every warehouse and any order reported lost because of inadequate delivery is held for analysis by the system at intervals.

The company subscribes to a nationwide site index on which are recorded all plants or other industrial properties that come up for sale. When a likely piece of real estate comes up, WSI's system simulates its acquisition and the consequences on the predicted routes, mileage, and traffic pattern of all the warehouses. After these consequences are converted, using standard cost projections, into money terms, the site price plus the cost of reorganizing, building, and conversion are subtracted. If the calculation shows a cost reduction after discounting appropriately the future savings, all the data needed are printed. A minor modification of the program simulates the elimination of a warehouse to see whether it could be closed down.

AUTOMATED WAREHOUSES

The WSI warehouse system is based on a system developed in the mid 60s by the medical-supply firm of Johnson & Johnson. As each item is produced and packaged it is accompanied by a card which appeared in the original manufacturing shop-order package and which followed it through assembly and now down to the warehouse. As the box is placed on a rack by the storeman, he notes the rack's position number. A micro-switch on the rack is opened by the weight of the package and remains open until the storeman places the card in a nearby card-reader and types in the 'address' of the package – that is, the number of the open switch. If the switch whose number is given is not open, the computer demands that a check be made; by the same token, if a switch stays open for fifteen minutes, a warning is given.

The racks are on a 30 degree slope and the packages are held by relay-operated gates. If these are opened, the packages topple onto moving belts which convey them to the loading bays. Each rack is a known distance from the bay in use and the belts travel at constant speed. Each truck is scheduled to deliver a particular set of packages to different customers in a known sequence, so each set must be loaded in the reverse sequence of deliveries. Given the details and sequence of loading, it is easy for the computer to actuate the relays at the right time – that is, the packages travel along the various belts in an order and at a frequency so that they merge properly on the transfer conveyor that takes them into the waiting truck. Packages must not be dropped onto each other from the racks, so the computer keeps track of everything on each belt and checks that there is nothing in the way of any drop; if there is, it can modify the sequence, or take a pack from another rack even though it has not spent the longest time on the shelf.

CONTROLLING THE DELIVERY FLEET

The WSI delivery fleet's disposition, loading, routes, performance, manning, depreciation, and maintenance are all handled by the computer system.

Knowledge of the order and delivery routes for each day is not sufficient to produce the truck-loading plan. A few dozen trucks leave each day to visit perhaps half a dozen customers. Each morning the computer calculates the weight and physical volume of each customer order, and then the capacity of the vehicles available. It must then choose which truck is to go to which customer in which sequence in order to minimize time and mileage, allow adequate unloading time, conform to legal restrictions regarding size, unloading hours, and driver shifts, stay within limitations on vehicle type on some premises, and provide the operators with variety. The program uses drivers' log sheets, weather reports, and the latest details of highway-construction projects to maintain a permanent, up-to-date inventory of potential routes from customer to customer or from warehouse to customer. When this program was instituted, only grid reference and map mileages were available, along with estimates of vehicle performance. Establishing feedback to improve operating data required major efforts in winning drivers' confidence and improving standards of log-keeping.

Getting the orders to the warehouse is the first link in the chain that leads to delivery. Here WSI has built on the experience of Westinghouse whose Tele-Computer Center was completed in 1962. This system was then described as follows:

A processing system handling around 90 per cent of all industrial orders sends shipping instructions by teletype in just three seconds to the warehouse nearest the customer that stocks the wanted item ... An updating system keeps inventory records on a real-time basis, reporting every transaction and permitting immediate answers to inquiries about stock levels. This has cut the company's inventory from $36 million to $18 million, permitting closing of eight out of 35 warehouses and reduced processing costs from $12 to $2 per order.[1]

The reduction in inventory and order costs result from the computer's ability to watch all items with the attention that men could devote to only a few. The initial requirement is that the computer analyse demand for each item and estimate the cost of placing and receiving an order, spoilage, pilferage, and the capital that is tied up and therefore not available for other uses. For all WSI's stock items, the computer calculates the amount of each order so as to balance the cost of frequent small orders against the costs of a high average inventory; it then ensures, by calculating each time a disbursement occurs, that in the light of experience orders are placed early enough for the stock rarely to run out entirely before the new batch is in.

Data for these continuous calculations are provided by the storemen as goods arrive and are stacked on the rack. Every time stock is depleted below a certain level, a demand for a physical check is generated, thus monitoring the accuracy of the system at minimum cost in manpower for counting. Random counts are also called for, the computer drawing parts for auditing each day.

PURCHASING

Not everything held in WSI's warehouses is made by WSI. A lot of items for its own production processes have to be 'bought out'. The purchasing officers, contrary to the belief of their colleagues in other departments, do not spend most of their time being wined and dined in nightclubs by the vendors' salesmen. The computer now plays a major part in the procurement operation, again in close cooperation with the men who use it.

The purchasing officer for WSI in Indonesia – where farmers produce specially mutated crops for the organic processors – has a terminal with a typewriter and screen. When he arrives at work each day, he types his name and a special security code, and hits the start key. The day's agenda then appears on his screen: special requirements to be met urgently, orders that have failed to arrive, quotations to be sought, invoices to be approved and paid, batches that were damaged on delivery, and quality specifications to which a supplier failed to adhere. Each item is assigned a departmental priority, although the buyer can ignore it. Some items occasionally belong in another man's workload; these he can direct to the right place.

This particular buyer's speciality is farm produce. When he has to place a contract he calls up details of all those farmers on his files who produce what he needs. One of WSI's strict rules governing purchasing is that no one farm should be dependent on WSI's order for more than one-third of its crop; equally, no order will be cut by more than 50 per cent from one harvest to the next unless quality or delivery is inadequate. Borrowing from the techniques of credit rating, these policies have been coded into the computer, so that when a vendor is sought, all the relevant details can be made available. Besides corporate policy, these details include performance on price, quality, and delivery, and analyses of financial stability and productive capacity. Major suppliers are analyzed just as their smaller counterparts are.

There is next to no paperwork in WSI's procurement operation. Salient facts are always fed straight into the computer by the purchasing officer. Supplier's documents are filed, as they arrive, in numbered folders and placed in numbered drawers. These numbers are told to the computer. When anyone wishes to refer to original material, the computer tells him where the folder is. Bills and payments are generated on the appropriate date to take advantage of terms offered; any printing is, of course, automatic – most payments are made by 'computer-to-computer' banking.

PROCESS-CONTROL

Because WSI has processes that require control and job shops whose operations require scheduling, it uses machine tools that utilize numerical control.

Because its raw materials are bulky and costly to ship and its output is small and cheap to transport, WSI's processing plants are set up near sources of raw materials. The absence of skilled local labour has been an incentive to minimize operating staff. In any case, since irradiation at high pressure and low temperature is required, men cannot approach the system to control it.

When the processing plants were set up, the move from pilot operation to full-scale plant had to be rapid. Systems scientists, even in the late 70s, were still arguing whether control of a plant could be achieved without a full understanding of the relationship between all the variables in the process. The purists maintained that it could not; the pragmatists said it would have to be, and that the computer could be trusted to learn the effect of the control with which it was provided. The purists returned to their laboratories, and from their research came the theory necessary to design the current processes that effect an economical partnership of process and controller.

The pragmatists, armed with their pilot-plant experience, placed an array of sensing devices to measure temperature, pressure, radiation, viscosity, flow, conductivity, and any other property of the processed material that seemed useful in every conceivable place. They then used a large digital computer as a *data-logger* to record and analyse all instrument readings. All the information from these *sensors* was converted from analog to digital form for transmission to the main computer. The computer investigated the relationships between all the variables, determining thereby how each influenced the others. Unfortunately, chemical theory describes reactions by relationships between variables which cannot be directly measured. The values of these, therefore, have to be inferred from what can be measured. (For example, the frequency of collision of gas molecules near the surface of a catalyst must be deduced from the temperature of the gas and from the weight changes of the catalyst as it gets dirtier.)

After a while, it became apparent that some instruments could be removed, as the variables they measured changed directly with others, and so any one of a set would indicate the reading of the rest. Also, there were fluctuations in some measurements not attributable to known causes and which could not be correlated with any other readings. During this phase the positions of all the controls manipulated by the engineers who had *learned* on the pilot plant were noted by the computer, which began to *learn* the effects of each adjustment as time went on. At this stage, the only automatic control was that used merely to make small localized adjustments – say, to maintain the rate of flow through a single valve or the pressure at one point. A series of small local feedback loops were busy 'suboptimizing' their respective sub-systems – for instance, a thermostat might be maintaining the temperature under its control. The best procedure to correct a variation in one part of the system might well be to make a change elsewhere. This would be particularly true regarding a change in input quality – the most economical solution may be to adjust the process to the material, not vice versa.

Determining how to adjust the process requires analysing all the instrument readings describing the state of the system, and then altering the controls on all the actuators effecting the condition: valves could be opened wider, voltages reduced, or pressures raised. The computer systems must then find the best running conditions and, having found them, maintain them.

Since the operating environment of the process plant is continually changing and is not under the control of the process operator, the best running conditions are also subject to continuous change. The speed of these changes is vitally important. The slowest may be changes in market prices of raw materials or products. A catalytic cracking plant in an oil refinery may have 100,000 tons of Persian oil one week and 250,000 tons of Venezuelan oil the next week; the plant must thus be adjusted to handle different quantities of, say, sulphur and naphtha. Even with the same inputs, market prices could reward the company better for a different ratio of fuel oil to high-octane gasoline. These variables can be included in the process model, as Mobil demonstrated during the 60s at their Paulsboro, New Jersey, refinery, where a computer responded automatically, 'on-line', to feedstock changes and varying prices. In a paper mill, temperature, humidity, and pulp quality may change from hour to hour and require resetting controls to maintain constant paper thickness and surface quality. In a nuclear generating station, increased power demands require the withdrawal of control rods and their restoration into place at a higher flux density of neutrons; and the control must be capable of shutting down the system in a tenth of a second if necessary. The time lag between controlling action and measurable response can be very long, so the system must learn to be patient as well as to ignore random trivial fluctuations.

Since computers never become bored, they are good at this sort of job, and, unlike men, do not cause fluctuating behaviour in the systems by impatient action.

DIRECT DIGITAL-CONTROL

Widget Spleeters Incorporated now employs *direct digital-control*; that is, all instruments are read and all controls operated by a single team of computers. Although the theoretical study of the system greatly improved the ultimate integrated design, most of its benefits were achieved without first developing a mathematical model. The sensors and controllers may be simple meters and switches, or they may be microcomputers acting as servomechanisms maintaining local, preset conditions. If we think of these as workers, then we may consider the computer a manager which reads their reports – after they have been converted to his language – and sends back instructions. The managing computer takes orders locally and reports to the man on the spot, and sends data to the 'boss' computer in the corporate network for analysis and recommendations on future operation. The system not only maintains operating conditions, but adapts to changes and controls start-up and shut down. This procedure not only increases production, holds quality tolerances, and cuts down on waste, it also reduces maintenance, accidents, and manpower requirements. Furthermore, because of these continuous measurements WSI is learning more about the process and, above all, costs.

NUMERICALLY CONTROLLED MACHINE TOOLS

At WSI, as in many engineering companies, short runs of complicated parts are needed. An integral part of all widget spleeters is a pressure vessel which is smaller than a man's thumb and very contorted. It is made from special steel and has ribs for strength with low thermal capacity. Using traditional techniques – assuming it could so be made – would require drawings from which a craftsman would make a model for the development of special dies by the machinists. At WSI the parts are produced on two machines; there are no jigs and fittings, and little scrap.

Among the by-products of the automated design process are a punched paper tape and a magnetic tape. The latter controls a *continuous-path* machine which sculpts the part by moving both a cutter – or milling wheel – and the tool-bed to which the component is fixed; either can be rotated about any axis and moved in any direction. The paper tape is fed into a special drill control which can place smooth or tapped holes precisely where required as the component is twisted and stuffed under a battery of drill bits. In effect:

Electronic or hydraulic controls, operating from instructions fed into the machine tool from paper to magnetic tape, replace the numerous handwheels and levers that need setting by the operator on conventional machine tools. This removes much of the metal-cutting skill from the shop floor into the program and planning offices where the tapes are prepared.[2]

Even where computers do not aid in design, programming can be taught to suitable men who have a knowledge of tool speeds and metal characteristics and who know how to read mechanical drawings. Special programs allow preparation of control tapes for normal machining operations in a few minutes, merely by entering data on a chart of coordinates. Such techniques are applicable not only to very small, close-tolerance parts but, more importantly, to large components which could be ruined by a single mistake in a long operation. (The first problem ever tackled by numerical control was the cutting of sections of aircraft from solid metal to insure smoothness, strength, and lightness. One result is that all parts match perfectly when the aircraft is assembled.)

The major difficulties of numerical control involved achieving accurate positioning of the work without the trial-and-error method used by craftsmen: the dimensions of the drive mechanism and tool body had to be more precise than the parts being made.

PRODUCTION-CONTROL

Anyone who has attempted to cook a four-course meal with only an oven and one gas burner has met with the problems of production-control.

In the typical *job shop* at WSI a great variety of parts are made on drilling machines, milling machines, presses, and grinders. The production orders for this work are determined by the computer. Customers' orders and warehouse records are compared to determine what must be found from outside the warehouse. Using the resultant determination, the computer analyses each product into its components, checks the stock of each component, and places orders with the purchasing department or the manufacturing plant for whatever else is needed. We have already looked at the procurement operation, so let us now follow what happens in manufacturing.

The order for each batch of components is broken down into a series of manufacturing steps: drill here, mill there, press somewhere else, and so on, in accordance with the procedures laid down by the designer and stored in the computer's data bank. No firm – least of all WSI – can afford many idle machines waiting for work, so jobs have to wait until the machines are free. At any one time there may be a backlog of a few hundred jobs waiting for several dozen operators and machines to be ready for them. At this point the computer not only has to decide which job to assign to a free machine, it must also know where the job is and be certain that the last operation has been completed.

Scattered all around the machine floor are terminals that read plastic badges which operators carry in their pockets. There is a similar card for each machine, and a punched card accompanies each batch of parts. When work starts on a batch on a new machine with a new operator, badges and card are placed in the terminal, the 'attention' button is pressed, and the computer records the start of the job and displays the number of parts in the batch in lights. If batch, operator, or machine is in the wrong place, a warning is flashed. When the job is finished, the operator 'clocks' it off the same way, noting any spoiled parts. All the details are recorded for subsequent cost-accounting, machine-maintenance, and payroll records.

Many things can go wrong in any system, not least among which are: machine breakdown, operators getting sick, parts getting damaged. The advantages of WSI's system are that nobody has to worry about what has not gone wrong, and new schedules can be produced rapidly. The actual scheduling technique is complicated, since there is in fact no established mathematical method of finding the best sequence of jobs. At WSI a small set of possible schedules is produced each of which will get the work out on time while avoiding such clashes as assigning two jobs to one machine at the same time; the best schedule is then chosen. Schedulers know from past experience how many such feasible solutions must be generated to achieve a reasonable approach to the best.

COST-ACCOUNTING

All products cost money to produce, and so do services. The cost-accountant is concerned with the contribution of each activity to the total, so as to establish control of costs.

· The essential difficulty of cost-accounting is properly associating a unit of work with a unit of output. What is the connection, for instance, between the five minutes a storeman takes issuing a box of nuts and bolts and the cost of the final widget spleeter? Traditionally, cost-accountants maintain the fiction of knowing the answers to such questions by a process known as 'allocation of the overhead burden'. The major contribution of cost-accounting to industrial efficiency over the last half century shows how valuable the technique is. Yet, it suffers, as does ordinary bookkeeping, from the presence of many assumptions which, because they are not spelled out in reports, are credited with greater validity than they warrant, and which are neither challenged nor tested.

With a computer at hand, test can follow challenge easily. The essential operation is the recording of each identifiable unit of work and component as it is used and classifying it not into a single category but into many. The computer is then set to analysing these cost elements and combining them into a wide variety of meaningful sets: by machine type, product, labour class, geographic region, payment method, and so on. Performing such an operation clerically would not only be uneconomical but would take longer than the time available for making decisions based on such analysis.

QUALITY-CONTROL

At WSI, as in many firms, statistical quality-control is routine for all production. Whatever processes are used, chance factors will contribute variations in the output; ball bearings will never be pure spheres, although the variations in radius may well be less than one part in 1 million. In a well-designed process, the variations are within the tolerances imposed by the designers, and the process is 'under control'. The output of WSI's big process plants is a very valuable organic compound which must be analysed for purity. A small amount is withdrawn at random time intervals, set and recorded by the computer, and passed through an automatic analyser whose results are collected after a few minutes by the computer. If the proportions found on analysis are well within tolerance, a new sample is taken only when due; if the sample is only just satisfactory, a faster (but more expensive) rate of testing is used; if the sample is unsatisfactory, the quality- and process-control engineers are summoned.

Whereas individual measurements are important, their trend can be even more significant: imagine discovering that all readings at 2 p.m. are just within an upper tolerance level, those at 3 p.m. near a lower limit. There may be a single direct cause (the restaurant dishwasher drawing high-pressure hot water?) which needs investigating, and its origins determined by engineers freed from watching a satisfactory operation well under control.

A familiar problem of managerial accounting is that people regard as important that which *is* measured, instead of measuring what they regard as important. A sales manager who is assessed on sales volume, say, and not on contribution to company profit, aims to sell more rather than to sell more efficiently.

Management should be in a position to set the objectives and standards without which cost-accounting and quality-control are wasted. It is of little value to know what your costs are if you do not know what they ought to be. Planning these targets, which could merely be 'the best that we have achieved in the past', 'as good as the competition', or 'within 1·5 per cent of the theoretical maximum utilization of the machine' is management's concern; so is determining whether the standard of the work is wrong when the variance between target and achievement is excessive. But the actual measurement is drudgery fit only for computers.

FINANCE

Widget Spleeters Incorporated, which pirates good ideas from anyone, lifted its cash-management system straight from Westinghouse.[3] They have funds for ready use in many hundreds of bank accounts, and have arranged for messages to be tele-processed straight into the corporate computer complex whenever receipts or disbursements occur in any one of them. They then treat their stocks of 'cash' like any other commodity in their inventory, keeping enough to meet demands but not tying up too much capital, and also maintaining the level above that needed to minimize bank charges. The central account is used to maintain account levels by transfer in and out of the regional accounts; and when it has a surplus, the Controller invests in short- or long-term securities and notifies the computer system.

The decision on the length and amount of the investment is made with the help of a display of the cash flow anticipated over the next few months. The display brings together many factors – the sales trend and the income that follows sales by a few weeks; manpower forecasts, particularly overtime; all dividends due and interest anticipated; major and regular purchases and sales of fixed assets – and shows the resultant cash position. The Controller can also call for a display of the maturity dates of his securities and use the chart to determine which to sell when.

Balancing a national network of bank accounts is only one part of the work. With profits (and losses too) in various countries, determining where to reveal profits and where losses is a major task. The income and expenditures in each country are not the sole determinants of the tax that will be paid. The tax laws vary from one country to another, as do the accounting disciplines. Charging costs to running expenses or capital investment, depreciating assets quickly or slowly, valuing stocks by cost of purchase or value if sold – and so on – can all alter the pattern of profit over the years. Even more important, when WSI moves its product from one national company to another, it must change the recipient at its inter-company transfer price. In some countries this price is regulated by law, more or less in line with port-of-entry cost to WSI's customers; others have no relevant legislation. With a knowledge of the various laws, WSI simulates on the computer all legitimate methods of distributing costs and prices between repatriating earnings and profits as dividends, through management contracts (the hire of managers from one country by another), and by adjusting transfer prices. This results in a determination of the minimum taxation legally attainable.

The same methods can be applied to raising funds for further investment. Bank and discount rates vary continually from country to country, devaluation threatens here, revaluation there, and so on. Again, simulation of possible future events reveals the best present action.

Once the money is raised, there is always the problem of what to do with it – a problem not of lack of opportunities, but an excess. The future is never certain, but some eventualities are more probable than others. In particular, early returns are more valuable than delayed ones, since the former can at worst be invested to earn a few per cent. Projects with different patterns of returns over the future are difficult to compare: investment in research has a low probability of a high but far-distant pay-off; a new machine will make small but certain savings starting the day it is installed; a training course will slow down present output and improve next year's quality. At WSI, thousands of good projects are put forward each year. The Finance Department scrutinizes them all and then generates for each an assessment of the probability distribution of investment and returns. The whole set of projects is then fed to the computer, which uses a company-policy directive on the tradeoff between risk and return to choose a good portfolio, some risky but potentially very profitable, others certain of a modest return. The groups chosen do not all rely on the same future events. Some are indifferent to a change in tax, others are strongly affected; some assume the success of a marketing strategy, others do well even if it fails. The choosing procedure is highly iterative, and therefore lengthy. It starts with an estimate of the cost of raising various amounts of capital and, using this, eliminates all projects whose probable rate of return will not pay the interest on the capital. It then classifies the projects by risk and return and drops all those whose risks are too high for the expected returns. It then tries many combinations of projects and selects that set which offers the best results. Managing this assessment manually would be impossible; the standard method in the past has therefore been to use some rule, such as the 'payback-period rule', and choose those schemes which promise to recover their costs most quickly. Arbitrarily applied, such a scheme would choose to invest $500,000 now for a return of $1 million after one year, rather than for $2 million after two years.

Another tedious but useful money-making proposition is the careful choice of discounts to be taken, refused, and offered. A company that can make 25 per cent per annum on any cash it invests does not take a 2 per cent discount for payment in 30 days, but could well offer to its customers 5 per cent for payment in 10 days. In any case, paying accounts before they are due is bad for the cash flow: WSI programs its computer to pay all '2 per cent/30-day' bills on day 27; anything less advantageous they pay net on the due date. The accuracy and reliability of the machine are essential to the maintenance of supplier goodwill under these conditions.

Their suppliers, after all, have the same sort of credit-rating system as does WSI, which contributes to and draws on the nationwide credit-rating service through direct access to the agency's data bank. The payment profile of each customer kept on file naturally includes his attitude to discounts.

PERSONNEL MANAGEMENT

Dealing with people is not the kind of job usually delegated to machines. Nevertheless, the staff of WSI's Personnel Department are enthusiastic about the help they get from the computer system.

When a new post has to be filled, the job description is fed straight to the computer. This 'profile' is compared with every employee in the company and a list of the most suitable selected, along with details. These are displayed on the terminal in the manager's office so that he can prepare a short list. The consequences of each potential appointment are worked out ('If we put Smith into this vacancy, then there are twenty-seven people who can do Smith's job as it was described when he was appointed; but if we give it to Schmidt, then we must hire outside,' etc.). All the details are, of course, held in a data bank accessible only to a limited group.

Accessible to its many subscribers, however, is the data bank of Head Hunters Incorporated, which has classified and cross-referenced listings of many millions of active and quiescent job-seekers. Since the unfortunate occasion when a certain senior manager was asked 'in strictest confidence, and without revealing his name' whether he wished to apply for his own job when he did not know his company sought to replace him, security checks have been more thorough.

Not only do such personnel-search companies save much of the expense of finding and selecting the right people, they also provide valuable statistical analyses of mobility, related benefits, salaries, and other inducements such as location. For example, WSI now chooses new plant sites where people are prepared to work, rather than paying exorbitant salaries to make them move.

UNIVERSITY RECRUITING

Many universities have adopted computerized résumés for visiting recruiters. Before the annual visit of the recruiters is due, students may submit themselves to interrogation by a terminal which collects basic details for the computer. The computer then abstracts relevant facts from its own files and assembles the résumés. It asks the student questions about tastes and ambitions, and inquires after any specific requests or exclusions. Recruiters insert company profiles and job opportunities (and any special inclusions or exclusions), leaving the computer to match individuals with companies and to schedule interviews. Computerization neither prevents 'private' approaches nor removes the need for interviews and short lists. Data banks of qualified people become nationally accessible.

One aspect of using such data banks caused a flutter in the academic dovecotes. This was the rating of universities and of the schools within them by correlating salaries and achievements with academic alma maters. The files also contained members of professional associations. It was found, when cross-checking was first done, that the number of impostors claiming to be professional engineers and certified public accountants was quite large. Now recruiters normally look up prospective employees (and their wives) in the data banks, using their terminals.

THE EFFECT OF COMPUTERS ON MANAGEMENT

By the early 1980s WSI could not conceivably have been run without computers. The progressive uses of data-processing unquestionably made possible its prime position in the marketplace. There were, however, a number of bad effects which were increasingly worrying top management.

The first of these was an inflexibility in the methods of operating. It took a long time, often two years or more, to change the way a computer system functions. Because of the difficulties involved, there was great resistance in the data-processing management to some types of changes. A comment occasionally heard at the board meeting was, 'We cannot do that – the data-processing system will not handle it.' The more complex the machines became, the more difficult it was to change them.

Exception-reporting techniques also had their problems. The information needs of management change continuously, and so most measurable facts were captured in the data base at WSI.

To avoid the unreadable deluge of paper that was earlier associated with computers, reports were produced only on request or when predefined exceptions arose. Unfortunately, this sometimes caused managers to overlook important happenings in the world outside the data-processing system's purview.

The advent of the computer, with its appetite for data and its propensity to be upset if not fed on schedule, has forced managers to adjust to the pace of a machine in much the same way that process operatives adjusted with the arrival of production lines. Naturally, this has improved managers' productivity when the *anticipated* continued to happen, but it reduced their capability to respond to the unexpected and in some instances destroyed their curiosity about the world outside.

This concentration of interest on measurable, recordable, and local facts is a common psychological phenomenon best observed in bureaucracies, which demand certificates, diplomas, reports, identification papers, and a host of such material – and which are not in the least interested in the facts of any case presented verbally and in person. Bureaucracies are human data-processing and decision-making systems with restricted capacity to handle variety in their input.

The psychological problems of managers in an era of almost complete information are increased by the resultant changes in organization. Every manager's performance can be measured by his superiors, and the measurement could appear in a report on the boss's desk even before the subordinate knows himself how he has performed.

'Decentralization' had been the catchword of the 50s. By breaking huge corporations into one-man-manageable enterprises and evaluating the one man on his performance every few years, the diseconomies of scale in large corporations could be counteracted. These diseconomies were principally due to lack of communications within geographically dispersed organizations. (Geographical dispersal had been the answer to the diseconomies of scale caused by drawing labour and materials from ever wider areas.) Utilizing integrated data-processing and telecommunications, corporations can once again be run from a centre, since all the data necessary to decision-making are there.

When computers were first installed in on-going enterprises, they achieved cost savings – if at all – by displacing clerks. In large firms where they were installed in many divisions, they petrified the systems and procedures, and in particular the nomenclature and codes, used by each division. When the divisional systems came to be linked, it turned out that no one set of procedures was common to even two divisions.

By starting later, WSI avoided this problem. Every item of data has the same definition across the corporation, all procedures are identical, and everyone uses the same language. Because WSI has yet to merge with anyone, it still faces the potential problems of integrating strangers into itself. The cost savings arising from the use of computers at WSI come not from displacing staff (whom they never hired) but from avoiding duplication of effort between divisions and, above all, from being able to make decisions which are best for the whole corporation, not just part of it.

The high degree of centralization made possible by computers and data-transmission has been a key to WSI's profitability; but now, with the computer systems so immensely complex and difficult to change, and the corporation becoming so large, it has also become a cause of inflexibility.

Some of the younger staff fear that the mobility that made WSI such an aggressive growth company for the last ten years is in danger of being lost. Management, they think, is becoming too dependent on the computer systems, and there is too much resistance to changing these. The computer methods are becoming an institution. Perhaps like some aspects of society itself, they think WSI is becoming locked in the grip of the machines.

REFERENCES

1. *Business Week*, 'How Computers Liven a Management's Ways,' 25 June 1966.
2. Robin Sanders, 'The Numerical Control Nightmare', *Management Today*, October 1966, p. 70.
3. *Business Week*, loc. cit.

10

DOCTORS

'Come into my parlour,' said the computer to the specialist.
– Marshall McLuhan.

In medicine as much as in other fields the computer promises revolutionary changes. Many doctors are not entirely happy about this, and some express the fear that computer usage is 'dehumanizing' and thus has no place in medicine.

There are two very powerful reasons for wanting to use computers in medicine. First is the serious shortage of doctors and medical staff in most countries of the world. President Johnson said in a speech on 7 May 1966, 'The nation faces a critical shortage of doctors, nurses, and other health personnel.' In Great Britain it has been estimated that 44 per cent of interns are imported from medical schools in underdeveloped countries,[1] while many of Britain's own doctors are drained to North America. The underdeveloped countries themselves are desperately short of medical staff, many having less than 1 per cent (per capita) of the doctors employed in the United States. Automation will help this situation somewhat by relieving medical staff of some of their more routine work and thus increasing their overall effectiveness.

Second, computers are likely to extend considerably the frontiers of medicine. It seems likely that with their use diagnosis and special care of patients will improve dramatically and that medical research may conquer some of today's diseases. The role of the doctor may eventually become substantially changed.

Let us examine some of the uses of computers in this field, starting with the least controversial ones. To some extent they are analogous to the types of computer usage in industry and other fields. They include the following:

1. Hospital Administration

Computers in medicine today are most widely used in straight-forward applications: handling administrative functions, inventory, payroll, and the financial operations of large hospitals. Typically the computer keeps track of the allocation of beds in the hospitals; records data concerning each patient on admission in an extensive patient file which is updated by the machine; schedules tests and consultations to minimize waste of time; controls the inventory of drugs and other items used, in a way similar to inventory-control in a factory. (In the optimum maintenance of drug inventory alone a large financial saving is possible.) As this corresponds broadly to administrative systems in other fields, we shall not describe it in more detail.

2. Handling Medical Information in Hospitals or Clinics

Details of tests conducted, patient conditions, clinical information, special diets, and so on are fed into the system. A typical system today prints out ward summary reports every two hours or so and patient summary reports every day. This is broadly analogous to a data-collection system in a factory. Eventually, as in factories, it will have inquiry terminals for hospital staff and will print instructions, reminders, and exception reports to nurses. This is a natural extension of the hospital administration system. The same machine is normally used for both sets of functions. Here, as in other fields, the more work that can be given to one machine, the easier is the economic justification.

A BMA report in 1969 found that direct cost savings from such computerization were seldom shown but the reduction in errors was a 'very real benefit'. 'Machine methods permit a great reduction of the very high error rates of traditional hand methods of identifying patients, making requests, linking specimens with labelling devices, labelling devices with request forms, request forms with reports and reports with patient journals ... up to one third of results of pathological investigations may never appear in a clinical record.'

The emphasis with such systems will shift away from the

present self-contained operations supporting a single hospital task, such as laboratory work-flow and record keeping, to a central overall system containing the basic data needed by any department of the hospital. At the Karolinska Hospital in Stockholm, the patient's record is started before he arrives; a questionnaire goes out with the letter of admissions, and the medical history thus collected is entered into the system; treatment can begin when he arrives, and need not be delayed while routine data are collected. To free the doctor from non-medical work, records are handled by clerical specialists using terminals. The doctors and nurses get the information when they want it, on-line if necessary but usually in batches later, in response to a request entered at any time.

In contrast, at London's King's College Hospital, 'data secretaries' are regarded not only as too expensive, but also as contrary to medical and legal ethics and the principles of clinical accountability. The conversational language has therefore been designed for ease of use by nurses and doctors.

An example will illustrate the technique. Suppose the ward sister wishes to order a routine nursing procedure to be performed as a part of the patient's routine treatment. She would identify herself, and the patient. She would then be presented with a screen displaying the types of conversation she can initiate, as follows:

History
Physical exam
Vital signs
Investigations
Diagnosis
Nursing procedures
Follow-up notes
Interrogation

The sister would type the choice number (6) and transmit. Information for Nursing procedures would then be presented:

Nursing Procedures
Bathing
Mobility
Care of pressure areas

Feeding
Oral hygiene
Lavatory
Dressings
Special treatments

These are separate categories of nursing care, each of which may be the responsibility of a nurse. Let us assume the sister takes Feeding, the system response is as follows:

Nursing Procedures   Feeding
To be fed
May feed self
To be encouraged to feed self
Needs help with feeding
Needs food to be cut up
Nothing by mouth

With this process the sister continues with detailed instructions for feeding the patient. As each choice is made, the text associated with the choice is added to the instruction. Occasionally specific parameters are required which cannot be conveniently handled by choices, and a 'form' is displayed for the sister to complete. This could occur in a later point in the conversation illustrated above; as follows:

Nursing Procedures   *Feeding
To Be Fed   *Fluids Only
Give]   [Millilitres
   At]   [Hourly intervals
        [Times a day]

The sister then types the required parameters between the square brackets, to complete the form.

The system then displays the total message to the sister, for her to verify:

Nursing Procedures   *Feeding
To be fed   *Fluids only
Give 25 Millilitres—At 2
Hourly Intervals

The sister may indicate her acceptance (in which case the message is filed) or back-track and restate part of the message. The message is now available for interrogation. It will be printed on a daily work sheet which will be used by the nursing shifts. It is evident that the response of the system must be apparently instantaneous to all users during periods of relatively heavy traffic. Without fast alphanumeric visual display units, the technique would be impractical.[2]

The hospital of the future will abound with computer terminals. When you first arrive the receptionist who greets you will use a terminal, probably with a screen, for entering data about you into the hospital system. If you have been there before, this process will be speeded up because the machine will be able to display your past record on the screen. In some hospitals and clinics the interviewer will not be a receptionist but the computer itself. Patients will sit at the screen of a computer which directs questions to them. The computer will decide its question sequence on the basis of the patient's responses, recording information about the patient and preparing a preliminary report for the doctor he is to see.

A doctor might, say, order a blood test for a patient in the hospital. The sample can be taken only after the patient has fasted for a period. The computer would instruct the kitchen to stop feeding the patient, arrange for the blood sample to be taken at the appropriate time, and print out blood-test instructions for the laboratory. The laboratory would feed the results into the computer, where they are stored in the patient's record. The computer than instructs the kitchen to start feeding the patient again. The doctor visiting the patient would inspect the results on his display screen.

The computer would function similarly with much more complex operations.

3. *Direct Patient-Monitoring*

Processes in industry that require continuous monitoring and control by computers are heavily equipped with such instrumentation as devices for measuring temperatures, pressures, movements, and chemical characteristics. In the hospital the analogue of this 'process-control' involves placing sick patients 'on-line', directly connected to the computer.

The computer continuously monitors patients' conditions by scanning instruments attached to them. Data will also be entered 'off-line' by nurses. Like industrial processes, different patient conditions require different response times from the computer. In intensive-care units where there is a chance that a patient's heart will stop, for example, the computer must watch the patient second-by-second and take fast action when an emergency occurs. A large and fast computer can have many patients connected to it.

One of the early installations of this type is at the Los Angeles County General Hospital Shock Research Unit.[3] The unit is concerned with circulatory shock, which results in low blood pressure, low blood flow, and an extremely unstable circulatory system. The proportion of circulatory-shock patients who die is large. The computer was installed to improve the monitoring and care of these patients. A patient in this ward has many instruments attached to him: six thermistors measure the temperature at various parts of his body; pressures are measured in the circulatory system; catheters are inserted into an artery and a vein; and a urinometer measures liquid waste. The instruments are all directly connected to the computer and other details about the patient are given to the machine by the nurse. The computer prints out a log about the patient's condition, and if any measurement falls outside critical limits set by the doctor for that patient, the computer will notify the ward staff.

As in other fields, the computer when first installed may be limited to collecting data in this way and instructing human staff when action is needed. After these procedures have been deemed successful, it may be decided to program the computer system to take action itself, thus 'closing the loop'. For instance, the Shock Research Unit administers certain potent drugs by means of manually controlled 'infusion' pumps, and the unit's directors are contemplating putting these pumps under direct computer control to maintain desired levels of blood flow and pressure. The ward staff could, of course, override the computer control at any time. Computers may be used in a similar way during surgery. In heart surgery, for example, the machine would notify the operating group of any computed parameters that fall outside the specified critical limits.

The computer is not replacing the nurse in intensive-care units: there will always be a nurse on hand, and others can be called when needed. It is, however, taking away some of her most tedious work and permitting her increased time for the more human parts of her job. The machine is also improving the patient's chances of survival. A cardiac-arrest patient normally needs a nurse at his bedside full-time. When his heart stops, immediate action must be taken by a team trained to massage the heart, administer oxygen, and so on. The nurse must first determine that his heart *has* stopped and then summon assistance. Automation decreases the time interval between stoppage and action being taken. In a San Francisco hospital a computer can signal 'low cardiac output suspected', alerting physicians to a decline in pulse and pressure before any noticeable physical signs have appeared.

A secondary objective of the Shock Research Unit system discussed above, and of other such systems, is to gain more complete and accurate data – both for the doctors involved and for later fundamental research.

The doctor may take certain action with the patient and wish to monitor its effect closely while the patient is not in the hospital. Again automation may assist. In the future certain patients are likely to have miniature recording instruments taped to them, much as do the astronauts, to record certain measurements. The tape (or wire) would later be played back by a doctor or transmitted to a computer.

4. *Electrocardiogram Analysis*

An electrocardiogram is a plot of rhythms, picked up from the heart by a number of small transducers (like microphones) attached to the chest, ordinarily analysed by a heart specialist, who can tell from inspecting the waveforms whether the heart is healthy. Computer programs have been written to analyse waveforms for both normal and abnormal conditions.[4] These programs have generally been highly successful, although with certain less common abnormalities the skilled eye of an experienced specialist is desirable, at least at the present state of the art.

Using modern instrumentation taped to the patient's chest, recordings for analysis can be made not only when the patient is lying on a bed, but also when he is pacing up and down, or doing all manner of exercises.

Electrocardiogram signals, like other such medical signals, can be transmitted over the world's telephone lines. Distance is no deterrent. Hospital computers in the United States, for example, have analysed heart rhythms transmitted from Europe. In the future a doctor will be able to dial such a machine, and with a small portable device send a patient's heartwave over the telephone for which the computer will provide a spoken diagnosis – a technique that has been successfully demonstrated already.

The computer's immense capacity for processing and storing data allows the collection of vast quantities of valuable electrocardiogram information for research into heart problems. It is possible that computer analyses of heart signals will eventually become more detailed and more sensitive than those of today's heart specialists, and will thus become a more elaborate and better documented guide to heart troubles. The same techniques may also be applied to such other highly complex signals as electroencephalograms (from the brain) and electromyograms (from muscles).

Computer technology opens up the possibility that mass screening of hearts could be conducted in much the same way as today's chest X-ray and tuberculin tests. If this could significantly cut heart illnesses, which presently cost the United States an estimated $30 billion per year, it would be worthwhile.

5. Physiological Modelling

Physiological systems generally perform in a manner more complex than physical systems. The ear is more complex than a microphone; the lungs are more complex than an airpump.

Their behaviour and ailments cannot be described in such simple mechanical terms. Models of their behaviour have been constructed in the form of computer programs, however, in much the same way as models representing the flow of work through a factory. Such models use mathematical or simulation techniques, and represent the system in question with varying degrees of accuracy depending upon how carefully and how elaborately they have been constructed.

Models have been programmed which simulate, among other parts of the anatomy, the heart, the lungs, and the cardiovascular and blood-chemistry systems. The models imitate the behaviour of the original closely and can be experimented with endlessly under both healthy and sick conditions. They currently help in understanding the effects of diseases, and are likely to help in therapy – for testing the effects of treatment before it is given to the patient. For example, in the therapy for some plumonary disorders the patient breathes in different gas mixtures. A model of the lung system which dynamically represents the gas exchange and blood flow in the lungs has been used to simulate the lung operation of patients with these pulmonary disorders and tested against data from these patients. Physicians using such a model will thus be able to predict the effect of different gas mixtures on their patients and the effect on the oxygenation of their blood.

Such models will be particularly useful in teaching future doctors and nurses. The medical student studying the workings of the kidney, for example, may be able to sit at a screen on which are displayed various pictures of the kidney, possibly moving. With a keyboard or a light pen he can make all manner of changes and the computer, using a programmed model, will show him their effect on the organ.

6. Medical Records

Currently the medical records of most people, if they have been kept at all, reside in a cardboard folder in their doctor's office. Some doctors and some clinics have many shelves full of such folders recording patients' past ailments. When a person moves, his records more than often do not move with him. He may have many sets of records, none giving a complete medical history, scattered among different doctors.

Diagnosis and care of his ailments could be facilitated, first, if the doctor had all available records of his complete medical history and, second, if this history was kept in much more detail than are most such records today.

Patient records are more and more likely to be kept by computers in the future, even legal documents, X-rays, and other items external to the machine will be referenced by the machine record. The machine records may contain computer electrocardiogram analyses and information recorded during hospital stays. A system will probably develop in which doctors normally have a terminal screen, and can examine the patient's records, past and present. Some parts of a record will be kept in such a form that they can be processed by the machine automatically; others will be in a verbal form, not capable of being processed but for giving certain information to a doctor. When the patient moves, his records will be transmitted to a new computer. The beginnings of such a scheme can be seen today in large hospitals and clinics. We expect that most large hospitals and clinics will have such a system in ten years, and eventually the private practitioner will also be able to dial the relevant data bank from a screen unit in his office.

Perhaps in the future if you have a near-fatal car accident your identity card will be inserted into a roadside telephone, or police transmitter, when the ambulance is summoned. The ambulance staff will thus immediately have relevant medical details about you, and will be able to bring a transfusion of the correct blood type, an electrocardiogram report, and drugs to which you are not allergic. The ambulance will have automated equipment to keep you alive until you reach the hospital. Perhaps, in some countries, your credit rating will have been checked also.

The British National Health Service ought to be in the forefront of such developments. It is a symptom of the questionable management of the service that the opportunities have been neglected. Managerial and financial responsibility is so split between regions, central and local authorities, doctors and administrators that effective cooperation and efficient use of resources is almost prevented. The medical profession is, quite properly, so concerned with maintaining and improving medical standards that it has no time to give to management; but it refuses to be managed by professional managers.

The main computer applications in Britain are directed to the *ad hoc* automation of what is done manually at present. Instead of tackling each problem area in a concerted manner, by allocating the development of different applications to separate units, with a view to national adoption when success has been achieved, parallel and ineffectual efforts have proliferated.

The basis of automation is standardization, but the basic medical record card is not even universal, let alone applicable to health department, clinic, hospital and general practitioner. Nor, of course, is it a suitable document for computerized record-keeping. Between a quarter and a third of the cost of the NHS is involved in record-keeping – a collosal £600 million per annum. A central medical records system for the country, with an on-line terminal for each doctor, would cost about 5 per cent of this to run each year.

7. *Medical Statistics*

Some doctors would say that a more important reason for keeping detailed medical records by computer is that such records provide extremely valuable information for medical research. For some questions that frequently arise, the scanning of a large quantity of past experience would be valuable. In heart research, for example, large numbers of electrocardiograms are being stored along with records of associated symptoms. Future computer analyses of these may reveal new facts about the causes of heart attacks, perhaps even making it possible to spot impending attacks. Again, investigations of possible side-effects of drugs would be possible using the medical data bank.

Considerable fear has been expressed concerning the desirability of storing very detailed medical histories of individuals; such records, it is argued, might constitute an invasion of privacy. We shall discuss this aspect further in Chapters 14 and 15.

8. *Information-Retrieval*

The quantity of medical information that a doctor might need to know is growing at a furious rate – a situation that also exists in other professions. The specialist cannot possibly read all of the reports on his specialty. Computer information-retrieval systems in use, however, let him know what work has been done in various areas. Most such systems are today 'off-line'. He can request a computer to scan available information on a certain topic and print out a list of relevant abstracts. He can register his requirements with the machine, which then sends him listings of new reports or articles as they are written. In the future, the search for relevant information may be carried out at the same terminal screen which the physician uses for other purposes.

9. *Patient Counselling*

In some clinics data-processing has been used experimentally to 'interview' patients before – or after – they see a doctor as part of the diagnostic process. The Mayo Clinic in Rochester, Minnesota, now plans for its patients to sit at the screen of a computer terminal and answer, in a simple manner, questions which the machine asks them. There are a number of reasons for doing this. First, the machine can collect information for the patient's records. Second, the machine can make a pre-liminary excursion into assessing the type of illness the patient has, and then direct the patient to an appropriate specialist. Today many patients visit one doctor only to be passed on to another who specializes in a given area. Much of this time-wasting process could be eliminated. Third, the machine could collect certain routine data needed for diagnosis and make a report for the doctor, thereby saving some of his time. Doctors perform much routine work that, if computerized, would allow him to spend more time on vital areas of his work that cannot be computerized.

The Mayo Clinic has for some years used a computer for *psychological* assessment of its patients. The machine prepares a report on patient psychology which is valuable in diagnosis. The basis of this report is a well-established form of psychological test, the Minnesota Multiphasic Personality Inventory (MMPI).[5] In this test, the patient checks off a large number of 'true-false' statements about himself on punched cards, which are then, at Mayo, processed by the computer.

The computer cross-checks to test the truthfulness of the responses and produces a personality report for the clinician. The figures printed rate the personality on scales such as hypochondriasis (HS), depression (D), hysteria (HY), psychopathic deviate (PD), and so on, and certain types of adjustments are made to the ratings.

This type of test could be more effectively conducted at the screen of a terminal: not only would the computer be able to use many more questions, but it could ask only a portion of them by selecting questions on the basis of responses, exploring doubtful areas in more detail. The conversation between a Mayo Clinic patient and the computer terminal in the near future will look something like this:

COMPUTER: Do you like mystery stories?
PATIENT: Yes.
COMPUTER: Would you like to be a lion tamer?
PATIENT: No.

and so on.

The gathering of data about patients is likely to extend far beyond psychological testing. There are many areas in which preliminary fact-findings could be accomplished in this manner and an appropriate report prepared for the physician by the computer. In this way much of the physician's valuable time would be saved.

In the majority of clinics or hospitals the doctor, rather than patient, will answer the computer's questions. The doctor, sitting with the patient and talking pleasantly with him, will enter the data the machine requests. Where the questions are of a detailed medical nature rather than simple ones such as those above they must be answered by a doctor. IBM is developing a computer system to assist doctors in diagnosis and to record data about a patient, called the Clinical Decision Support System (CDSS). The use of this system begins with a screening process in which the computer obtains data about the patient and completes questionnaires available for each of the following functions: symptoms and complaints, physical examination, laboratory findings, radiology reports, patient history, and pathology.[6] A physician may select a questionnaire and, beginning at any point, answer questions which appear on the screen.

10. *Mass Screening*

Doctors are far too busy to spend much time with the apparently healthy. But preventing disease is as valuable as curing, and that means locating it early. The British United Provident Association has a screening clinic in London. A ninety-minute examination by computers and medical technicians, assisted by X-ray, cardiograph and automatic chemical analysis equipment, produces a comprehensive report on the 'non-patient's' health for his local general practitioner. Again, many of the basic data are collected by terminals from those who are being screened. This all helps to keep down the cost of the system, which relies on using expensive skilled personnel very intensively. BUPA expects to screen 20,000 people a year for £25 a time (less for BUPA members).

Mass X-ray and cervical cancer tests are valuable methods of detecting conditions which become serious if neglected but are cheaper and easier to treat if spotted early. Persuading people to attend a mobile clinic for a chest X-ray is possible merely by local advertising. But it takes a direct personal invitation to persuade most women to have a simple and painless cervical smear taken. So a computer has been programmed to address a personal invitation to women most at risk, giving a specific time and place for the appointment.

11. *Diagnostic Assistance*

The application of computers to diagnosis has aroused great controversy. Many doctors claim that the diagnostic process is empirical, vague, based on ill-defined experience, and above all *intuitive* – that is, not a process than can be defined in terms of mathematics and logic. The old country doctor is said to have been able to smell a disease as soon as he walked into the patient's room.

On the other hand, a *New York Times* headline proclaimed, 'Computer Program Seeks to Diagnose All of Man's Diseases'.[7]

The eventual use of computers in diagnosis may prove to bear out neither of these two extreme views. As with managers, teachers, architects, and other professional people we have discussed, the doctor has certain human qualities which we should not *attempt* to replace. On the other hand, he also has human limitations. The work of diagnosing illnesses is so complex that a mere human being cannot hope to make a first-rate job of it – and diagnosis is rapidly becoming more complex. Furthermore, medical knowledge is increasing at a rate far exceeding the general practitioner's ability to keep up with it. It has been estimated that, at best, one quarter to one third of clinical diagnoses are incorrect. In many instances the ratio is worse than this.[8] It seems likely, however, that the doctor's experience can be extended in range, and the chances of correct diagnosis improved, if he works closely with a computer system. The computer and its data bank aiding the physician will again represent the optimum combination of man and machine: computer-aided diagnosis, not computer-diagnosis, is the aim.

Certainly, if computer diagnosis consisted of feeding a list of symptoms to the machine and then letting the machine use these to determine the illness on the basis of probability calculations (for example, using Bayes' theorem or a modified version), then the sceptics would be justified in their doubts. Not only is such a scheme far too complex to be workable in the foreseeable future, it would also fail to account for psychosomatic factors. The number of possible symptoms that a doctor can recognize is enormous, and today's computers are singularly poor at recognizing patterns in complex input. It would be very difficult to program a present-day computer, for example, to recognize a human face under all lighting conditions; and if we did manage to program this, the machine might still be outperformed by any human being. Recognizing patterns in medical symptoms presents a similar problem. Different patients with the same disease rarely exhibit the same sets of symptoms.

Inductive logic plays a major part in diagnosis. The doctor first assumes that he knows what the disease is; his medical knowledge then tells him that he must gather certain types of data to check that assumption. If the data do not verify his hypothesis, he tries something else. This trial-and-error process continues until he forms an opinion which appears to be backed up by the symptoms.

The computer can assist in this process in a number of ways. First, it can make suggestions to the doctor, on the basis of symptoms he has fed into it, about the identity of the disease. Second, for any one of the identifications suggested it can also suggest to the doctor what additional information he should collect. Third, it can analyse certain types of data with the precision that only a computer – and not a doctor – can muster. Fourth, if the machine has access to the patient's medical history, electroencephalograph analyses, and so on, it can bring these to bear on the case.

Computer diagnosis, then, is more likely to take the form of a conversation between the computer and the doctor. The doctor's bedside manner and human insight will still be all-important, but will be backed up by the vast, precise memory and analytic powers of the machine. The diagnostic process will follow a trial-and-error sequence in which the machine requests certain data as needed, and doctor and machine together steadily narrow down the possible list of diseases.

Such a system may encompass the whole field of medicine, which is the intention of certain research in progress today, or it may concentrate on certain specialties to which the computer is best suited. In either case, it seems likely that the physician of the future will have to have a computer screen; the visiting doctor may eventually carry a portable computer terminal. The years ahead will see much research in this area, although we doubt whether computer diagnosis will have filtered down to the general practitioner in the next ten or fifteen years.

The main purpose of IBM's Clinical Decision Support System, discussed earlier, is to aid doctors in diagnosis. During the patient-screening process, as the doctor enters data the computer eliminates many diseases and puts others on a list of possibilities. The computer normally asks for more information relevant to the diseases it has implicated. The doctor may stop the process at any time, ask the computer for a display of the diseases implicated, and, if he wishes, ask its reasons for listing a particular disease. The computer, on the other hand, may request laboratory examinations, recommend X-rays, or other time-consuming investigations. The diagnosis may be a long continuing process.

The computer may also recommend types of treatment, and for this purpose make available details of a large number of drugs and treatments from its data base.

Such a system as IBM's certainly does not eliminate the need of a doctor's skill; it merely enlarges his available knowledge and speeds up certain processes. If he wishes, the doctor can delegate some of the screening process to an assistant, thereby gaining valuable time. The system does not *order* the doctor to make examinations or take certain actions. The relative success of the system will not be known for some years, as it will take a very large amount of work in several medical institutions to build up the necessary data base of medical knowledge. Indeed, its success will directly depend upon the participation of the medical profession. Over the years, however, the body of knowledge that is built up in such a system will grow and grow. Hundreds of man-years from dedicated doctors, specialists, and university staff will be poured into the construction and improvement of the data base, and eventually the machines that work with a doctor will store incomparably more knowledge than he could ever hope to acquire. Furthermore, they will be capable of analysing certain symptoms with a power that far exceeds human ability.

THE CHANGING MEDICAL ART

One effect of the computer on medicine will be pressure for greater precision. If medical textbooks are steadily translated into computer logic, they can become much more elaborate. Physicians have immortalized themselves in the past by recognizing the association of a group of symptoms with a disease and attaching their name to it. With more precise analysis of symptoms there are probably many more such groupings yet to be found. Some of these may emerge from computer-scanning of large numbers of stored medical histories.

Thus although the computer will be introduced merely to assist in diagnosis as it is known today, over the course of decades it may change the whole nature of diagnosis.

Doctors, like the members of other professions, will come to terms with the computer. However, when discussing the impact on managers in Chapter 9, we mentioned the tendency to place too much confidence in the machine and to ignore factors beyond its ken. Let us hope that doctors never do this. When human beings become bored or bewildered and leave the machine to decide everything, then danger lies ahead.

The skills required to bring about this revolution in medicine are formidable. The medical discipline has traditionally been descriptive, empirical, and intuitive – but extremely complex. What we are now postulating would link medicine with the disciplines of mathematical statistics, systems analysis, electronics, and telecommunications. A complete course in medical training is already very long. The men who can bring into being medical computer systems will need some years of computer training and experience in addition to their medical training. Once the machines are set up, however, we are not demanding any greater talent of the practicing doctor than we do today.

How quickly this medical revolution comes about depends upon how quickly the medical profession accepts the role of computers. The main difficulties will involve not the engineering, but the translation of medical knowledge into computer terms. This can only be accomplished by persons with medical training, and it will require large expenditures of money. Because the pay-off is likely to be great, however, it would be unfortunate if fear and distrust of computers delayed it.

REFERENCES

1. *New York Times*, 16 September 1966.

2. N. J. Crocker, 'On-line Patient Records', *Data Processing*, July–August 1970, p. 343.

3. See R. L. Patrick and M. A. Rockwell, Jr, 'Patients On-Line', *Datamation*, September 1965.

4. See Theodor Sterling and Seymour Pollack, *Computers and the Life Sciences*, Columbia University Press, New York and London, 1965, Chapter 11.

5. W. A. Dahlstrom and G. S. Welsh, *An MMPI Handbook*, University of Minnesota Press, Minneapolis, 1960, and T. R. Sarbin, R. Taft, and D. E. Bailey, *Clinical Inference and Cognitive Theory*, Holt, Reinhart and Winston, New York, 1960.

6. *CDSS manual* (if made public).

7. *New York Times*, 25 September 1966.

8. R. H. Gruver and E. D. Freis, 'A Study of Diagnostic Errors', *Annals of Internal Medicine*, Vol. 47, 1957, pp. 108–120.

11

CITIES AND TRANSPORTATION

In the next 40 years, we must completely renew our cities. The alternative is disaster. Gaping needs must be met in health, in education, in job opportunities, in housing. And not a single one of these needs can be fully met until we rebuild our mass-transportation systems. – *Lyndon B. Johnson*

It is impossible to spend any time on the study of the future of traffic in towns without being appalled by the magnitude of the emergency that is coming upon us ... We must meet it without confusion over purpose, without timidity over means, and above all without delay. – *The Buchanan Report for the Ministry of Transport, Traffic in Towns*

The ills of large cities are likely to get worse before they get better. The population of the United States may increase by 100 million before the end of the century, and the drift from the country to cities will continue. In two decades' time, 70 per cent of the population will probably be living in metropolitan areas. Sixty per cent, it has been estimated, will be living in four presently emerging 'megalopolises' – 'Boswash', stretching from Boston through New York and Philadelphia to Washington; 'Sansan' from San Diego to San Francisco; 'Mijack' from Miami to Jacksonville; and 'Chipitts' from Chicago to Pittsburgh. Rising worldwide affluence means that more people will want to have cars, especially in Europe and Japan, where the great public desire is to catch up with Americans – but where the cities are already blocked with too much traffic.

USE OF COMPUTERS

In the running and planning of cities the computer will help in at least three ways. Each of these is characteristic of computer usage in other areas and together they provide an interesting illustration of its potentialities. First, the computer will be used in a real-time fashion to control air and rail traffic, regulate vehicle flows, plan and book journeys, organize cargo movement, and perhaps arrange door-to-door transportation facilities. Second, it will collect and disseminate the vast quantities of information needed for planning and control in cities: City Hall, like the Pentagon, will have its computer screens. Third, computers will be used to build 'models' of city facilities which will enable planners to observe the effects of alternative plans for very expensive projects.

CITY TRAFFIC

Let us examine first the immense problem of traffic congestion. It is being increasingly realized that transportation is a major cause of many of the ills of modern cities. In one way or another it is destroying the attractiveness of most towns and, in addition to pollution, congestion, noise, death, and maiming, it is causing the urban sprawl which isolates so many people; it is also a major cause of frustration and tension.

In the years immediately ahead, the numbers of private vehicles will increase sufficiently to choke present road systems in most cities of the world (see Fig. 11.1). Piecemeal improvement does little good. J. M. Thomson[1] in an article in *The Times* compared central London's traffic before and after the many traffic-management schemes that were carried out in the 1960s. He doubted that any increase in capacity had resulted. Average travel was a mere ½ mph faster, but routes 5 per cent longer. Because traffic volume seemed to be dictated by road capacity, between 1960 and 1966 the primary effects of one-way systems and of turning and parking restrictions were negligible. It is now even harder for motorists to find a particular route or a parking space. All that was accomplished, it seems, was a shunting of bottlenecks and a ripple of side-effects in neighbourhoods adjacent to the schemes.

Standard methods having failed, Scotland Yard decided to

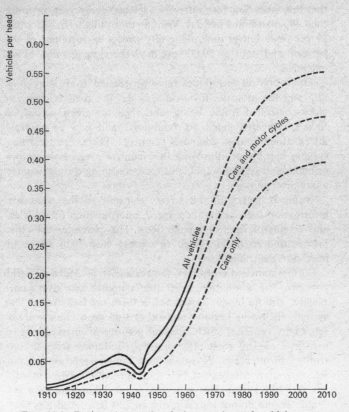

Fig. 11.1 Projected growth of the number of vehicles per capita in Britain. *From the Buchanan Report for the British Ministry of Transport, Traffic in Towns. Her Majesty's Stationery Office, London, 1963.*

experiment with a computer to control the hundred sets of signals in a 6½-square-mile area of West London.

In some cities computer control of traffic has worked well. San Jose, California, has used a computer to control the traffic lights on a three-mile stretch along which 35,000 cars move per day. A. R. Turturici, the city's director of public works, claims that this saves San Jose's drivers $250,000 every year, or more than 400 man-hours per day. Waiting time at lights has dropped 14 per cent. Travel time along city streets has dropped by 10 per cent and stops by 50,000 per day. The system is now being extended.

Traffic-control techniques have progressed from stop signs allowing the motorist to advance when the road is clear, to clock-controlled traffic lights, and then to 'green waves' of linked, staggered lights, to 'tidal-flow', and now to multiple intersections under computer control. There are learning systems capable of improving their control over whole sectors of cities, and systems capable of recommending through-routes along diversions.

Traffic detectors in the streets transmit to the computer information such as number, speed, and direction of vehicles, and lengths of lines at intersections. The computer uses this information to decide when to change each light so as to maximize the traffic flow.

Europe's ancient cities lack the regularity of America's grid patterns. The temptation to let the computer take over completely, with no human assistance, is therefore less strong. The systems in use in London and other European cities employ sets of TV cameras which transmit pictures of intersections to men in a control room. If abnormal conditions arise, a controller can intervene. When the new London scheme was put into operation, Ministry of Transport specialists advised motorists that 'the system works best if drivers cooperate with the computer and do not pit their wits against it ... And it is very much up to the motorist to obey rigidly the traffic lights and diversion signs that come under computer control.'[2]

A similar scheme in Munich has been highly successful, perhaps because of the discipline for which the Germans are famous. There, two computers control 400 intersections. The traffic flow has daily, weekly, and seasonal patterns, and the crossings are not all simply two-way. At the complex intersections, the computer can modify its program in the light of its experience, as recorded on reels of magnetic tape. Based on the current situation, it can also select appropriate traffic-control programs from its library. A controller, observing TV pictures of the intersections and a large illuminated map showing traffic-light colours, can intervene if necessary. A series of road signs can be selectively lit up to lead drivers quickly through a maze of sidestreets, thus relieving congestion on the main arteries. The success of the scheme probably depends on the drivers' willingness to be guided – and on the reactions of the residents of erstwhile quiet backwaters to the surging tide of redirected traffic.

All traffic-optimization schemes can be confounded by an individual motorist who has goals different from those of the system. Those who park in forbidden areas, double-park, stop 'just to pick up a parcel', or otherwise indulge in 'suboptimal' activity can snarl up the best-laid plans and programmes. This is also true of other applications of computers and gives rise to some interesting reflections on regulation and conformity in the computerized society – reflections in which we shall indulge in the second part of the book.

RESTRICTIONS ON TRAFFIC

Computerized traffic-control systems can be made to work and they do help to speed up traffic and relieve jams; but alone, they are not enough to deal with the increase in traffic that is forecast. There are many other possibilities, however.

The city centres are the areas that suffer most. Piccadilly Circus, in the heart of London, was once a pleasant place to stroll; now it is a swirling congested mass of noisy vehicles, and couples no longer sit on the steps of the statue of Eros; it is merely an inaccessible traffic obstruction. The Buchanan Report argued with unassailable logic that traffic access to certain parts of cities will have to be restricted, in some cases partially, in others totally.[3] (Julius Caesar, in 46 BC, banned all wheeled vehicles from Rome during daylight hours to relieve congestion.) Certain locations, the Buchanan Report said, should be turned into traffic-free precincts. These would be shopping areas, market squares, cathedral closes, and so on. This plan is particularly desirable in the older cities, where such areas were not designed for the automobile and where it is destroying their beauty. Even in modern cities, if the public can be made to walk 400 yards from their cars, then more pleasant environments can be built by separating shopping and walking areas from thundering traffic.

In many current cities it would be extremely difficult to increase traffic capacity by road-building or -widening. Thus as the number of vehicles increases, more must be kept out. This may be done by charging vehicles for use of the roads – perhaps charging extra during peak periods – using either licensing arrangements or toll booths. It could also be done automatically by computers.

AUTOMATIC VEHICLE IDENTIFICATION

One feasible method of identifying individual vehicles for road charges is to fit each with a radio transmitter which would operate on a common frequency and transmit a code representing the vehicle's registration number. Local receivers detecting the signals would feed them to a central computer, which would record the date, time of day, and location, and debit the car-owner's account accordingly. The road tax a car-owner paid when licensing his vehicle each year would have to be sufficient to keep the account in credit; presumably, the owner would be advised of its status monthly.

The police would also have detectors listening for 'wanted' vehicles. They could detect cars breaking the speed limit and automatically debit the fine from the owner's account. The fine for illegal parking – possibly even the fee for legal parking – could also be debited automatically.

NEW TYPES OF TRANSPORTATION

If it will not be possible for everyone to drive his car into the city whenever he wants to, alternative forms of urban transportation will be needed. Travelling on a well-planned public-transportation system is often quicker and more relaxing than driving a car in heavy traffic and finding a parking space. A variety of new forms of transportation offer advantages over those in today's cities. Furthermore, the city of the future is likely to have several different types of transportation which interlink systematically.

All urban transport systems are very expensive to build. Moreover, it is difficult to assess exactly what their effects will be. With so many differing proposals for the future, it is an extremely complex task to decide how best the vast sums involved should be spent. This is where the computer can help. Planners can build a model of the city in question in the form of computer programs and superimpose different transportation systems on it. The variables involved, of which there can be an extremely large number, can be changed easily in the computer model. We can thus 'experiment' with the city to find out what will be the best for it. Before discussing this in more detail, let us describe some of the new, promising forms of transportation.

In 1968 the United States Department of Housing and Urban Development (HUD) submitted the results of their transportation study to Congress.[4] After studying many proposals and designs, they recommended the following as 'the more promising of these new systems'.

1. *Continuous Conveyor and Capsule Systems*

In cities of the future various types of conveyors are likely to be used in the major activity centres, perhaps in precincts where private vehicles are not allowed. We have seen prototypes of these in various World's Fair exhibitions. They will include continuously moving belts of seats and capsule-transit systems – some on guideways, some suspended above the city streets.

Figure 11.2 shows one of many proposed designs for such conveyors.[5] A system of belts moving at successively greater speeds enables pedestrians to sit on the seat conveyor, which

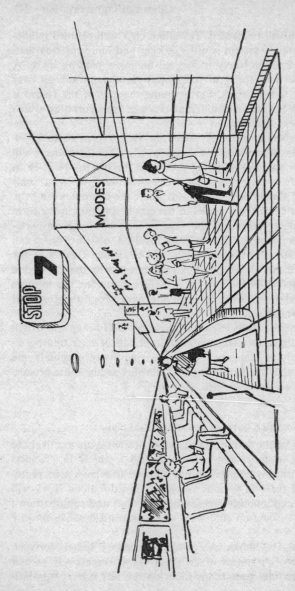

Fig. 11.2 For high-activity centres of future cities, continuously moving pedestrian conveyors have been proposed. This is one of many such designs. *From New Movement in Cities by Brian Richards, Studio Vista, London, and Reinholt, New York, 1966.*

might travel at a speed of about 8 miles per hour. Another system uses automatically controlled capsules which follow each other quietly on continuous tracks above the street level. Each capsule holds two people, with room for parcels. They travel at 15 miles per hour, only a few feet apart, allowing capacities of 8,000 vehicles per hour. A person enters a capsule at one of many sidings, pushes the start button, and the capsule accelerates into the main traffic stream; it can then be stopped at any siding.

2. 'Personal Rapid-Transit' Vehicles for Guideways

Current urban mass-transit systems are slow because, except for 'express' vehicles, they have to stop at every station (or bus-stop). The Underground (also called the Metro or Subway in different cities) are expensive to build and often inconvenient for passengers. Moreover, they lack the comfort and privacy of private automobiles.

Many designs have been proposed for what HUD terms 'Personal Rapid-Transit' systems, which would consist of guideways for small vehicles, each with about the same passenger space as a private car. In a typical vehicle, one or more passengers would enter, each would key his destination, and the vehicle would travel there on a network of 'guideways', possibly a monorail system. The average speed would be 50 to 70 miles per hour, and the vehicle would not stop at intermediate stations. Passengers entering a station would normally find a vehicle available, and so there would be no waiting. The network of guideways would be under computer control, and the computer would transfer vehicles whenever necessary to maintain an adequate number of available vehicles at each station.

This system would provide faster (and certainly safer) transportation than an automobile for most journeys. Because the vehicles are light, the guideways would be much less massive than today's subway installations and consequently much less expensive. Impact on the environment would be minimized: rubber wheels or air bearings could make the system relatively quiet. It would perform economically with a travel demand ranging from 1,000 to 10,000 persons per hour in the travel corridor. This is less than the load of today's subways. These densities prevail in almost all but the largest metropolitan areas, and today's subways cannot handle them economically. The guideway network could economically reach out into the suburbs. Such a system, HUD estimated, could operate at costs below 10 cents per vehicle mile if capacity were 6,000 riders per hour and if the demand were sufficient to generate 15,000 riders per day.

3. Dual-Mode Vehicle Systems

The lower the population density, the less encouraging are the economics of mass-transit systems. The Personal Rapid-Transit systems described above would economically reach into areas where subways are inconceivable. In low-density areas, however, guideways would still be too far apart for convenient walking access. They would be unsuitable for local neighbourhood trips. A *dual-mode* vehicle has therefore been designed which could both be driven along the roads and coupled to the automatic guideways. Such vehicles could be privately owned or, rented for one-time use and turned over at the destination point for redistribution to other users.

New forms of propulsion have been proposed for the dual-mode vehicle.[4] Its development presents more problems than the completely automatic system, but it could be more economical.

4. Automated Dual-Mode Bus

There are various designs for buses that run both on roads and on tracks. These would be particularly useful where there is high-volume travel between an urban core and diffuse outlying areas: the buses would collect passengers in the low-density residential areas and bring them to the rail lines for fast travel to the metropolitan centre.

5. Pallet or Ferry Systems

The advantage of dual-mode vehicle systems is that vehicles can be removed from the roads in congested areas. An alternative to dual-mode vehicle systems are the *pallet* or *ferry* systems in which conventional vehicles board high-speed trains. A pallet may be used on various types of transport mechanism, and may carry a car, a minibus, or freight. The pallets may be 'parked' in computer-controlled warehouses from which they can be automatically retrieved.

On a rail 'ferry' system, automobiles may be carried on flatcars with driver and passengers still inside. Loading and unloading may be automatic. Cars are presently carried by British Rail and by certain European airlines, allowing a person to take his car to a distant location without driving it. Similar high-speed radial car-transportation systems may speed up cross-city travel and lessen congestion, but still enable a person to use his car where he needs it in low-density areas which have inadequate public transportation.

6. The Dial-a-Bus System

Urban transportation today is unable to provide convenient and economical services in low-density areas. Bus service is too infrequent – if it exists at all. In many areas today, it is infeasible to operate *any* form of public transportation.

Some authorities believe that public transportation will never successfully compete with the private automobile until it can provide the convenience of comparable door-to-door travel; if it could, the need for a second car in many families would be removed.

The 'dial-a-bus' scheme would provide such door-to-door service, and be economically feasible in very low-density areas. The buses, holding perhaps between twelve and twenty-four passengers, would be routed by a central computer. Customers desiring service would telephone the controller from their home, or office, or from a theatre, and so on. The computer would collate details of each request for service with information regarding the location of each bus, how many people are on it, and where each passenger is going, and then dispatch the most suitable vehicle to the caller according to an optimal routing program. The bus would thus be continuously collecting and dropping passengers, and routing would be chosen so that no one passenger journey takes an excessive amount of time.

Such a system is highly flexible. It could provide different classes of service for different fares, ranging from scheduled-route pick-up, such as today's buses provide, to immediate individual service similar to taxi service. It might possibly be part of such dual-mode systems as those above. Some vehicles could operate on scheduled routes during rush hours and handle door-to-door travel at other times.

The communication links and computing technique needed for operating this system are available today. The dial-a-bus service is the only one of the new forms of transport discussed here that could be brought into operation in an average city at a cost of well under half a million pounds. It could be operating within three years after such a scheme being initiated.

7. *High-Speed Transit Links within Urban Areas*

High-speed intra-urban links are needed to move high passenger volumes between city and suburban centres. These could be built in the future with systems of a much lighter design than today's subways and railroads. High-speed point-to-point links could couple into the other forms of transportation discussed. Besides rail links, short-takeoff-and-landing aircraft, helicopters, hovercraft and gravity vacuum tubes are under consideration. New types of propulsion have also been considered. High-speed intra-urban rail links have been designed with capacities of twenty passengers per coach, which could move 6,000 passengers per hour. The comfort and privacy of the coaches would approach the automobile's. A computer-controlled single track would not take up excessive space in cities and might be installed along the edge of existing freeways.

SMALL CITY-AUTOMOBILES

In addition to the new forms of public transportation recommended by the Department of Housing and Urban Development's 1968 report and described above, a variety of small 'city-automobiles' has been designed. These could be parked in a very small space and, in some cases, in computer-controlled parking 'warehouses'. Figure 11.3 shows one such design.

It has been suggested that these might be rented at the outskirts of a city area. A person driving his car to work or coming in by public transportation might stop at the perimeter of a city zone and travel into the zone in such a vehicle.

AUTOMATED HIGHWAYS

There has been much discussion of automated highways, and probably within the next fifteen years construction of these will begin. Automated highways offer at least three potential advantages. The first, and perhaps the most important as road carnage increases, is that accidents could be reduced. The second is that driving on such highways could be more relaxing: the driver could read the newspaper, listen to his stereo, or, as one sick joke would have it, mix himself martinis. Third, comparable highways could probably be built more cheaply than today: vehicles could be packed closely together on automatic lanes, which would bear five times the capacity of today's lanes for fast travel.

One General Motors design for an automated highway proposes three lanes: an outer lane for manually operated vehicles only; an inner, automatic lane; and a centre, transitional lane (see Fig. 11.4). Vehicles on the automatic lane would travel close together at a speed of 70 miles per hour. This lane would have a capacity of 9,000 vehicles per hour past a given point. This is the equivalent of *five* additional lanes on current highways, and the cost of the automatic-control equipment in the roadbed would be much less than the cost of the equivalent extra lanes.

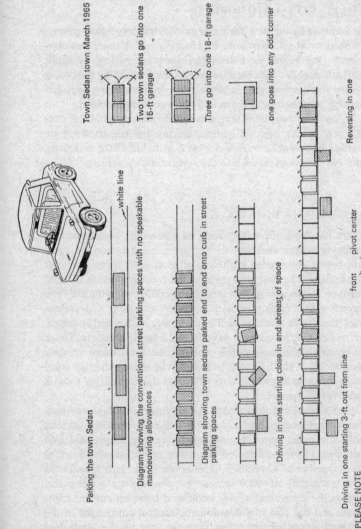

Town Sedan town March 1965

Parking the town Sedan

white line

Diagram showing the conventional street parking spaces with no speakable manoeuvring allowances

Diagram showing town sedans parked end to end curb in street parking spaces

Driving in one starting close in and abreast of space

Driving in one starting 3-ft out from line

Reversing in one

front   pivot center

Two town sedans go into one 15-ft garage

Three go into one 18-ft garage

one goes into any odd corner

PLEASE NOTE
SIGNALS POSITIONED ON TOP OF SEDAN
CLEARLY INDICATE INTENTION OF DRIVING

Fig. 11.3 Town sedan with 13-foot turning circle. These cars will fit within one 20-foot by seven-foot-six parking space. *Courtesy Eric Roberts Associates.*

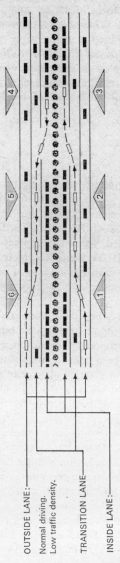

OUTSIDE LANE:-

Normal driving.

Low traffic density.

TRANSITION LANE

INSIDE LANE:-

'Autoline lane.'

Traffic density equivalent to a five-lane highway.

Speed 70 miles per hour.

1. A motorist travelling in a normal lane but wanting to enter the Autoline lane would move into the transition lane and signal his desire to enter the Autoline.

2. By putting his car on automatic control, his speed and position would be monitored and adjusted.

3. The car would be automatically guided into position at the end of the first available group on the Autoline lane.

4. To leave the Autoline lane, the motorist would first signal his intention to the system.

5. His car would move automatically into the transition lane at the first safe opportunity.

6. He would return his car to manual control and then move into a normal lane.

Fig. 11.4 An automated highway. *Courtesy General Motors.*

The electronic equipment in the car might cost somewhere between $100 to $200. General Motors proposes that before a vehicle may be permitted to drive in the automatic lane, a check of the car's equipment should be made – automatically. An instantaneous check of the equipment would be conducted while the car is in the transitional lane via a transmission link to a computer. A fail-safe system of handling emergencies has also been devised.

SIMULATION

Engineers generally agree that these new types of transportation systems could be made to work in cities within a few years. The question remains: *Should* they be built? They are all very expensive, although less expensive than a major subway system. The United States Department of Housing and Urban Development (HUD) issued contracts for studying this question by means of simulation. Using a computer 'model', the effectiveness and economics of such systems can be investigated in detail.

Simulation models may be either large-scale models of movements within an entire city or small-scale models for investigating one aspect of a proposed system. On a small scale, for example, models have been built to investigate sets of rules for changing traffic lights at a series of intersections. The models investigate the rules for selecting the light phases or patterns so as to maximize traffic flow depending upon the density, speed, and lines of traffic.

Experiments with real traffic are expensive, exasperating, and often inconclusive. Models built using computer programming allow investigation of a much wider range of conditions so that curves can be drawn and formulas produced. They thus provide deeper insight into the effects of different factors than confused and limited 'live' experiments.

The modelling of entire cities was carried out by the General Research Corporation of Santa Barbara under a HUD contract. Some 200 models were built and their results summarized in a 500,000-word report.[6] Four cities were selected to provide a representative cross-section of transportation problems in America: Boston because it is a large city strongly oriented to the private automobile, and two smaller cities, New Haven and Tucson.

A great amount of demographic detail of the cities is needed for such models – population density and average family income in each area of the city, locations of businesses and numbers of people working there, and so on. It is necessary to evaluate the effects of variations in cost, crowding, and accessibility of transportation on where and how people travel. For instance, would a worker earning £1,500 a year and living in such-and-such a location walk on a winter's day to save a 10-pence fare? The accuracy of the model's performance depends upon how accurately such factors are assessed.

The General Research Corporation tested the validity of its results by simulating contemporary conditions in the four cities before it introduced proposed future changes into the models. It found that the computer printout corresponded well with actual measurements of present-day traffic movement. It was therefore judged to be a realistic representation of future effects of building new transportation facilities. Adjustment of the model also introduced into the calculation such factors as air pollution, the intrusion of the car into city life, the accessibility of key areas, and the mobility of ghetto residents, in recognition of the fact that the most economical and fastest system is not necessarily the 'best'.

Two approaches seemed practical in the simulated cities: either a gradual improvement of present-day facilities or a radical introduction of some form of the 'new technology' described above. City authorities, hard pressed for immediate relief of traffic problems, tend to seek immediate means of improving present facilities – widening roads, one-way traffic schemes, more or better buses, extension of subway lines, and so on. The models developed by General Research indicated that this is generally not good enough. Money in many cities would be better spent on a new-technology approach. The Houston model suggested that in most large cities developing

COMPARATIVE PLANS OF TRANSPORT SYSTEMS (drawn to the same scale)
Scale in feet: 10 0 50 100 150 200

COMPARATIVE SECTIONS FOR 2 DIRECTIONS
Scale in feet: 10 5 0 10 20 30 40

Transport system	Comparative sections (radius)	ECONOMIC DISTANCE BETWEEN STOPS OR STATIONS	PASSENGER OR VEHICLE CAPACITY PER HOUR ONE WAY	AVERAGE SPEED	ECONOMIC RUNNING COST PER CAR OR PASSENGER MILE
PRIVATE CAR ON SURFACE STREET IN CITY		as required	700–900 v.p.h.	up to 30 m.p.h.	1p–1.25p
			500–2,000 v.p.h.		2.9 cents
PRIVATE CAR ON AUTOMATED MOTORWAY (AUTOLINE) SYSTEM	40' radius	interchange points 2 mile intervals minimum	7,200–9,000 v.p.h.	40–70 m.p.h.	as above plus possible toll
MINI CAR ON SURFACE STREET IN CITY (no other traffic)		as required	2,000 v. p. h. One 8ft wide lane	30 m.p.h.	2.1p
MINI CAR ON ELEVATED AUTOMATED ROAD (STARRCARR) SYSTEM	6' radius	interchange points 5–25 mile intervals	3,000–5,000 v.p.h.	15 m.p.h. (city use)	2.1p
EXPRESS BUS ON GRADE SEPARATED ROAD (one lane)	66' radius	1 mile	1450 v.p.h. 60,000 people	35 m.p.h.	1.6p 30 cents per car mile
DOUBLE DECK BUS ON SURFACE STREET IN CITY	70' radius	2 mile	120 v.p.h. 7,200 people	8–15 m.p.h.	18.3p–19.1p per car mile 0.8p–2.1p per pass mile
TELECANAPE (non-stop system)	47' radius	2 mile	8,300	8 m.p.h.	10.5p per pass mile

System	Turning radius	Length	Capacity	Speed	Cost
CARVEYOR 4 SEAT (non-stop system)	12' radius	½ mile	5,000 seated 10,000 seated and standing	15 m.p.h.	
MINIRAIL	50' radius	½ mile	5,000 seated	8-15 m.p.h.	10.5p
NEVER-STOP RAILWAY	15' radius	½ mile	12,600 seated 18,000 seated and standing	15 m.p.h.	10.5p
OPEN BUS TRAILER (automated robotug tractor)	26' radius	as required	100 v.p.h. 7,500	8 m.p.h.	
CARVEYOR 10 SEAT (non-stop system)		25 mile	11,000 seated 23,000 seated and standing	15 m.p.h.	32ft belt = 3,000 people
TELEPHERIQUE GONDOLA CAR (4 seat car)		1 mile or over	500-1,000	6-10 m.p.h.	2.1p
PEDESTRIAN CONVEYOR OR MOVING BELT	straight only	100 feet-800 feet	48ft belt = 10,000 people	1.5-2 m.p.h.	0.10p

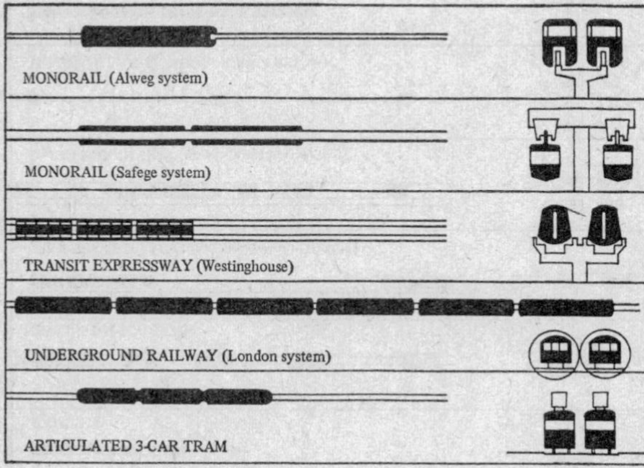

MONORAIL (Alweg system)		5 mile min.	16-20,000	50 m.p.h.	2.1p-1.6p
MONORAIL (Safège system)	100' radius	5-3 mile	8 coach = 48,000 2 coach = 12,000	50 m.p.h.	
TRANSIT EXPRESSWAY (Westinghouse)	150' radius	16 mile	8-20,000	75 m.p.h. 23 m.p.h. (5 mile stops) 39 m.p.h. (2 mile stops)	
UNDERGROUND RAILWAY (London system)	330' radius	5-2 miles	40,000	20-30 m.p.h.	28 cents 15-1.6p per car mile
ARTICULATED 3-CAR TRAM		25-5 mile	20,000	20-30 m.p.h.	23p

Fig. 11.5 A comparison of transportation systems. Actual running cost is shown in the right-hand column, excluding construction and staff costs. *Reproduced with permission from New Movement in Cities by Brian Richards, Studio Vista, London, and Reinholt, New York, 1966.*

'personal rapid-transit' systems would be preferable to building a subway. In smaller cities – those with populations of less than half a million people – these systems are harder to cost-justify. Congestion in these cities is less, however, and the use of the Buchanan Report's precincts and of dial-a-bus systems may be the best approach.

Whatever the results, the important fact is that these models now give us a way of 'experimenting' with a city, of measuring the various effects of alternate plans and optimizing its effectiveness – before the large expense is committed.

DESIGN OF NEW CITIES

New technology in transportation will probably not be seen in the old, established cities. It is certainly difficult to imagine the above schemes being implemented in New York, London, or Paris in the next few years. They may be built first in those parts of the world where the level of innovation is highest. Perhaps we will see them in Japan before they appear in America. There are schemes in many countries for building entirely new cities. Free from the forces of existing politics and vested interests, these can be designed to be cleaner, with transportation systems a major feature of the design.

The following paragraph describes transportation in Romulus, a subcity being considered for construction on two islands in Boston Harbor:

A transportation plane extends under the entire central portion of the city. On this level, we find an automated guideway system, a conventional street system on which cars and trucks can be driven manually, and ample space for parking. Above the transportation plane is the ground level, which, throughout this area, is freed for pedestrian malls, shops, and a specialized low-speed passenger-transport system. This last system provides movement both along the ground and vertically up special outside elevator shafts on all of the major high-rise buildings in the area. Residents can thus carry bundles directly from the shops to their apartments with no need to transfer or walk long distances. Several areas in the central city

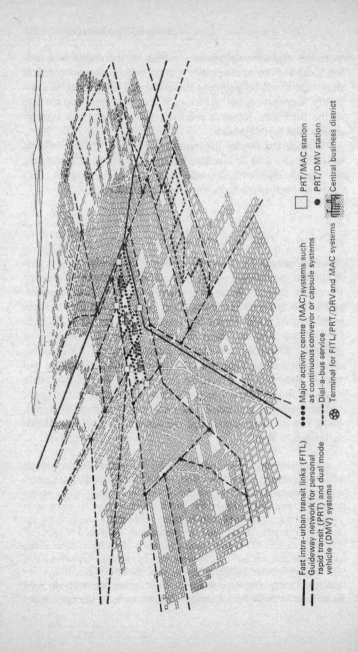

Fast intra-urban transit links (FITL)

Guideway network for personal rapid transit (PRT) and dual mode vehicle (DMV) systems

●●●● Major activity centre (MAC) systems such as continuous conveyor or capsule systems

--- Dial-a-bus service

⊗ Terminal for FITL/PRT/DRV and MAC systems

☐ PRT/MAC station

● PRT/DMV station

▓ Central business district

are domed to permit pedestrians to roam about freely during inclement weather. During the spring and summer, portions of the roof are rolled back to take full advantage of nice days. Since this city is built on a pair of islands, extensive advantage has been taken of its water location for seaside parks and marinas.[7]

Computer models of new cities will of course take into consideration many factors other than transportation. The city should be designed as a total system in which all of the necessary facilities intermesh. It should attempt to attract a balanced mixture of people with a suitable blend of shops, factories, offices, theatres, and recreational facilities. Planning this is systems analysis on a large scale with much social research providing input to the computer models. With such planning techniques, we may well design cities that become truly thriving and pleasant to live in, rather than barren shells like the glamorous Brasilia.

Fig. 11.6 New systems of urban transportation offer many alternatives for travel throughout a metropolitan area. Fast intra-urban transit links can serve travel needs between major urban nodes. 'Personal rapid transit' can automatically serve low-to-medium population densities over a network of guideways. Dual-mode vehicle systems can augment personal rapid transit with small cars or buses that can be driven on city streets and travel on automatic guideway networks. 'Dial-a-bus' service controlled by computers can collect and distribute passengers in low-density areas on demand. Through combinations and efficient use of these new systems, people can be provided with greater accessibility to the opportunities of urban life.

The efficency and economies of such systems will be explored by building computer models of cities that simulate the movements that take place. In this way, experiments can be made with different combinatons and configurations of transportation systems before any system is actually implemented. Preliminary models indicate that novel systems such as 'personal rapid transit' and 'dial-a-bus' systems will often be more economic than building more conventional highways and subways. *Redrawn from US Department of Housing and Urban Development study report, Tomorrow's Transportation, Washington, DC, 1968.*

CONTROL OF TRAFFIC FLEETS

Another use of computers in transportation is for the control of fleets of vehicles such as taxis, buses, and trucks. It is now practicable to pinpoint the position of a given vehicle by means of radio signals. Figure 11.7 illustrates this. The vehicle in this diagram may be many miles from the transmitter. The transmitter sends out coded signals addressed to each vehicle in turn. The vehicle responds only to the signal addressed to it, and the response is picked up by three or more roadside receivers. The vehicle's location is computed from these signals by a method known as 'triangulation'. Such a system could keep track of many thousands of vehicles and pinpoint the location of each to an accuracy of within 100 feet. Its use may be shared by private and public users such as police, taxis, and trucking companies.

AIR-TRAFFIC CONTROL

If land-vehicle control is desirable, air-traffic control by computer is becoming absolutely essential. Day and night, at large airports today, controllers sitting at radar screens guide the large numbers of planes coming in to land. It is a highly exacting and often tension-filled job. All aircraft within a certain range of the airport appear as green points of light on the screen and are identified with small plastic markers known as 'shrimpboats'. The controllers move the shrimpboats by hand as the aircraft change their position on the screen. Newer systems dispense with the shrimpboats; instead, labels on the screen show aircraft identifying codes and altitude. These systems, however, are not yet in wide use.

The controller gives orders by radio to the captains of aircraft that are landing and taking off. The controller needs to muster concentration in order to retain a mental picture of all planes in his sector – identifications, positions, speeds, altitudes, types of aircraft, and landing sequence. If he lose the picture, a crash could result. It is a job that few people are capable of.

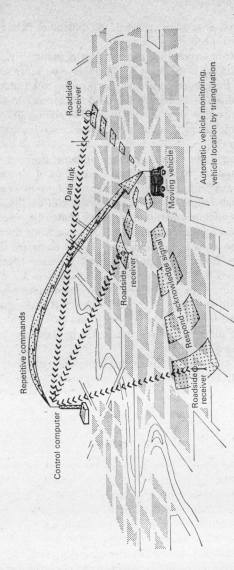

Fig. 11.7 Automatic monitoring of vehicle positions. *Redrawn from US Department of Housing and Urban Development, Tomorrow's Transportation, Washington, DC, 1968.*

Repetitive commands

Control computer

Data link

Roadside receiver

Moving vehicle

Roadside receiver

Respond-acknowledge signal

Roadside receiver

Automatic vehicle monitoring, vehicle location by triangulation

As the skies become more crowded, the difficulties of manual air-traffic control become more severe.

The number of aircraft will increase greatly in the 1970s, and supersonic transports (SSTs) will contribute even more to the difficulties of the ground controller. Eurocontrol, the European upper-air-traffic-control agency, has used a computer to simulate the consequences. The agency found that adding 200 SSTs to 700 subsonic aircraft would increase the controller's workload many times.

By 1970 air-traffic controllers in southern England had Marconi computers for analysing all flight plans of aircraft in their airlanes. A computer system such as this must provide immensely high reliability. Three computers will do the same calculations simultaneously and compare results; the objective is to avoid more than one failure of 30 seconds or more in five years. The plans of incoming aircraft will be passed on to the computers from neighbouring control systems. Those of outgoing planes will be typed into the computers by pilots before departure. The machines will then provide air-traffic controllers with warnings of converging aircraft and continuously check the locations of all planes in the vicinity. Humans will still have ultimate control, but it will be based on exact up-to-the-second information.

This and similar systems are the beginnings of a network that will constantly monitor the increasingly crowded skies of the 1970s.

COMPUTERIZED RESERVATIONS

While some computers keep track of the planes, others will keep track of the passengers. Most of the world's major airlines now employ real-time systems for handling flight reservations. Most of them make spectacular use of communications. Some have about a thousand terminals, all obtaining very quick responses from a far distant computer. Some have communication links that circle the world. Pan American, for example, maintains a large room full of agents in London, near Piccadilly, who answer telephone calls about bookings from most of England. While they talk to a passenger on the telephone they are also carrying on a 'conversation' with a computer in the Pan Am Building (over Grand Central Station) in New York. As one stands in the room in London watching them, it is difficult to believe that the replies they are receiving, with a response time of a second or so, are coming from the other side of the Atlantic.

Information concerning a booking made in London or any other part of the world must be stored in the data bank in New York. First, the record of which seats are available on that particular flight must be changed. Second, a new record must be created for the passenger in question giving details of his journey, where he can be contacted, when he will pay for the ticket, and any special requirements he may have. The records are then made immediately available to all other terminals; thus if the passenger telephones another agent to make a change in the booking – perhaps an agent in a different city or even a different country – this agent will be able to display immediately all details he needs about the passenger's reservation. There are no longer filing cabinets full of paper records.

As the bookings received – possibly from around the world – fill the flight, the current availability of seats on a given flight can be ascertained by any agent with a terminal. If one flight is full, the computer indicates the best alternatives. Hotel bookings and car rentals can be handled similarly if desired.

On a typical computerized booking system a passenger wishing to make a booking on a given airline may walk into its office and talk to an agent operating a terminal with a screen and keyboard. The agent does not use a timetable, but obtains information about possible flights from the computer. For example, the passenger might want to arrive in Los Angeles on 25 March at about 5 p.m. (Pacific Time). The agent keys in this fact in a coded form, and a listing of the best four flights that have available seats and meet the customer's requirements will appear on the screen. If for some reason none of these flights is suitable, the computer can be instructed to scan outward from this time and indicate the four next-best flights, and so on. The screen indicates to the agent the appropriate fares and all the other details he needs to know.

When it has been agreed which flight (or flights) the passenger will travel on, the agent then enters details about the booking into the system: the passenger's name, his home and business telephone numbers, ticketing arrangements, and other pertinent information which are sent to the computer and stored on its disk files.

One by-product of such a system is worth noting because it has important implications for our lives in a world of computers. An airline which prides itself on giving *personal* service to its passengers can use such a system to advantage. Once the files, the computers, and the vast network of communication lines are set up, it costs little extra to use them for attending to simple personal details as well as the basic facts of business. For example, if a passenger will require a wheelchair at the airport, this fact can be recorded. If he is flying on his birthday and he would like a cake on board, or if he speaks only Italian, or if he wants a window seat with a good camera view, or if he is a cousin of the airline's president, or if he wants to take his pet Dalmatian on board – these and other personal items can be recorded in a 'comments' section of the computer record. The comments can then be made available to other agents handling this passenger, to the check-in girl at the airport, and to the hostesses and captain of the plane on their passenger list if so desired. Thus a lady with arthritis can be assured a seat near the bulkhead where there is more room for her stiff leg, or the captain can greet certain special passengers.

The point is that a computer system, contrary to much public opinion, need not be something impersonal which treats people as if they were numbers. Before the computer was installed, agents found it too much trouble to record and convey these personal details which can make the journey more enjoyable. Now it is no trouble. In the employment of other computer systems, too, we may find – if we use our imaginations – that an inexpensive by-product of automation can be an improvement in the human aspects of efficient service rather than depersonalization.

BOOKING MULTI-MEDIA JOURNEYS

Many airline-reservation systems currently also handle hotel reservations and car rentals. In the future, perhaps entire journeys on different transportation media may be booked in this manner. One can presently book train seats, in this way in Japan, for example. A journey in the future by train, plane, and 'dial-a-bus' may be entirely planned and booked by computer, making possible minimal waiting and fuss between connections. If one commutes to work, a comfortable seat will be reserved and connecting transportation arranged.

FREIGHT

Many forms of transportation, starting with the birchbark canoe, began by conveying passengers but eventually carried more and more freight, until freight dominated their business, as it does today's railroads. Air-freight transportation presently is growing at a much faster rate than passenger air travel. As such planes as the enormous Lockheed C-5 Galaxy come into increasing use, freight will eventually predominate in the airline business too.

'Containerization' is presently the favourite solution for cutting cargo costs. Its advantage lies in the ease with which steel or aluminium boxes can be switched between air, sea, rail, and highway carriers. The resultant reduction in turn-around time is great – for example, a ship can spend less time in port and more time at sea earning revenue. Stevedoring and cargo-handling contribute a large proportion of a ship's operating costs, costs which will be reduced if the container is packed by the consignor away from the dock. The containers, however, are themselves expensive and so must be kept moving. To do this, the container-owners will increasingly use computers to track and optimize their movements, and to direct them to waiting customers. Computers can also be used to calculate how best to position containers in a ship's hold, to keep customers informed of their consignment's whereabouts, to locate overdue and lost units, and also to check for space on specific vessels for new orders.

It has been suggested that worldwide container pools of completely interchangeable units will be a necessity within a decade. Such units would be interchangeable among ships, planes, trains, and highway vehicles. Automated warehouses will handle them. International lists of container movements and availability will be required as an aid to world freight distribution. Internationally linked computers for this specific purpose will become commonplace, with communication satellites providing the link instead of Telex systems as at present.

It is possible that containerization will come into use for people also. Costs and congestion might be cut if aeroplanes were designed for automatic loading of small, light, comfortable passenger 'containers'. Passengers would board a container at a city terminal, and the container would then be transported by bus, rail, helicopter, or hovercraft to the airport (which may be a very long way from the city terminal by the late 1970s). The doors of several containers aligned end-to-end would then open to form an aisle in a plane or train. Their windows would position against the plane's (or train's) windows – although some planes, especially supersonic transports, may have no windows in the late 1970s. The movements of the passenger containers would be scheduled and controlled in an optimum fashion by computers. A passenger today can go to sleep in London and wake up in Paris, his bed having travelled with him by rail and ship. Tomorrow the bed may fly also.

Paperwork is currently one of the major causes of delay in cargo-handling; the delay not only exasperates customers, but also causes piles of freight to collect in expensive warehouses. Swissair has brought in a computer to achieve same-day clearance of goods at JFK Airport in New York. When a plane takes off from an airport in Europe, the cargo list is transmitted to the New York computer and fed to terminals which print carrier certificates, release orders, and transfer- and transit-cargo manifests. The papers are ready in an hour or two, several hours before the plane's arrival, and in time for either same-day customs clearance of the shipment or immediate onward transmission for inland clearance.

At London's Heathrow Airport, customs and the airlines propose to cooperate by means of a computer system capable of processing the huge volumes of freight. The expected air-cargo boom would otherwise swamp current procedures. The system will operate like Swissair's New York system, except that it will be at the service of any airline, using BOAC's computer network to channel appropriate data to the customs machine. The latter will then record the arrival of goods, spot discrepancies and help to clear them, and calculate duty, tax, and so on. Customs officers will use terminals to inquire about items they wish to inspect, and the airlines will have the same ability (with regard to their own cargoes, not their competitors'). On the export side, the computer can check freight lists for embargo and other restrictions, ensure that cargo regulations are met, and prepare all the paperwork for the forwarding agents. The system will be run by the Post Office's National Data Processing Service and will collect statistics regarding imports and exports not only for accounting and for charging the participants, but also for trade returns to the Board of Trade. By 1975, 1 million tons of freight per year – the equivalent of 1000 DC3 loads a day – will be passing through the airport's automated warehouses. Freight will be physically handled on pallets, with containers guided automatically to the right part of the storage and transit area.

HOW TO RUN A RAILWAY

Railways are beginning to plan their operations by computer too. A computer keeping track of all the rolling stock can plan its movements so as to conform to an optimal pattern.

Railways in the United States operate at a disadvantage because of the boundaries – determined by historical accident, not deliberate policy – which divide their routes. Because railway operation is most economical over long distances, trucks ordinarily leave the jurisdiction of their owners and travel on other systems. One line's railway truck is much like another's, and the rational thing to do with an empty one is to fill it with suitable freight and send it on its way. However, rules set up among the railroads require that empties be returned home as soon as possible by their recipients – often at some inconvenience, and frequently empty and hence unprofitably. Many railroads employ elaborate tracking systems which record the location of all their trucks while on their lines, using data teleprocessed from the freightyards where trains are made up. Once a truck leaves its home system, however, it is 'lost' until it reappears on the line. Unfortunately, it is unlikely that economic logic will prevail in the establishment of a single data bank for *all* rolling stock across the continent, coupled with an agreement to treat all similar trucks as interchangeable, like the container plan outlined above.

The capacity of the computer to perform central decision-making for large numbers of cargo units was dramatically demonstrated by the Louisville and National Railroad in 1966. A fleet of more than 1,000 of their wagons was used to move wood from forests in Kentucky and Alabama to pulp mills on the Gulf Coast. Originally, the railroad stationed an agent at each forest with instructions to keep his customer happy by maintaining a ready supply of empty wagons in the yard. Each of the many agents attempted to do this, but those farthest from the mills, where the empty cars originated, found that their colleagues nearer the source of supply were in a better position to grab the available woodracks as they passed. Thus certain customers were always short while large stocks of idle cars awaited loading in other yards. The L&N therefore had to obtain more cars to meet the demand, only to discover that the idle stock grew and the dissatisfaction remained. We have briefly mentioned 'suboptimization' before (Chapter 9, p. 208), and L&N's problem is a classic example. A computer was installed, therefore, to optimize the whole situation by meeting all the needs as they arose. The agents were withdrawn, and the customers themselves supplied the computer with their daily requirements. As cars were released from the mills, they moved up to freightyards from which trains departed for the forests. With a record of where every car was at any given moment, the computer could determine which ones should be sent from which yard or train to meet a particular requirement most easily. It was not long before every demand was met within one day, still leaving a number of idle cars which could be profitably leased to neighbouring lines. A simulation model was built to see how many could be spared at a given time without service being reduced significantly.

This case-history points up one of the fundamental problems faced by industry and society in an age of massive data-handling and communication ability. Centralization and optimization go together, and in some instances, planning proves to be more rational than competition. The question remains, however, whether man is competent to determine the ends to be sought by planning. Although competition is sometimes wasteful, it has the advantage of preventing the total and efficient achievement of the wrong objective. The present condition of our cities and transportation facilities epitomize the failure of market economies: regulation and control of them are found even in capitalist countries. Systems engineering and computer techniques offer an alternative to market competition as a determinant of certain human needs. The next decade may show to what extent this alternative can prove workable.

DATA FOR CITY PLANNING

Good planning – whether for transportation systems or for entire cities – requires accurate, up-to-date information organized for easy access and evaluation. City data banks are therefore being developed for the accumulation of departmental files in a form that allows all agencies to draw on them for their planning. The basis of the files must necessarily be geography, since the location of people, businesses, services, and facilities is the primary interest.

By means of display terminals or digital plotters, maps can be displayed showing densities of housing, crime, traffic, telephones – in short, anything measurable – using the most recent

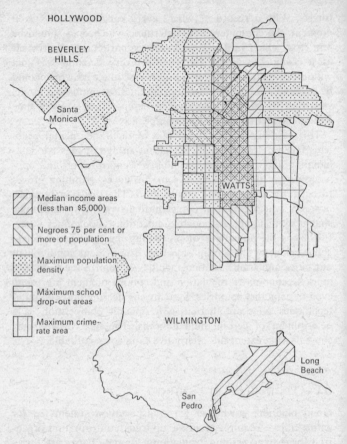

HOLLYWOOD

BEVERLEY
HILLS

Santa
Monica

▨ Median income areas
   (less than $5,000)

▨ Negroes 75 per cent or
   more of population

⬚ Maximum population
   density

▤ Maximum school
   drop-out areas

▥ Maximum crime-
   rate area

WATTS

WILMINGTON

Long
Beach

San
Pedro

Fig. 11.8 How to predict trouble. In 1965, Space-General
Corporation undertook a systems study for the state of California on the prevention and control of crime and delinquency.
In the course of that project, a company team made a special
analysis of Los Angeles County to discover whether there was
an underlying social pattern that determined which areas showed
the greatest potentials for trouble. The team selected five social
criteria (highest population density, lowest income level, etc.)
and constructed the demographic map above. The five sets of

available data. The location of a new school could clearly be better chosen if the future distribution of school children in the district could be projected and translated into a series of maps. Figure 11.8 shows a dramatic example: a demographic map of the type that could be produced automatically from data-bank information reveals potential trouble areas in Los Angeles County.

City planners are much better served by maps than they ever could be by tables and lists. A detailed streetmap is much more meaningful, for their purposes, than a street directory, for a map can simultaneously display not only physical shape and facility locations, but also the location of licensees, churches, general stores, fire outbreaks, gas mains, bars, restaurants, bus-stops, population, or most anything else. One map is worth thousands of lists.

Quite soon we can expect City Hall to be equipped with display screens on which a planner can throw out maps of any section of the city on large or small scale, entering demographic, topographical or administrative data of any kind. Ultimately, he may be able to 'add' a new facility and gauge by simulation the effects on people or traffic.

Backing the map with mathematical models allows authorities to make short-range decisions as well as long-range plans. For example, Connecticut's pollution-prediction program combines survey data about power-plant, vehicular, domestic, and industrial emissions to create a model which generates a picture of the state of the atmosphere at thousands of points within the state. This picture can be altered by 'closing down' different pollution sources in the model to determine the value of so doing before issuing a recommendation. Of course, a huge

criteria – identified by the shaded boxes at left of map – turned out to overlap most completely in the Watts section. Shortly after the map and report were completed, the historic Watts riots broke out in the late summer of 1965. The event bolstered the report's conclusion that analysis of better and more refined social data could pin-point the real danger areas in cities in time to institute preventive programmes.

quantity of pollutant is blown or drifts in from out of state. The system thus also measures the amount of 'imported' carbon monoxide, soot, sulphur dioxide, etc., but it cannot issue recommendations with any great expectations until the boundaries of its jurisdiction are aligned rationally.

Systems engineers would like to build a complete urban model that included all demographic, aesthetic, social, economic, and political data. It might then be possible to predict the consequences of total city plans, and hence to evaluate them. Appropriate management-science techniques exist for analysing such data – when they are available; but the ability to quantify social, as opposed to economic, benefits and costs has yet to be developed. Nevertheless, armed with quantitative information, one can at least argue about the human values without the facts being in dispute. A large flyover (one might conclusively *assert*) will cut the light from 740 dwellings for more than 50 per cent of the day, nine months of the year, and reduce journey time from Suburbanville to Downtown City by an average of 17 minutes for 14,000 rush-hour motorists. But (one might *argue*) is it worth it? What about the 'blight' of gas stations and diners that will grow up along its route, and the noise of heavy traffic on the approaches? Can we extrapolate the past? If the information had been collected in the past, predictions could have been made and the hypotheses tested.

In the next chapter we shall discuss further the sort of data banks that are needed to back up such decisions.

GOVERNMENT

The changes in our environment that technology has brought about are immense. It can clearly bring great changes in the future, probably even greater than those in the past. The changes in technology, however, have not been matched by corresponding changes in our political apparatus, which is still based on the communication potential of the horse-and-buggy age. In no other area of human life is the need for a 'systems approach' more evident than in dealing with the urban crisis: the city of Los Angeles draws some of its water from the Sacramento River, 450 miles away; the cities of five states have fouled Lake Erie; downtown Manhattan is inundated daily with commuters who arrive from three states by way of subways, bridges, and railroads *under independent management*. Only by seeing these problems as *one* can they be solved. Only a single agency or organization can command the necessary vision.

Although the various problems of the city are so interrelated as to be truly inseparable, solutions are frequently sought from a very limited range of possibilities which are limited not by logic or technology but by politics. Fresno County had a contract with Aero Jet General Corporation to examine 'solid-waste management'. When asked why his pollution-control study for Fresno County was for *solid* waste only, Aero Jet's Bowerman replied:

It's because the money was obtained by the State on a matching grant from the US Public Health Service's office on solid waste. For water-oriented studies, you have to go to the Department of the Interior's Federal Water Pollution Control Agency or the Office of Water Resource Research or the Office of Saline Water. For air studies, you must go to the Center for Air Pollution at the Department of Health, Education and Welfare. So you see, if you wanted to do a total waste-management study, you would have to get money from quite a few sources.

Of course, you cannot actually 'do' a study of solid-waste management without seeing how that relates to air and water. So we just have to do that anyway with our limited funds.

New York City has studied problems in a similarly isolated fashion. Britain's recent overhaul of its railway system ignored related forms of transportation, such as roads. The list of examples can be extended endlessly. It will be a long time before the ends of public policy catch up with the advances in the means of their achievement. The solution to traffic problems may not be in rapid-transit systems, but in the redesign of industrial and social activity, so that people are prepared to live and play near where they work and to make better use of new communication media. Democratic institutions change slowly, largely due to the difficulty of obtaining meaningful agreement on any particular one of the many alternatives to the status quo. We live, therefore, in a world of small piece-meal improvements, where most authorities can tackle only the problems within their limited jurisdictions.

As modern technology tightens its grip on our environment, this piecemeal, compartmentalized approach will prove increasingly inadequate. We now have the good fortune to possess the computer technology and the ability to conduct wide-ranging systems analyses. Clearly we *must* use it in our cities; yet there are two aspects to this mandate. First, development cannot be left *solely* to private enterprise inspired primarily by the profit motive. Energetic government planning and heavy spending are needed in many areas (often to make the appropriate actions of private enterprise profitable). Second, the narrow, compartmentalized viewpoints and actions of much of today's government must be broadened.

The authors showed an early version of this chapter to a high official in the British Ministry of Transport, and he wrote: 'There is a very serious conflict between the desire of the technocrat to have a large organization which can utilize the latest technological apparatus, and the need to keep local government on a basis which involves people and gives a true sense of community. There is no simple answer to this conflict.' It is a point well taken, which articulates a dilemma that exists in other fields in the computerized society as well. We have to devise means somehow to enable local government to use the immense power of the new technology wisely, just as we must also bring this power to the small businessman, the country lawyer, and the village schoolteacher.

REFERENCES

1. *The Times*, 22 January 1968.
2. *The Times*, 24 January 1968.
3. *Traffic in Towns*, Report of the Steering Group and Working

Group appointed by the Ministry of Transport ('The Buchanan Report'), Her Majesty's Stationery Office, London, 1963.

4. *Tomorrow's Transportation*, a report to Congress describing the results of a study on urban-transportation systems by the US Department of Housing and Urban Development, Office of Metropolitan Development, Urban Transportation Administration, US Government Printing Office, Washington, DC, 1968.

5. Brian Richards, *New Movements in Cities*, Studio Vista, 1966.

6. William F. Hamilton, II, and Dana K. Nance, 'Systems Analysis of Urban Transportation', *Scientific American*, July 1969.

7. William W. Seifert, 'Transportation – 1993', *IEEE Spectrum*, October 1968.

12

TOWARDS A NATIONAL DATA CENTRE

Wherever we are, it is but a stage on the way to somewhere else, and whatever we do, however well we do it, it is only a preparation to do something else that shall be different.
– Robert Louis Stevenson.

A significant part of the information explosion is the ever-growing quantity of records that is kept by the various levels of government. If we can accept civil servants' estimates of the amount of records necessary to run our increasingly complex society, then – *without* automation – in a few decades time we would all be employed governing and policing ourselves, and have no time to do anything else.

STATE AND LOCAL GOVERNMENTS

Table 12.1 shows some of the categories of data under active consideration for record-keeping, using automation, by local

TABLE 12.1
Typical Categories of Information Stored in Data Banks of Local and State Governments in the United States

1. *Personal-Identity Data*:
   Social Security number, recognizable physical characteristics, address, race, blood type, relatives;
2. *Personal-Employment Data*:
   present employer, previous employer(s), occupation, earnings;
3. *Personal-Education Data*:
   schools, courses, degrees, achievements;
4. *Welfare Data*:
   periods on welfare, aid received, basis for aid;
5. *Health Data*:
   physical deficiencies, reportable diseases, immunization, X-ray results;

6. *Tax Data*:
    details of income-tax returns, dependents, investments, life
    insurance;
7. *Voter-Registration Data*:
8. *Licence and Permit Data*:
    type, number, date issued, issuing agency, date of expiry;
9. *Law-Enforcement Data*:
    records of offences, outstanding warrants, ex-convict registra-
    tion, suspects, missing persons;
10. *Civil-Action Data*:
    plaintiff and defendant, court, date, type of action, result;
11. *Probation/Parole Data*:
    probation number, agency, court, offence, terms, conditions;
12. *Confinement Data*:
    type of confinement, period, place, reason, treatment, escapes;
13. *Registered Personal-Property Data*:
    cars, motorcycles, boats, firearms, dogs, ambulances (in USA),
    elevators;
14. *Vehicle-Registration Data*:
    owner, licence number, make, engine type, body, number of
    axles, wheel base, whether registered in other states;
15. *Landed-Property Data*:
    zoning, uses, assessed value, taxable value, deeds, water and
    mineral rights, drainage, productivity, soil type, details of
    building, owners, sales;
16. *Residence-Owner/Occupant Data*:
    names, address, licences, occupation group, race, place of work,
    children in school, income group, vehicles owned, police
    records;
17. *Street-Section Data*:
    name and class of street, length, boundaries, intersections,
    surface, drainage, traffic and parking information, streetlights,
    kerbs, sewers, signals, accidents.

and state governments in the United States. Many other items
are currently stored by government bodies in addition to these.
A more complete list is given in the Rand Corporation study, *A
Data Processing System for State and Local Government*, by
Hearle and Mason.[1] Each of the categories in Table 12.1, if
listed in full, would be quite lengthy. Table 12.2, for example,
is Hearle and Mason's listing of category 1, 'Personal-Identity
Data'. The other categories in Table 12.1 could similarly be

TABLE 12.2

Expansion of Data in Category 1 (Personal-Identity Data) in Table
12.1

1. *Name.* In addition to last, first, and middle names, the following items should be included when applicable: 'also known
   as' aliases, maiden names, former names, 'doing business as'
   designations, and similar categories which identify a person
   by name. Soundex code designations when used would constitute a separate file which could serve as an index to the
   System file.
2. *Social Security number.* Social Security number, as the currently and potentially most comprehensive number system
   relating to persons, can be the basic numerical identifier.
3. *Sex.*
4. *Date of Birth* and birth-certificate number as filed in the
   appropriate governmental office.
5. *Place of birth*, including city, county, state, and country, if
   necessary.
6. *Date of death* and certificate number is a useful item since
   inquiries are often made about persons whose demise is unknown to the inquirer.
7. *Place of death*, including city, county, state, and country, if
   necessary.
8. *Race / descent.*
9. *Religion* is an item of little interest to most functions, but is
   recorded by others as part of the basic information describing a particular individual. Since it is collected by very few
   functions, it would not be an item filed for most of the persons described in the System.
10. *Current address*, showing the street address where the person
    maintains his domicile, together with
    a. *date moved to current address*;
    b. *permanent address*, if current address is temporary.
11. *Previous address(es)* for, say, the past 10 years.
12. *Date came to state*, if not evident from above.
13. *Date came to county*, if not evident from above.
14. *Height.*
15. *Weight.*
16. *Colour of eyes.*
17. *Colour of hair.*

18. *Complexion of skin.*

19. *Marks and scars.*

20. *Physical handicaps* specifying any handicaps that clearly relate to the identification of an individual, such as blindness, amputations, and continuous wearing of glasses.

21. *Blood type.*

22. *Marital status* indicates whether the person is married, single, divorced, widowed, or separated.
    a. *date and place of marriage(s);*
    b. *name of spouse,* and spouse's Social Security number;
    c. *names of former spouses* and Social Security numbers;
    d. *data describing any divorce decrees,* such as court, decree number, and date.

23. *Citizenship,* listing the country of which the person is a citizen.
    a. *alien-status number* is given if the person is an alien;
    b. *naturalization number* is shown for naturalized citizens.

24. *Fingerprint classification* lists the code-numbers for the classification of the person's fingerprints. These classification code-numbers are a series of digits based on the loops and whorls of the fingers. Classifications are assigned by local law-enforcement agencies, state criminal-identification bureaux, and the Federal Bureau of Investigation.

25. *Military Service.* Specific items describing military service should include:
    a. *period of service;*
    b. *branch of service;*
    c. *type of discharge;*
    d. *service serial number.*

26. *Selective Service number* is the number assigned to persons registered by the Selective Service System.

27. *Parents' names and Social Security numbers* refers to the parents of the individual. Mother's maiden name can also be recorded here.

28. *Children's names and Social Security numbers* refers to the children of the individual and their relationship, whether normal, adopted, foster, etc.

29. *Siblings' names and Social Security numbers* refers to brothers and sisters. Detailed data about parents, children, or siblings are available in the records for the persons referenced.

expanded to considerable length. Many such records are being or have already been stored in data-processing systems.

DUPLICATION OF DATA

The data collected today by national and local governments contain much duplication. This was true before computers were used and with the early computer systems. There is duplication, indeed, when we fill out government forms. How many times do we write on such forms our address or a physical description of ourselves? Perhaps in the future we shall only need to write in a National Insurance number* – or National Insurance number and name for checking purposes – and the computer, with a data link to the relevant file, will provide the rest. Before the systems analysts started their work, the various government departments kept independent files – and most still do. (It takes a very long time to change most government departments.) Departments of motor vehicles collect one set of data, local police departments another, the public utilities another, and so on; and to a large extent the information in these sets overlaps. Filed information has thus been organized according to its functional use rather than its data content.

Proper use of data-processing can lessen this duplication by making common information available to any department that requires it, thus lowering not only the cost of storing the information but also the greater cost of collecting it. Figure 12.1 illustrates the use of an *information centre* shared by various local government departments.

An example of an interdepartmental computer centre is Alameda County's 'People Information System' in California.[2] This system, still in its early stages as of this writing, consists of two separate sub-systems. The first is a system containing police information which can be interrogated over telecommunication links and gives information to policemen on patrol. This is described in Chapter 5. The second keeps 'people infor-

---

*Napoleon introduced identity numbers to Europe, and Britain may adopt the system, or issue National Insurance numbers at birth.

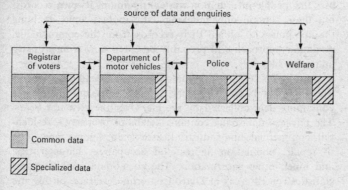

(1) EACH DEPARTMENT HAS ITS OWN FILES

source of data and enquiries

| Registrar of voters | Department of motor vehicles | Police | Welfare |

Common data

Specialized data

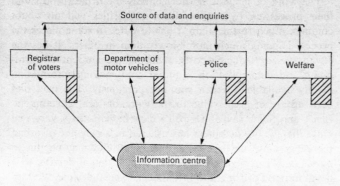

(2) COMMON DATA IN A COMMON DATA BANK

Source of data and enquiries

| Registrar of voters | Department of motor vehicles | Police | Welfare |

Information centre

Fig. 12.1 Interdepartmental use of centralized data bank reduces duplication of data storage and data collection. It makes existing data available to all departments. *Redrawn from the RAND Corporation study, A Data-Processing System for State and Local Government, by E. Hearle and R. Mason, Prentice-Hall, 1963.*

mation' for various social-service agencies in such areas as health, welfare, hospitals, and probation. Many government departments use the data-processing centre of which these systems are a part. The centre was created by combining the data-processing systems of several departments. Each agency which uses the 'people-information' system maintains its own records, but these contain only indexing data about people, enabling the computers to obtain their records from the common file. A similar approach will probably be taken now with property and other files.

Detroit has two such data banks: a 'social-data' and a 'physical-data' bank.[3] These are two computer-and-file systems containing data collected by the city over the past few years. The physical-data bank contains details of property – residential, commercial, and industrial; assessment figures; age; state of repair; population figures and occupancy characteristics; and much other information. The social-data bank contains statistics for all areas of Detroit on crime, welfare, births, and deaths; school truancy and drop-outs; diseases; and so on.

The work of integrating 'horizontally' both the records and their processing by various local authorities will no doubt continue much further than it has to date. In some places the process is going much more slowly than in others. It has the advantages of reducing costs, reducing the amount of form-filling and data-collection required, and making information readily available where it was not previously. The data files built up are of great value to the work of local government. Black and Shaw say of Detroit's data banks: 'In a very real sense the age of computers has ushered in a new age of urban planning,' and with their social data bank they may be able to devise 'an "early warning system" which could alert us to a neighborhood drifting into instability or social decline.'[4]

COMMUNICATION OF DATA

As well as 'horizontal' integration there is work on 'vertical' integration. A massive amount of information flows upwards from local to state government and from state to federal government. This flow is increasing. The federal government requires much information on crime, health, education, poverty, urban renewal, and so on. New requirements for information can place considerable strain on local organizations that must provide such facts and statistics. Well-planned data banks ease the problem; and when all necessary systems analysis and organization has been accomplished, we shall probably have a regular flow of information upward to the departments which need it.

The computer is an integral part of the government's internal communications as well as its relations with the public. The General Services Administration is installing an Advanced Record System to link up 1,600 federal offices in 600 cities so they can communicate by voice, teletype, punched cards, or magnetic tapes. Connection time, it is estimated, will only be a second or so for priority messages between any two points in the country. A message-switching centre will use computers to recognize and route messages as soon as lines are free. It can record delayed messages, alter codes and formats to suit the equipment in the receiving office, and send one message to many offices. Several million words a day are involved.

Information systems at various levels of government will be gradually connected to one another by telecommunication links in the 1970s. Figure 12.2 illustrates the types of interconnections likely to be implemented. An endless variety of such configurations is possible. We shall probably see the development of a vast network of interconnected computers and data banks, with terminals for interrogating them in many government offices.

Automation of the flow of information for government will have many advantages. One department will be able to use information collected by another, and readily find out whether a particular item of information is available. Relating all the information concerned with, say, a given taxpayer will be far less difficult than it is now. Statistics for all manner of decision-making will be available.

However, as we shall discuss in the next section of this book, increased centralization and availability of information has potential disadvantages as well. If the contents of a centralized data bank were destroyed, the result could be catastrophic. The thought of a tax-evading Luddite burning down a data

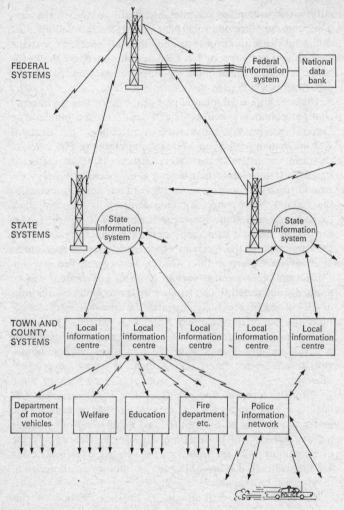

Fig. 12.2 The data-processing systems at different levels of government will be linked together in a variety of ways.

centre is the systems analyst's nightmare. Precautions can be taken, however. Files will be duplicated periodically and stored, it is hoped, in felon-proof cellars. Some real-time files might be stored in more than one computer centre.

More alarming in the eyes of the public is the issue of invasion of privacy. The more standardized, transmissible, and accessible the data become, the more caution is needed to maintain individuals' privacy and to prevent related abuses. Computer records will have programmed 'locks' on them so that only authorized personnel may have access to them; also, only that part of an individual's record which is cleared for an inquirer's use will be given to him. Many safeguards are possible, but this is a complex subject that we shall reserve for later chapters.

This vast government information network will increase in complexity and develop organically throughout the 1970s. Already as of this writing there are more than a thousand computers processing information in state and local governments in the United States, and many more than this in the federal government. In Britain the growth rate of computing power for central government administration was growing fairly steadily at 65 per cent per annum in the late sixties, and is expected to increase tenfold in the seventies.[5] And even this is only a preliminary first step. Most of these computers now in use are isolated machines churning through data on a single aspect of government administration. In the next decade many of them will become interlinked, both 'horizontally' and 'vertically'. Communication lines will flash information between the various data banks. Terminals and display screens will make facts immediately available to whoever needs them. The network of data links will grow and multiply, like a giant spiderweb, across the face of America.

NATIONAL DATA BANKS

There has been much discussion concerning the possible establishment of a federal data bank and data-processing centre in which statistics concerning the entire nation would be made available to government and research bodies. The proposal to set up a national data bank in the US has been greeted with howls of protest. In other non-communist countries the progression toward such a system is slower, and as yet less vehemently protested. Certainly it would be easier to set up a national data centre in such a country as Britain, which is smaller, has shorter telecommunication links, and is not faced with the problem of a profusion of differing state laws. On the other hand, the communist nations are probably those most in need of national computer networks; the inefficiency in running such massive and cumbersome centralized systems without the aid of computerized data-processing must be appalling.

In some countries – Holland, Sweden, Germany, and Canada, for instance – the national government's statistical record-keeping system is completely centralized. In others, the advantages of making statisticians readily available to policymakers in each department outweigh the confusion that occasionally stems from poor coordination.

TODAY'S MULTIPLE AGENCIES

The statistical record-keeping system of the US federal government is now highly decentralized. Unlike many other countries, the US has no central statistics office responsible for the nation's record-keeping, but instead maintains a large number of uncoordinated administrative and regulatory agencies.

The Bureau of Labor Statistics collects information on wage rates, prices, employment, and much other data related to labour. The Department of Health, Education and Welfare collects data and compiles statistics in its areas of concern. The Office of Business Economics collects and digests data on business activities, national income accounts, and other matters concerned with the functioning of the economy. The Federal Trade Commission and the Securities and Exchange Commission collect quarterly financial reports on manufacturing corporations. And there are innumerable other departments, agencies, sub-agencies, commissions, and so on – all collecting their own statistics.

Most of the available statistics are produced by the various agencies as a by-product of their main activities. Only the Bureau of the Census has a primary purpose of providing information to others. An attempt to coordinate all federal statistical work is made by the Office of Statistical Standards, while the National Archives try to preserve important records.

These agencies' methods, however, are primarily geared to the pre-computer era, and to the immense task of issuing *printed* reports. But the purpose of *reading* reports is to abstract the information they contain for subsequent analysis and as a basis for decision-making. Thus it makes sense to provide the data in machine-readable form, since no one will wish to reduce large volumes of data by hand.

Essentially, therefore, a national data centre must include a large organized bank of data and a battery of abstracting programs that can provide summaries of reports as required. This data bank must be well indexed and the format of the output of its programs made clear to researchers.

The 1965 'Ruggles Report' – on 'the Preservation and Use of Economic Data' – indicated that this is presently far from being the case :

At the suggestion of the Committee on the Preservation and Use of Data, the Bureau of the Budget and the National Archives jointly undertook a survey of machine-readable data held by various government agencies. The survey covered some 20 agencies in the Departments of Agriculture, Labor, Interior, Treasury, Commerce, and Health, Education and Welfare, and the Board of Governors of the Federal Reserve System. Over 600 major bodies of data were listed in this preliminary survey. The data are stored on approximately 100 million punchcards and 30,000 computer tapes. The decentralized nature of the federal statistical system makes it extremely difficult for users outside the government, and even in other federal agencies, to find out what data exist on various topics and how to obtain access to them.

In addition, agencies often do not provide sufficient information on the layout, classification, and definitions of data contained in a tape. As a result, even for the agency's own purposes it becomes very difficult to go back after a few years and make use of the information, unless it happens to be in the same format and classification system employed for current data-processing. The turnover of personnel within federal agencies often makes it impossible to trace back precisely what was done in the original coding of the schedules or programming. In view of these circumstances, what is needed is some system which will insure that for important data all federal agencies will provide clean, edited data with accompanying information describing layouts, coding and programming, so that these tapes can be served by both the agency itself and by other groups.[6]

Because researchers require data in much the same form, one might anticipate that certain tapes could be bought from a central agency, and would contain standard summaries found to be generally useful.

The essence of any such scheme must be the integration of the efforts of many departments. At the moment, files pertinent to the following areas (listed on the left) are held in the agencies (indicated on the right):

population:	Census, Public Health
housing and real estate:	Census, Labor Statistics
labour force and wages:	Census, Social Security, Labor Statistics, Employment, Securities and Exchange
education:	Education, Census
health:	Public Health Service
consumer behaviour:	Labor Statistics, Agriculture, Census
agriculture:	Agriculture
business and industry:	Census, Federal Reserve
government, financial and tax:	Census, Internal Revenue, Securities and Exchange, Communications
foreign trade:	Census, Treasury, Federal Reserve

UNIQUE IDENTIFIERS

To bring these many thousands of files together, the data must be organized into records with common *identifiers* that point to individual reporting units, which might be companies, farms, or persons.

Suppose we wished to relate illness with income for a selected section of the population. The Internal Revenue Service (IRS) tax data will indicate income, the Public Health Service (PHS) data the morbidity. The two records might look as follows:

IRS No.	Income ($)	PHS No.	Days Ill
278142	27,894	A41610	7
278143	6,310	A41611	91
278144	12,471	A41612	4
278145	4,076	A41613	0

The computer tapes of which these form a part may contain a few million records. But how would we know that the individual known to the PHS as A41611 is the same individual known to the IRS as 278143? Suppose the records were in the following form:

IRS No.	Income ($)	Soc. Sec. No.	PHS No.	Days Ill	Soc. Sec. No.
278142	27,894	999 136 531	A41610	7	704 680 401
278143	6,310	777 137 532	A41611	91	777 137 532
278144	12,471	412 139 538	A41612	4	681 432 901
278145	4,076	381 130 501	A41613	0	314 159 267

Then we could sort the tax records in order of Social Security number, repeat the same process for the health file, and then match identical numbers to give a set of records comprising incomes and a corresponding 'days-ill' records.

Without the link provided by the Social Security number (or other *unique identifier* such as name and address), the two files could only yield such unconnected items of information as: 10 per cent of total incomes exceeded $11,000, and 10 per cent of total employees had less than seven days' sickness. We could not determine that, say, the top 10 per cent of earners averaged fewer days' sickness than their poorer fellows.

This highly simplified example illustrates the vital role of the unique identifier which will link together all the myriad government statistics so that they can be used in combination. The provision of the identifier means that the identified unit has been *classified* into a class containing only one member. When we come to use the information, our first operation is to aggregate individuals into such larger classes as 'blue-collar workers', or 'one-man businesses', and so on.

An ever-present problem in dealing with government statistics is achieving uniform classification of industries, occupations, regions, commodities, and so on; these categories form the bases for different treatment by taxing and regulatory agencies. Perhaps the most problematical aspects of classification are the indexes that ostensibly reduce a whole range of economic behaviour to a single number. Changes in social habits play havoc with the 'cost-of-living index' and the 'price level', just as the growth of new industries does with Standard & Poor's indexes of stock-market trends.

CROSS-CHECKING

The pressure for more cooperation has led to cross-checking of information, particularly in the area of taxation. In 1963 it was reported that 'for one reason or another, be it oversight, ignorance, or deliberate omission, nearly $4 billion in taxable interest and dividends have not been reported on tax returns each year. This is often referred to as the dividends and interest gap.'[7]

The adoption of tax-account numbers led to the reporting of $2 billion of this amount within one year, but the procedure involves getting Social Security numbers from shareholders, not all of whom cooperate. Under new IRS procedures, tax returns are transferred onto magnetic tapes at Regional Service Centers; the tapes are then sent to the National Computer Center, where the Master Files of taxpayers' records, identified by Social Security number and employer-identification numbers, are updated. Failure of an individual to file a return is spotted promptly and payers without corresponding payees start investigations. The computer can cross-check recipients of interest, dividends, and wages against reported payment. The results are expected to insure comprehensiveness and uniformity and, hopefully the honest citizen will have to pay less tax.

Changes in tax policy can be tested on these files to see their effects on revenue and workload. Uniform administration is easier, since rulings can be consistently applied; however, major problems of coding the law have still to be solved.

AUTOMATING THE COLLECTION OF DATA

National statistics can help both businessmen and policy-makers – by providing the former with meaningful predictions of future events and the latter with predictions of the effects of alternative courses of action. But the raw material on which the econometricians operate must be good – that is, accurate, comparable, timely, and complete.

There are numerous methods of collecting data, including survey of intention as well as past action; covering a sample of all relevant sources; by-product information from tax, unemployment, customs, and excise data, and so on. The key to the use of the information is its classification, which must be both flexible and consistent.

The task of collecting the desirable data from around the nation is enormous. Today, although it involves a staggering quantity of paper-work and clerical effort, the information collected is still inadequate for many planning purposes and is badly out-of-date. A 1963 inquiry into the US 'Federal Paper Work Jungle' estimated that between one and two billion reports were made annually by the public.[8] Large organizations frequently employ special staffs to prepare these; usually reports are a by-product of the firm's computerized accounts, but at least it gets some return in the form of useful statistics

296 · Euphoria

for planning. It is the little man who suffers, whether he is the owner of Joe's Bar and Grill reporting his trade, or the taxpayer spending five dollars for advice or two days of his time to fill in his form, or the airline traveller kept waiting by immigration clerks.

The census taken in the United States is probably the largest and most complex in the world. As early as 1890, the Bureau of the Census started to use punched cards. Herman Hollerith, who invented the technique, was then a Bureau employee. After World War II a UNIVAC computer was used for thirteen years following its development for the Bureau of Standards. For the 1960 census, the Bureau of the Census developed FOSDIC – Film Optical Sensing Device for Input to Computers – which can transfer information from microfilm to magnetic tape at the rate of 2,000 personal census records per minute. That census resulted in 141,000 pages of tables, and contributed to a Bureau output of 1 million lines of print per day by the mid 60s.

Even so, all over the world information collected is inadequate for many planning purposes. Britain's National Economic Development Council (Neddy) and its subordinate 'Little Neddies' for various industries are constantly plagued by lack of data for planning. Many otherwise elegant planning efforts have been sadly based on a census many years old and inadequately detailed. To make good use of the simulation techniques and econometric models discussed in the next chapter will require very much more refined and up-to-date data.

The widespread use of computers in industry and of datatransmission and real-time data-collection techniques should change this picture. The government may in the future ask for all information in machine-readable form from its departments and from industry. The process of conversion of figures from manuscript is slow and error-prone and needs to be avoided wherever possible. In contrast, machines can convert any type of card, magnetic tape, or paper tape quickly and accurately. Organizations too small to be using any form of automatic data-processing can only afford the time necessary to make a few returns anyway. Information from them can be obtained best from tax returns or by other 'indirect' methods. Large firms spend several man-years making government returns, but reckon to get their money's worth in planning information.

By tapping the computers of industry and public administration and bringing the findings together with reasonable speed, continuously valid census, production, export, input-output, employment, consumption, stock and earnings data could be available on request to any organization with a terminal or computer linked to the government central statistics bank. A company economist, or a law-maker preparing a speech, would call for the statistical index on his typewriter terminal, call the appropriate analysis program, indicate the name of the statistics he wanted, and get the details he required. Also available would be regression analysis, smoothing and projecting programs which he could instruct to operate on the figures.

ON-LINE DATA COLLECTION

Accurate input of the large quantities of data that are needed will always be a problem. Ideally the moment at which data are created is the best time to capture them for later use. In some cases this is possible. The cash registers in a shop can punch paper tape to be fed into the computers. The computerized telephone exchange can record whatever is desired about the numbers subscribers dial, and these data can be analysed to determine traffic patterns and, in Sweden, are used to trace malignant anonymous callers. Terminals in some factories gather information on every manufacturing step of each job and this enables tight production planning and control.

On-line capture systems are the ultimate. No doubt these will be used where possible for government information in the future. The computer will monitor an operation, record what interests it, but also check at the time what it notes is reasonable and consistent. At London Airport (Heathrow), for example, a real-time computer was installed in 1971 to control and account for the importation of cargo. The system, designed jointly by the airlines and Customs and Excise, is seen as the only way to cope with the massive increases forecast for air-freight traffic. The airlines and forwarding agents feed information to a computer terminal which converses with them by means of a screen; the computer checks the validity and completeness of their input and indicates duty payable. The desired government data would be obtained directly from the computer.

Passengers entering a country may in the future fill in a card which is machine-readable instead of filling in today's immigration form. The immigration officer who checks their entries will feed the card into a terminal, keying in additional data if required. The machine will quickly check its list of wanted or inadmissable persons and check that the data given are valid. The machine may check a voice print or fingerprints. With such a system the police and government could gather the information they needed in greater detail and with less delay than today.

We shall probably find computer-terminal keyboards and screens in many government offices. Sooner or later the public also will be confronted with the terminals. Offices where they fill in forms of various types may well contain cubicles with terminals before the 1980s. A man applying for a licence will sit down at a terminal, the screen will instruct him exactly what to key in, reacting in a 'conversational' manner. If he has difficulty, clerks will be available to help him. The occasional person who may be completely incapable of using the terminal will have all his information keyed in for him by a clerk. The computer operating the terminal will have access to the records of the person filling in the form. If, for example, he is renewing a car licence, the machine will have stored in it all the data associated with his present licence. This would make the renewal process very quick. The machine may directly debit his bank account. A more detailed description of this is given on page 416.

In a similar way, the easiest way to fill in one's tax return in the future may be at a terminal, attached to a government computer, operating in a conversational mode. The machine will already have many facts, which will ease the process. It will do calculations for you, look up stock dividends and appreciations, and so on. You may have stored information throughout the year in a computer file record allocated to you, so that when the day of reckoning comes the process will be relatively easy.

Much census and other information may eventually be collected on-line, or at least in a machine-readable form. If the data-collection and -processing become easy enough, censuses may be taken more frequently, at least for some of the facts. The decennial census may even be done away with altogether and replaced by a continuous flow of information that will keep the data banks up to date.

The key to obtaining information for government planning without excessive clerical labour lies largely in the cooperation between government departments and between government and business sources of information. Information from US tax returns obviated a census of 1 million small businesses. How many other such rationalizations could there be? Much duplication of effort could be avoided. A good place to obtain motor-vehicle information, for example, is the insurance companies. Many insurance companies are installing terminals to assist in selling the optimum policy and reducing clerical work. Details of a person requiring motor-vehicle insurance will be keyed into a terminal – often by an agent talking to that person on the telephone. The computer will check the details, assess the risk, and if it is acceptable display details of the premium and terms. The agent could issue the policy within 90 seconds after typing in the data, and the computer will have recorded the details of his commission. Ideally the required government information on vehicles could be collected at this point; the licence cost could even be included in the insurance payment. The government could pay the insurance companies for this service, or give them the facilities outright. Why is further duplication of effort needed?

With clever use of automation and with cooperation between different authorities, we could collect all the information desirable for government without drowning in our own red tape.

REAL-TIME RETRIEVAL

The information, once collected, digested, and statistically summarized, will perhaps be available in a variety of different forms. Today, for example, census information in the US is available as fourteen or so shelf-feet of large books. These tabulate selected combinations of data that someone at the Bureau of the Census had thought useful. The problem is that many of the questions that people have regarding the census data are not answered by these books. Such questions represent a combination not tabulated; to answer them one must process the original raw data. The Census Bureau will attempt to answer these questions, but it takes time and is expensive. Alternatively, one can buy a computer tape containing a representative sample of the census information with identifying information removed. Loading this tape on a computer, a person can browse and tabulate correlations at will.

In the future, in addition to such tapes, government data or samples of data may also be available in random-access files containing multiple indexes, tabulations, and summaries for answering many of the questions that might be asked. Some users will then be able to answer some of their questions in an on-line fashion with a fairly fast response time. A trained individual will be able to explore the data further in an inter-active manner. He will have a variety of statistical routines available to him for answering his question in different ways. Only when his question involves scanning a large quantity of data in a way that has not already been done will he have to wait a while for his answer.

How far such an information system will be taken in the next decade or so is uncertain. We may see no more than a development and centralization of the data facilities presently used by government agencies. If so, even then vast libraries of magnetic tapes containing facts and statistics would be accumulated and made available to research groups, planning committees, and other agencies for processing on their computers. At the other end of the range of possibilities is the growth of a vast real-time system comparable in some ways to the systems presently used by the American military. In this case the President would have screens in the White House on which his data specialists could display facts that he requests. The results of differing courses of action could be explored in advance using mathematical economic models based on the vast quantity of data that would then be available. Government department offices in all parts of the country could dial up the central system and connect their terminals and screens through the public telephone network in order to interrogate whatever data in the system they are authorized to use.

The Department of Defense is perhaps a decade ahead of other departments in the ways it uses computers, but we can expect that the techniques of war games will be extended to peace games. The computer screens used today in the Pentagon will be used tomorrow in Town Hall.

REFERENCES

1. Edward F. R. Hearle and Raymond J. Mason, *A Data Processing System for State and Local Government*, Prentice-Hall, Englewood Cliffs, New Jersey, 1963.

2. Gordon Milliman, 'Alameda County's People Information System', *Datamation*, March 1967.

3. Harold Black and Edward Shaw, 'Detroit's Data Banks', *Datamation*, March 1967.

4. ibid.

5. *Computers in Central Government; Ten Years Ahead*, HMSO, 1971.

6. 'Report of the committee on the Preservation and use of Economic Data to the Social Science Research Council, April 1965.' Reprinted in *The Computer and Invasion of Privacy* – Hearings before a subcommittee of the Committee on Government Operations, House of Representatives, 89th Congress, 2nd Session. US Government Printing Office, Washington, DC, 1966, Appendix 1, page 201.

7. *The Federal Paperwork Jungle* – Hearings before the Subcommittee on Census and Government Statistics, House of Representatives Committee on the Post Office and Civil Service, 88th Congress, 2nd Session, May 1–June 24, 1964. US Government Printing Office, Washington, DC, 1964.

8. ibid.

13

GOVERNING WITH COMPUTERS

It is an illusion to think that, because we have broken through the prohibitions, taboos and rites that bound primitive man, we have become free. We are conditioned by something new: technological civilization. – *Jacques Ellul.*

The collection of data has little value unless the data are used to understand the world and prescribe action to improve it. A mass of disorganized facts and figures is of little use – to governments or anyone else. The computer is the tool *par excellence* for collecting enormous quantities of otherwise indigestible data. The central problem of much data-processing is extracting from a mountain of facts an essence of value to human users.

Often, to make sense of data, hypothetical *models* embodying the relationships of the various measurements are needed. Economists have been building such models since before Adam Smith, and they are still a long way from understanding how all the factors operating in an economic system are connected. But the art is improving rapidly, and we are now beyond Smith's day, when he could rightly say:

The natural effort of every individual to better his own conditions ... is ... so powerful a principle, that it is alone ... not only capable of carrying on the society to wealth and prosperity, but of surmounting a hundred impertinent obstructions with which the folly of human laws too often encumbers the operations.

Current models are more complex, more sophisticated, and more useful than Smith's. Moreover, the manipulation of economic variables is within the competence of governments and the results are reasonably predictable.

The data on which models depend still require a lot of processing to be useful. To predict the effect of a particular tax policy, one cannot simply feed the computer the names of all taxpayers and details of their yearly pay cheques. A useful measure of the effects of a direct tax requires a massive reduction of the raw information and a reduction of the *dimensions* associated with individual items. We may know a taxpayer's age, occupation, residence, income, and number of dependents, and yet for the purpose of budgeting our interest would be limited to the total tax paid by people in his income group. We therefore 'project' the mass of data onto the income 'axis' and come out with a single figure for the tax paid by each income class. (See Fig. 13.1.)

In this 'projection' process there will be a loss of information; the same problem with 'dimensions' occurs in the description of objectives. Most of us find it easiest to think in terms of maximizing some particular measure of merit, such as the growth rate of the national economy or the speed of a car. However, the achievement of such a maximum is usually accompanied by side-effects – such as increased imports or shorter engine life in our respective examples – which have to be kept within reasonable limits. Methods of handling these complex goals, or 'yes–buts' – '*yes*, we want growth, *but* we do not want balance-of-payments crises' – have been developed by operations researchers. Several 'component yes–buts' (i.e., one 'yes', a few 'buts') can be handled by clerical methods, but analysis of real situations with many 'buts' requires a computer, whose ability to bear all aspects in mind without over-emphasizing any one is essential. Tests at electronic speeds can be made of the effect of relaxing the constraints slightly, measuring the cost of so doing and the gains made toward the primary target.

GOVERNMENT STATISTICS

Statistical techniques provide a special way of recording data whose dimensions have been reduced by projection. Frequently, instead of losing the whole of a dimension of the information by projection, the characteristics of the dimension are summarized into, perhaps, a mean value and a measure of how

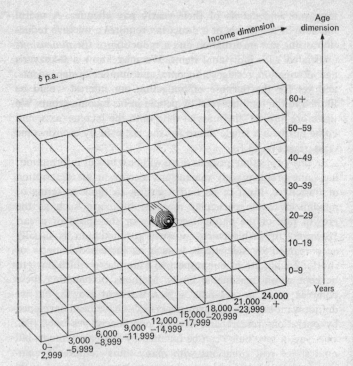

Fig. 13.1

closely an individual value is likely to approach that mean. We could, for example, record that American men average 5 ft. 9 ins. in height with a *standard deviation* of 2½ inches, meaning that a large representative sample of all adult male Americans would probably show that about two-thirds are *between* 5 ft 6½ ins. (i.e., 2½ inches less than 5 ft 9 ins.) and 5 ft 11½ ins. (2½ inches more), and that one-third is either taller than 5 ft 11½ ins. or shorter than 5 ft 6½ ins. Although this tells us nothing about any individual, it could help us design a comfortable automobile seat for most American men.

Government statistics are collections of such summary data, often associated with a date so that one can produce a *time-*

series. Some figures of this kind are publicly announced with much excitement, and people pay far more attention to them than warranted – for they are frequently averages without an associated probable error of measurement or historic standard deviation. To avoid distortions of this sort, the statistician's job is to interpret statistics and to demonstrate what degree of credence should be attached to them.

SIMULATION

The computer comes into its own when handling the time path traced by large, complex systems. With a knowledge of how certain economic variables are related, analysts can make changes in just one variable, over a simulated period of time, in a computer model and follow the resultant changes in all the other variables.

Interest in the dynamics of economic aggregates goes back to the 'business-cycle' theorists of the Depression and the 'macro-economics' of the Keynesian era. The first dynamic economic models were based on a set of equations relating the change in a variable in the course of one time period to the values of all the variables at the beginning of that period. Adding the calculated changes to the original values produces new values for the end of the period, a process which can then be repeated. By hand, the operation is tedious, and there is little scope for experiment. The validity of its conclusions is tested against real-life time-series data of such variables as aggregate consumption and income of the household sector.

Building economic models is much more difficult in practice than in theory. We must start by defining the economic system we wish to analyze and decide which variables are internal to it and which external or given. For instance, a model of the US economy might treat all other nations' economies as unchanging (i.e., 'constant') over the time span to be simulated. The proportion of personal income that is saved may be regarded as a constant in one model, but as dependent on total income and interest rates – in other words, as a variable – in a more complicated one.

Having decided what *exogenous* (given) and *endogenous* (dependent) variables to consider, we must next determine their relationships in order to provide the 'works' of the model. These relationships form the cruxes of such politico-economic issues as the impact of tax hikes on international balance of payments. There is a marked tendency for politicians to substitute ideology for scientific investigation in this area.

Each variable in the model depends on a large number of others, and the rate of change of each is frequently influenced not only by its own absolute value but by the others' rates of change as well as their absolute values. Furthermore, these relationships are seldom simple; tax revenue typically increases faster than total earnings, and capital investment swings more widely than corporate income. To add to the complications, most economic action depends on individual decisions, and these depend in turn on psychological factors. To make confusion worse, there is the factor of psychological feedback, which is manifested most dramatically in the stock market. The great crash of 1929 was precipitated primarily by financiers and speculators who paid more attention to each others' opinions than to the true facts. The prevalence in this decade of 'chartism' – a technique that prefers to determine future stock prices from price histories instead of concentrating on the intrinsic worth of a stock – has obligated all market operators to take into consideration the behaviour of chartists themselves, who thereby influence the cost of capital and hence the performance of the firms whose equity they trade.

However large the computer, it will never be able to cope with all the variables in an economic system. Some must be ignored, treated as constants, or lumped together. This last process – *aggregation*, as it is called by economists – is the bane of their lives. Information is inevitably lost in the process of projection that yields an aggregate – information that may have been vital.

If, eventually, a sound theoretical model of an economic system *is* built, there will still be problems in determining the values of the model's *parameters*: those elements of the mathematical equations that one claims to know. To obtain them, one must analyse the past using sophisticated statistical techniques which are themselves prone to esoteric errors of method and theory.

Finally, the analyst must provide the working model with a mass of accurate figures on which it can operate – data which are themselves subject to errors in collection and, more seriously, in classification.

Nevertheless, realistic forecasts of economic performance can be of immense value to business and government. Here, for example, are but a few descriptors of economic performance: national income, Gross National Product (GNP), total employment, unemployment, consumption, savings, investment, taxes, welfare payments, pensions, imports, exports, price levels, inventories, government expenditure, technical skill and education, family size, house-building, labour mobility, and growth and declining industries.

Typical of these models is the one developed by Professor Richard Stone and his associates, starting in the early 60s, at the Department of Applied Economics of Cambridge University. Nicknamed the 'Rocket Project', after the first steam locomotive, it is intended as a prototype computing 'engine' for the British Economy.

The capacity of economists to perform the necessary calculations for this project is appreciably greater than the ability of its creators to collect and adjust the data and to refine the model. The highest claim that has been made for its predictive ability is that 'the trends calculated here are at least as likely to be true as those produced by any other means'.[1]

The model starts with SAM – a Social Accounting Matrix – which shows the inputs and outputs of each identified sector of the economy. With a big enough machine, 253 such sectors might be distinguished. In early versions only 31 could be used.

The exogenous variables include population and balance-of-payments figures, price levels and purchase taxes for different classes of expenditure, replacement and new-capital expenditures, and consumers' propensity to save or spend their income. From these, total expenditures and then total outputs of all sectors are calculated and, using SAM, distributed over the various sectors of the economy. To produce the output, each sector must receive the right amount of inputs from the others. Thus the outputs of each must be increased appropriately to provide extra inputs to other sectors and, conversely, all must be reduced proportionately to keep the total correct. Since each addition to output in turn requires inputs from other sectors, the whole operation has to be repeated until all inputs and outputs match throughout SAM.

When all operations are completed, the results show forecast demand, production, imports, capital expenditure, consumer durables, taxes, and so on, in each sector for the period following that for which the data were entered. By feeding *this* output back into the model and repeating the calculation, the next period can be simulated. Any errors in the model will be amplified as time goes on and more cycles are simulated.

Such figures are essential to government for setting tax levels, restraining inflation, maintaining employment, and planning its own expenditures. They are no less important to business for estimating potential markets and forewarning of declining current outlets, as well as predicting future price levels of raw materials and labour.

Most of the major exogenous variables do not change very quickly. Some, however, are under government control, and political philosophy regarding the degree and manner of such control can change following elections. This leads to uncertainty in the input, and above all to uncertainty in the minds of those whose actions influence the relationships expressed by the equations of the model. For instance, SAM may say that 26 per cent of total private consumption will be in the consumer-goods sector. But the prospect of a luxury tax on such goods following a change of the party in power could cause a run on some consumer goods, with reduced expenditures elsewhere. Ironically, therefore, the improvement in planning techniques has made business more dependent on consistency of government policy. Firms who tool up to meet demand corres-

ponding to 26 per cent of personal expenditure will find it difficult to meet an increase to, say, 35 per cent in the immediate
future and may have to lay off labour when the proportion
declines later to a mere 20 per cent, perhaps.

The use of economic models is not restricted to governments
with overt interventionist policies. The Brookings Institution
in Washington is currently working on the largest model of an
economic system in the world, one which subdivides the US
economy into more and smaller sectors than has any previous
simulator. Included among the variables are the banks' prime
lending rate, the rediscount rate, government-bond coupons,
and reserves, all of which help to describe the monetary sector.
Since fiscal levers are the prime tools by which the economy is
manipulated, it is not surprising that the American Bankers
Association is funding the development of this part of the
model. Although still in an experimental stage, initial results are
encouraging its builders to expect that within ten years they will
be able to predict the consequences of government fiscal and
monetary policies. The projections will appear as a time path
to be traced by the endogenous variables.

Input/output models developed by Leontief in the United
States emphasize the cross-sectional structure of the economy
rather than its time path. Outputs from various industrial sectors are the inputs into others, including the ultimate-consump-
tion sector. Tables have been drawn up showing where the
products of 100 industries go, and where their raw materials and
components originate. The consequences of a postulated change
can be traced through the industries it affects. The availability
in Britain of natural gas following exploitation of the North Sea
fields, for instance, will change the inputs to fuel-using industries, the cost of their products, and the contribution of other
fuels.

All the models discussed have limitations which are incorporated in the assumptions needed to make them manageable
and computable in a reasonable time. It is possible to break
down the components into smaller blocks – even, ultimately,
to break down industries into individual firms. But this is not
necessary. The assumptions are embodied in the relationships

between the variables. These variables can be altered and the consequences calculated. The extent of the change in the consequences tells us how much value there is in tightening up the assumptions. The input/output model assumes that each unit of output will use the same proportions of labour and other input materials at high-production levels as at low. A government analyst who must plan training facilities for skilled labour may 'run' the simulator for 'ten years' on the built-in linear assumption, and then rerun it with an assumption that capital intensity increases with output. A very great difference in answers would justify considerable research into the nature of the 'true' relationship.

This approach has led to a 'building-block' approach to simulation models.

PRESCRIPTION

Collecting data and describing and operating economic models are valuable for the insights they give into the economy and society. This big pay-off in efficiency comes when the conclusions drawn are acted upon.

In comparatively simple systems measurements can be made, calculations performed, and controls moved to set the system on a prescribed course and keep it there, all without human intervention under normal conditions. Autopilots, thermostats, and numerically controlled machine tools are examples of 'closed loops' in control systems. Not all of these operate without human supervision, and frequently one finds humans monitored by automated systems. For instance, when a nuclear reactor is brought up to power, the rate of nuclear reactions must be increased at a controlled rate and then held steady at the required level. A slight human error in positioning the neutron-absorbing rods which establish the rate of increase of power could cause an explosive rise in energy release in a few tenths of a second. Instruments capable of spotting this increase in time are linked to springs which fire the control rods into the reactor before damage can occur. No man could spot and correct this error fast enough.

Mechanized systems capable of operating to specified plans in economic systems can be similarly designed to take measurements, evaluate them, and operate controls to implement consequent decisions. (At the extreme, they may even fire a worker!) The virtues of this way of operating are also the same – maximum speed of response, complete vigilance without boredom, precision and accuracy, safety and predictability. The vices usually result from human failures to give the automaton the right orders.

The Soviets, despairing of the automatic decisions of an invisible hand operating through the free market to achieve the goal of optimal resource-allocation, aim to create a vast national computer network. Linked machines will receive economic data from thousands of sources and process it rapidly for expeditious application to the task of planning the large and complex Soviet economy; and 'planning' means directive planning – what we would call 'control'. Note that such a system is not different in principle, although very different in effect, from a system of data-collecting entrepreneurs, who measure profits, compare success, and change their business if necessary; or measure demand and cost, and alter supply and price under the control of a free market. One Russian academician reckons that 10^{16} calculations (the number equals 1 followed by 16 zeros) are needed to find a single optimal plan. British housewives make about 10^{12} decisions a year among them, thereby controlling national patterns of supply and demand. It is difficult to guess how many calculations each decision requires, but remembering 'it's cheaper next door' is a complex bit of reasoning.

A controlling computer can act very much like a highly skilled centralized bureaucracy, with most of the virtues and fewer of the vices of its human equivalent. It handles everything in its purview by recognition and application of its instructions or precedents. It passes unusual business up the line for attention; anything really extraordinary will upset it completely. Unlike a clerk, however, it will not be tempted to pigeonhole and conceal what it cannot understand, but will have a tantrum and scream for help. It will be completely consistent in dealing with identical situations, absolutely impartial, and completely accurate. It will not quarrel, insist on a carpet better than it's neighbour's, or take a coffee break with a queue waiting outside. It will not even protest when it is replaced by a better one. But it is unlikely to be original, creative, sympathetic, ambitious, or nepotistic.

When computers make routine decisions, the burden of responsibility is taken off 'middle management'; exactly to whom it goes still involves legal problems. The sort of decision one can expect to be automated is the assessment of a welfare recipient's subsidy based on a means test. By asking the Internal Revenue computer for information (which it would process according to local rules but not actually reveal to the operator), the local computer could assess the welfare applicant immediately. It would also be able to 'leave a note' with the tax machine indicating when or if the assessment needs revision.

Large-scale systems of this kind are already being planned by banks and credit institutions and will operate on these very principles. The technological problems have been solved, but the social and political problems are only just appearing.

Not least among these problems is the achievement of a balanced view of the role of the computer. There is a vast gulf between the theoretical potential of economic models and automated bureaucracies and the practical realization of their benefits. Given the political and financial backing, computer systems could serve the nation as they served the Apollo moon-landing project. There would be false starts, and false dawns, mistakes, muddles, and budget overruns and, of course, arguments about the purpose and value of the whole exercise. In particular, there will be those who make it all sound far too easy and who thus give hostages to fortune. It is their 'euphoric' view that we have reflected in Section I. In Section II, we discuss the views of those who say that much of what we can do with computers in government is undesirable, and should not be attempted. We have, however, through necessity, skated over the immense practical difficulties associated with large computer systems, since all our examples are of schemes that worked, or have yet to be tried. To maintain a true perspective, we would need to lengthen an already long book by including examples from that *majority* of computer systems that have failed to achieve their objectives. The difficulties that led to these failures are even more considerable with governmental systems, but we believe, nevertheless, that they will be overcome, not quickly, but ultimately.

A COMPUTER-CONTROLLED ECONOMY

In principle, computer techniques can be applied to economic-control systems just as they can to any other. A national economy is not a simple system, and it possesses an environment which includes other nations as well as its own political and social structure. Nevertheless, certain activities are essentially economic and within the jurisdiction of government, and these activities are characterized by *transactions* which result in changes in the credit of individuals, corporations, or institutions.

In Chapter 4 we discussed the 'cashless society' in which the credit card or its equivalent largely replaced cash and cheques. Suppose now that every non-cash transaction were registered in computers across the country. A certain degree of regulation of spending habits it thought necessary of a healthy economy is to be maintained. (What makes for a healthy economy is almost as fruitful a source of discussion as what makes for a healthy man – but that is not the point here.) For example, a merchant may buy a consignment of umbrellas from an importer, the purchase being completed by the transfer of credit between the appropriate bank accounts. Economic management requires that sales and income taxes be paid and import duty levied. In addition, to set tax and excise rates, all such umbrella purchases need to be aggregated to determine the impact, say, on the domestic counter-precipitation industry, the general price index, the balance of trade, rate of inventory-building, and the level of commercial credit balances.

If the participants and their transaction are properly classified at the time by the use of a comprehensive and comprehensible coding system, all the appropriate aggregates can be updated at the same time. For instance, the retail-stock total will be incremented by the value of the umbrellas, and the wholesale stock correspondingly reduced. Now, retail stockbuilding has a normal seasonal pattern onto which is superimposed merchants' readings of the market in the near future. If retailers know what they are doing, a rise in seasonally adjusted stock levels presages a consumer boom which will affect the money supply, balance of payments, and so on. The 'economic manager' will set boundaries on the levels of the aggregates he measures so that he will be given warning of any whose behaviour is exceptional enough to warrant his attention.

But the system of real-time economic control need not stop there. If it is possible to measure the excise duty payable on a transaction from its classification, then it is possible to collect the duty there and then by crediting the appropriate departmental account and debiting the payer, either immediately or after a set number of days.

The classification of transactions need not stop at determining to what kind of aggregates they should contribute. A comprehensive coding system would enable us to get more information about what is going on in the economy than could be obtained merely from knowing the bank-account numbers that had been debited and credited, which is the total information in a bank statement at the end of a period.

If you can classify the transaction and the parties to it, why not classify the 'money' too? Just because one banknote is identical with another of the same denomination, it does not mean that the alteration of *credit* in one account need be exactly equivalent to that in another.

A parameter of vital concern to government economists is the 'propensity to consume', or its complement, the 'propensity to save'. This is a measure of the proportions of disposable family income which are spent or saved, and naturally it varies with income. By multiplying every family's income by its propensity-to-save factor and then adding all the products, we get the total savings of the household sector of the economy. By various routes this ends up as new investments in industry and influences the future productive capacity of the economy. Many people spend more than they earn in the same period and finance the expenditure from savings – their own or other peoples' invested in finance corporations or banks. A government tries to control the drain on savings by various measures, including making banks deposit money in its reserves instead of lending it to consumers, altering interest and rediscount rates, and so on.

The computer-monitored economy offers another and more attractive possibility, for it can enhance investment as well as discourage consumption. Suppose we held all household-sector credit in two distinct types of account, a consumption account and a savings account, and regulated the flow of funds between them.

Consider first an inflationary situation in which consumer spending is running ahead of production, and so too much money is chasing too few goods whose price level therefore increases. Since rising prices hurt those on fixed incomes, the managers of the economy will wish to encourage people to save more of their income, which industry will invest to produce more goods to restrain the price level in conjunction with the lower amount of income being devoted to consumption. When an individual receives his income, he may pay part of it into his 'savings' account, keeping the rest in his 'current-consumption' account. By making 'saved' income attract a lower rate of tax, the government can encourage the build-up of balances in savings accounts. It can also discourage drain on savings by taxing the transfer of funds from savings to consumption at a rate that varies with the state of the economy and the income of the individual. If inflation appeared to be reaching 5 per cent per annum, the spending of savings might attract a tax of 80 per cent for someone of middle income, while a retired pensioner would pay only 20 per cent. At another time, with business confidence low, stocks running down, and unemployment rising, the 'transfer-tax' rates might be 20 per cent and 5 per cent respectively.

Within the respective accounts, credit circulates quite freely. Individuals can sell their Motorola stock and buy life insurance without crossing from 'savings' to 'consumption'. In this way the flow of capital to the best opportunities would not be stifled, and investors could choose how best to balance risk and return.

The transfer tax could be a very sensitive regulator if the government were left free to manipulate it at any time. Because it is exacted at the time of the transaction, it has an immediate effect; and because its rate is specific to an individual, it can provide a measure of income redistribution. Whatever regulators are used, they could be much more sensitive than those used today if savings and various categories of spending were monitored by computers, as will be possible in a world of computerized credit transfers. The transfer tax could be highly flexible and permit 'fine tuning 'of the economy. Compare this with the 10 per cent income surtax implemented by the Johnson administration after months of debate.

INSTANT PUBLIC OPINION

Economic management is not the *raison d'être* of government, despite its present importance in the political forum. Just as it is now feasible to monitor economic activity by automation, it is also possible to sound out public opinion similarly with comprehensive polls on a frequent or even continuous basis.

This could conceivably be done using the home touch-tone telephone. The computers would dial a cross-section of the public and pose questions with a 'human' telephone voice. They would ask the recipient of the call to respond by pressing an appropriate button on the telephone. One computer would be capable of conversing with many thousands of people at once, and vast quantities of public-opinion data could be collected in this way.

Eventually, voting from the home may become possible, although this poses the difficult questions of security and insuring that a person does not vote twice. (Persons without telephones would still have to go to a polling booth.) If computer-controlled polling does come into use, polls and referenda could be taken much more frequently than at present.

Voting from the home is a technical possibility, and perhaps will be implemented in some countries before the end of the century. Continuous computerized referenda on all matters of public importance may appear to be the logical extension of Western democracy, but if attempted, would almost certainly be its ultimate downfall! Could any political system survive the volatility of ill-educated public opinion? The computerized society must steer a narrow course between automation of democracy and automation of tyranny.

REFERENCES

1. Slater, Lucy J. 'Computing the State of the Economy', *Computer Journal*, p. 11, May 1966.

Part II: Alarm

The trouble with all scientific utopias
is that once that they are stated they
look remarkably like nightmares.
– *Kenneth E. Boulding, The Image.*

14

THE BATTLE FOR PRIVACY

The most precious commodity we have now is the few years of lead time before this problem grows beyond our capacity for control. – Professor Alan F. Westin.

If we are to believe the popular press, we will not have to wait till 1984 to meet Orwell's Big Brother. The machines are 'getting to know too damn much' and will tie 'humanity in the chains of plastic tape'. The National Data Centre is 'one of the most dangerous ideas ever to come from the bureaucratic mind', and 'an appalling nightmare'. Individual privacy is threatened with annihilation by the advance of a legion of all-devouring data banks.

The debates concerning this issue in the United States and Great Britain are fascinating examples of forward-thinking democracy at work. In this chapter we shall summarize the course of these public arguments, and in the following chapter we shall discuss the nature of the threat.

DISTANT EARLY WARNING

The American Economic Association lit the taper in 1960 when it invited the Social Science Research Council to form a Committee on the Preservation and Use of Economic Data. Richard Ruggles from Yale, a team from various universities, and the federal government spent from 1962 to 1964 researching agencies and departments and in April 1965 proposed, in the 'Ruggles Report', a 'Federal Data Center' which would bring together many thousands of files of personal information.

In May 1965 Edgar S. Dunn, Jr, looked into the proposal on behalf of the Bureau of the Budget, and made it clear that the proposed use of the centre implied identifying individuals so that their records from different sources could be collated for analysis. The 'Dunn Report' reached the Bureau on 1 November 1965, and was published the following month.

Soon Carl Kaysen, then of Harvard, was appointed by the Bureau's Director to lead a task force, made up mostly of economists, in an investigation of better ways of storing and retrieving US government statistics; their report was made in October 1966. Kaysen was later to state that 'the participants were moved by professional concern for the quality and usability of the enormous body of government data to take on what they thought to be a necessary, important, and totally unglamorous task. They certainly did not expect it to be controversial.'[1]

How wrong they were! In February 1966 Orville Brim attacked the 'dossier' as the forerunner of *1984*-type surveillance. Lawyers Charles A. Reich and Alan F. Westin indicated the threat that the plan posed to constitutional rights. In August 1966 Edward Shils voiced similar concern to the American Political Science Association. Influential newspapers followed with reports and lead stories. The computer press involved its professional readers with editorials, letters, and articles.

In June 1966 Senator Edward V. Long's subcommittee investigating the invasion of privacy heard testimony from Dr Dunn. The House was not far behind; at the end of July Representative Cornelius E. Gallagher (Democrat, New Jersey) called Vance Packard, Professor Reich, Paul Baran of the Rand Corporation, and several computer scientists to testify. Ruggles and Dunn defended their respective reports. The following spring Senator Long's subcommittee devoted two more days to computers and privacy, and Representative Gallagher sought to enlist the help and advice of the computer men at their own conference in Atlantic City.

As controversy warmed the net widened. Senator Long had started by looking into the issue of privacy in general, but soon focused his concern on the computer and the Bureau of the Budget's proposal. Later, the private and state- and local-government schemes were scrutinized as the law-makers examined their expert computer witnesses, both lay and professional.

By February 1968, *Electronics* magazine remarked:

Fresh from having just about killed the proposed National Data Bank, the Senate's Mr. Privacy is planning new runs on other information banks ... Senator Edward Long (Democrat, Mo.) ... has expanded his attacks on electronic invasion of privacy to include all sorts of computer data banks – private or government.[2]

Reaction in Britain was slower. The Society of Conservative Lawyers reported on eighteen months of study in December 1968. The author of the foreword suggested that 'Indeed it might be said that this Report is perhaps even a little ahead of its time ...' Within a year, however, Kenneth Baker MP and Lord Windlesham had introduced bills in the Commons and Lords respectively. A lively debate in the Upper House at the end of 1969 aired many of the problems.

Meanwhile the Government was studying its own use of computers for statistics, personnel records, social security, police, vehicles and so on, with particular attention to standardization as an aid to data exchanges between departments. In Parliament three Labour MPs, Brian Walden, Alexander Lyon and Leslie Huckfield, each introduced bills. The first achieved limited success in the form of a 'Committee on Privacy' under the chairmanship of Mr Kenneth Younger, set up in January 1970. The committee survived the change of government with its terms of reference intact, and its report was due in early 1972. To the consternation of all interested parties outside the Government, however, its investigations specifically excluded the Government's own data banks.

In November 1970 Malcolm Warner and Michael Stone launched their book *The Data Bank Society* at a 'workshop' organized by the National Council for Civil Liberties. The computer professionals outweighed the few others who attended in numbers, concern and practical remedies. The British Computer Society already had a Computer Privacy Specialist Group and its president outlined the code of ethics later to be adopted by the Society. Representatives of the computer industry, the banks and hospitals (as large users) all revealed their concern for privacy and the steps they were taking to safeguard it.

The media have been interested, but not crusading. After all, much of their living comes from invading privacy. The 1971 Census engendered a little excitement when Liberal party members gave a lead in refusing to answer some or all of its questions. The BBC prepared a fifty minute documentary and then mistimed the screening trying to anticipate a period of heightened public interest.

But all this activity has produced neither widespread public interest nor legislative action in Britain.

In the United States, however, since this book first went to the printers in 1970, some progress has been made. In April 1971 the Fair Credit Reporting Act became effective, restricting the use of reports and ensuring disclosures by the collecting agencies of the data they hold to the subjects thereof. Representative Edward I. Koch introduced a Bill (H.R.854) to check on government record-keeping. In December 1971 the Senate passed a Bill protecting the privacy of Federal employees. State law enforcement data banks receiving Federal financial help were obliged to take adequate measures to protect the privacy of their data.

The Senate Judiciary Subcommittee on Constitutional Rights, headed by Senator Samuel J. Ervin, held hearings on 'Computers, data banks and the Bill of Rights' in 1970 and 1971. They heard that,

At one period in late 1969 [the US Army] maintained dossiers concerning approximately 800 civilian organizations and individuals ... These dossiers were commonly called subversive files. [They] contained ... official military intelligence reports ... reports from other ... [state] investigative agencies and copies of photographs ...

Dossiers were begun on any antiwar, militant, radical, violent or non-violent group or individual of the right or left or who showed sympathy or friendship with them.

Among the files was one on Adlai Stevenson. Originally the Secretary for Defense denied that such a file had been kept, but later the Assistant Secretary confirmed that it did exist. According to a witness the file had been started when Stevenson was seen with a leader of the Southern Christian Leadership Conference.

The investigation had its lighter side. 'In mid-September 1969, 5,000 to 10,000 demonstrators were expected (at Fort Carson, Colorado) to protest against the war ... Of the 119 demonstrators who showed up, 53 were intelligence personnel or newsmen.'

When, finally, the order to stop spying was given, it failed to reach junior levels and some files were copied before destruction.

It is worth going back and looking in more detail at each of the major stages of the controversy.

The Ruggles Report

The following is part of the summary preceding the full 'Report of the Committee on the Preservation and Use of Economic Data to the Social Science Research Council, April 1965':

During the past four years the Committee on the Preservation and Use of Economic Data has met with a considerable number of Federal agencies concerned with the collection and use of data in machine-readable form. The prime concern of the committee has been the development and preservation of data for use in economic research ...

First, the committee urges that the Bureau of the Budget, in view of its responsibility for the Federal statistical program, immediately take steps to establish a Federal Data Center. Such a Federal Data Center should have the authority to obtain computer tapes and other machine-readable data produced by all Federal agencies. It would have the function of providing data and service facilities so that within the proper safeguards concerning the disclosure of information both Federal agencies and users outside of the Government would have access to basic data ...

Second, the committee urges that the Office of Statistical Standards of the Bureau of the Budget place increased emphasis on the systematic preservation in usable form of important data prepared by those agencies engaging in statistical programs ...

Third, the committee recommends that at an early date the Social Science Research Council convene representatives from research institutions and universities in order to develop an organization which can provide a clearing-house and coordination of requests data made by individual scholars from Federal agencies.

The report made virtually no mention of privacy, beyond recognizing 'that Federal agencies should not violate the confidentiality of their data by making them available to outside research workers or other agencies'.[3] However, when he appeared before the Gallagher committee on 27 July 1966, Ruggles testified that 'first and foremost, it is essential to protect the individual from an invasion of his privacy and the misuse of information which may damage or embarrass him'.[4] He insisted that techniques be developed that allowed research to continue while preventing disclosure.

The Dunn Report

Edgar S. Dunn, Jr, consultant to the Office of Statistical Standards, Bureau of the Budget, and a research associate with Resources for the Future, Inc., Washington, DC, evaluated the Ruggles report in the six months prior to 1 November 1965, on behalf of the Bureau. The following is part of the summary that leads into his 'Review of a Proposal for a National Data Center':

The Ruggles committee report recommending the establishment of a National Data Center is only one of the more manifest expressions of concern, dissatisfaction, and frustration that have been surfacing among the groups that use numerical records for research, planning, or decision-making at all levels ...

The central problem of data use is one of associating numerical records, and the greatest deficiency of the existing Federal statistical system is its failure to provide access to data in a way that permits the association of the elements of data sets in order to identify and measure the interrelationship among interdependent or related observations.

There are a number of characteristics of existing programs and procedures that stand in the way of an effective association of numerical records for purpose of analysis:

1. Important historical records may be lost.

2. The absence of appropriate standards leads to low-quality files that contain many technical limitations to effective association of records.

3. Many of the most useful records are produced as a byproduct by agencies that do not recognize a general-purpose statistical service functions as an important part of their mission.

(Points 4, 5, and 6 concern incompatibilities between records which make association difficult.)

7. The structural problems of concern of today's policy-makers and the effort to bypass problems of record incompatability force the utilization of data at levels of disaggregation that place severe strains upon regulations restricting the disclosure of information about individual respondents.

8. There are new possibilities for more efficient management of large-scale numerical files in terms of storage and retrieval. These potentialities require the expenditure of time and effort on system design and software development that few agencies can justify.

Recommendations

Accordingly, it recommended that a National Data Service Center be established with the capability to

1. manage archival records;

2. develop referral and reference services;

3. provide explicit facilitating services for users including:

a. file rearrangement, cost tabulation, and extended output options;

b. tape translation and file modification;

c. record matching;

d. disclosure bypassing; and

e. standard statistical routines.

4. develop computer hardware and software systems essential to above;

5. provide staff support to work in conjunction with the Bureau of the Budget to develop and establish and monitor standards essential to the system capability; and

6. establish a research capability directed to an analytical evaluation of user requirements for the purpose of designing and developing the system components essential to perform these services.

The National Data Service Center would perform these services
for

1. archival records under direct management control of the Center;

2. the current and accumulated records of administrative and
regulatory agencies;

3. as a system resource to be used in connection with the current
records of any agency not in a position to meet the needs independently.

For Dr Dunn, 1966 was a busy summer. Senator Long, fore-
seeing the creation of a Frankenstein monster and great potential for abuse, and some petty official pressing a few buttons
and obtaining a dossier on any individual, called on Dr Dunn
to testify to the committee on invasions of privacy. The following month Representative Gallagher was questioning him.

In his testimony before the Long subcommittee, Dr Dunn
started to examine the issue of privacy more closely than he
had in his original report, which had treated disclosure restrictions as a mere complication in the system. Both personally
and professionally, he was concerned about privacy. He noted
the inadequacy of the law on the subject, the problem – familiar
in other contexts of law and administration – of competing
interests, and the need for a balanced approach. He reviewed
the kind of information being discussed. There was the 'public
face', defined by demographic and economic data, and the
'private face', delineated by medical and lie-detector tests. For
intelligent public planning, the former was needed; the rest was
unnecessary. Furthermore, the only reason for individual's
records carrying identifiers was to bring the data together for
subsequent aggregation into statistical tables. Dossiers were
unnecessary.

Senator Long was not concerned with necessity. What, he
wanted to know, was *feasible*? By violating current administrative and legal practice, Internal Revenue Service files could
be abstracted. If everybody had a number, all information
from cradle to grave could be pulled out at the push of a

button. Dr Dunn thought that this was impractical at the time, although perhaps possible in the future, but he did not think it was useful. His interest lay in groups, not individuals, but he wanted to be able to select a group appropriately to answer a policy question. On the subject of safeguards, he felt that aggregation, coding, and care concerning which records entered the system would protect privacy.

INTELLIGENCE V. STATISTICS

Six weeks later, before the House subcommittee, Dr Dunn enlarged his point on the significant and basic difference between 'intelligence' and 'statistical' systems. He reckoned that a secure statistical system could be developed to provide aggregates, averages, percentages, and other descriptions of relationships. But it would not provide 'intelligence' about an individual if it were run like the Bureau of the Census. Furthermore, there are technical as well as legal difficulties. Holding up a computer tape before the committee, he said that even the experts in the room could not get at its contents without a machine, a codebook, a program, and a technician. The programs in the statistical system could be made to provide only legitimate analyses. Congress, however, would have to insure that it was not used as an intelligence system. Such systems did indeed have their proper role in dealing between agencies and the public, and the 'intelligence' could be aggregated into 'statistics'. But the derivation of intelligence from such a statistical system as the Census could be, should be, and has been prevented.

The Kaysen Report

Carl Kaysen had succeeded Professor Oppenheimer at the Institute for Advanced Study at Princeton when he produced, in October 1966, for the Bureau of the Budget, the 'Report of the Task Force on the Storage of and Access to Government Statistics', from which the following are excerpts:

The committee was originally charged with the task of considering 'measures which should be taken to improve the storage of and access to US Government Statistics'. It is the best judgment of the committee that it can answer this question only in a much broader context, namely, by looking at the question of how the Federal Statistical System can be organized and operated so as

1. to be capable of development to meet the accelerating needs for *statistical* information, needs that are increasing in quantity, in variety, and in degree of detail with the developing character of American society, and the changing responsibilities in it of the Federal Government;

2. to develop safeguards which will preserve the right of the individual to privacy in relation to information he discloses to the government either voluntarily or under legal compulsion;

3. to make the best use of existing information and information-generating methods and institutions at its disposal; and

4. to meet these needs for statistical information with a minimum burden of reporting on individuals, businesses, and other reporting units.

The focus of the committee's concern is the Federal *statistical* system. Although different government agencies may require information about *specific* individuals or businesses as part of their legal operating responsibilities, the committee was unanimous in its belief that Federal agencies or other users should not be able to draw on data which is available within the Federal statistical system in any way that would violate the right of the individual to privacy. Organization and legal safeguards should be developed to prevent the use of data which is brought together for statistical purposes as a source of information concerning individual reporting units.

A body of data can provide useful *statistical* information only to the extent that it is live, in the sense of corresponding to a clearly defined and currently comprehensible system of identifying the sources of information, definitions of quantities being measured, classifications on which groupings· of units are based, and the relations of all these categories to those for other information collected on similar units, or the same units at different times.

The reference to privacy (in item 2, above) is now prominent and specific. Annexed to the report, in fact, was a discussion of 'The right to privacy, confidentiality, and the National Data Center':

After this committee was convened and well into its work, Congressman Cornelius E. Gallagher, as Chairman of a Special Subcommittee on Invasion of Privacy of the Committee on Government Operations of the House, raised questions about the possible threats to privacy and freedom that a National Data Center might present. These are serious questions that deserve to be met squarely.

In general, our committee believes that the problem of the threat to privacy can be met best by Congressional action, which defines a general statutory standard governing the disclosure of information that is collected on individuals....

The problem of disclosure of confidential information about individuals and businesses is not new. It has long been recognized that the information which individuals and businesses provide under law to the Bureau of the Census, for example, is confidential. This means that no other Federal agency is permitted to see or use the individual records, and even Congress itself cannot obtain Census information on any individual or company. The disclosure rules are meant to safeguard individuals so that they can feel sure that information which they give to the Census Bureau will never be used against them for such purposes as tax enforcement, antitrust, or Congressional investigations. The disclosure rule has not been interpreted, of course, as preventing the use of Census information for analyzing policy or providing information about specific groups, regions of the country, performance of industries, etc. In making tabulations of data, however, the Census Bureau carefully omits those classifications which might enable anybody to figure out information about individual firms or persons.

The enforcement of a statutory obligation as the primary method of dealing with the problems of safeguarding privacy can work excellently, as the experience of the Census Bureau shows. Indeed, the present situation, in which there exist a variety of different disclosure standards, some statutory and some executive, is much less conducive to protection of individuals' privacy than would be a situation in which, as our report suggests, the Director and the Data Center would have the obligation of enforcing a uniform standard over the whole system.

The Gallagher subcommittee has also raised the question of the creation of a vast file of individual 'dossiers' incorporating police and FBI information, Armed Service and government personnel records, and the like. This it not the purpose of the proposed Center at all, and it is clearly within the power of Congress to distinguish between the collection and organization of general economic, social, and demographic information of the sort that Federal statistical agencies have traditionally collected – much of it on a sample basis – to which our proposed National Data Center is directed, and assembly of the sort of personal-history information on named individuals that is contained in a personnel file or police file.

Finally, the subcommittee has raised certain questions as to the technical security of data stored in machine-readable form, and accessible through machine operations. Here again, this is not a new problem, and both organizational and technical means are available to control and limit the risks. Though bank robbers have not been totally eliminated, we have not on that account abandoned banks and banking, and the analogy seems to us perfectly appropriate. We think that the maintenance of privacy against both unwitting and illegal disclosure of information made available to the Government are real problems, to which our proposed new Center must direct attention and effort. However, they are neither insoluble problems, nor ones of such magnitude as to make the organization of effective functioning of a National Data Center possible only at the expense of significant inroads on liberty and privacy.

Note that Dr Kaysen believed there to be great advantages in uniformity of disclosure, and that policing a centre would be easier than policing a set of distinct agencies. When Dr Kaysen had to defend his views before the Long subcommittee in March 1967, like Dr Dunn he stressed the distinction between dossier- and statistical-data files. He proposed the Census as a model, and burdened Congress with the duty of defining which information could be assembled. He also believed that technical barriers could raise the cost of 'penetrating' the system even more than the benefits accruing to successful penetration due to the amount of information so neatly pulled together. He agreed with Senator Long that unscrupulous people could illegally gain access, but that the effort, skill, and teamwork necessary were greater than with manual systems.

The Lawyers and Law-makers

While the proponents of the centre were acquiring ever-wider white bands for their 'black hats', the posse of 'white hats' was growing.

Leading the pursuit, determined to apprehend Big Brother before he could do further mischief, was Senator Edward V. Long, the Democrat from Missouri. In his book *The Intruders*, the hearings of the Subcommittee on Administrative Practice and Procedure of the Senate Judiciary Committee were distilled. 'The invasion of privacy by government and industry,' Long wrote, 'is conducted with wiretraps, personality tests, cameras and lie detectors (polygraphs), by private eyes, government sleuths, and corporate peeping toms.' The computer might still have been one of their tools, but not if Senator Long and his subcommittee had their way.

But Representative Cornelius E. Gallagher was no less concerned, and he presided over the Special Subcommittee on Invasion of Privacy of the House Committee on Government Operations. His interests extended from polygraphs in 1963 through personality tests in 1965 to the computer dossier in 1966. Like most lawyers, he was not satisfied merely to prove something wrong. He had to make sure it was *unconstitutional* as well.

In this exercise he had the active support of many professors of law, among them Professor Charles A. Reich of Yale Law school, a constitutional lawyer with expertise on the collection of data into individual dossiers, who testified before the Gallagher Committee in July 1966. Reich was concerned about defamatory material entering the dossier unknown to its subject to errors of fact and vagaries of opinion. He goes right back to the authors of the Bill of Rights and tries to reconstruct their opinions concerning privacy in the hope of applying them to the present problem. (The problems of feedforward control systems are severe enough over short time spans and simple systems. It seems naive to assume a troublefree regulator of a nation operating after two centuries.) He went on to discuss the inability of computers to forget the past, and their embargo on beginning anew – the second chance. Believing that protection comes not from people's good intentions but from law, he wanted some laws on the subject of gathering information. In particular, certain facts should, in his opinion, be uncollectable; others should be limited to collection in a single instance and for a single purpose, and destroyed when the purpose is served; any that are retained should be open to challenge by the individual concerned.

Kenneth L. Karst, Professor of Law at UCLA, was also concerned with the accuracy of the files, but was less than optimistic about the possibilities of restricting their contents:

One need not believe, as I do on my gloomier days, that we are marching steadily to the Anthill in order to share the view that our main concern about 'the files' should be directed to their accuracy, not to their accessibility. Any society, whatever its degree of freedom, has a great deal to gain from unimpeded access to information. Any society risks serious losses in the quality of low-level decision-making if its leaders decide before the event that certain classes of information must be kept secret because they will not be relevant to future decisions. In the society of the near future, much more personal information will be relevant to predicting individual performance. Concurrently, much more will be thought to rest on predicting accurately. Whatever we do today to restrict access to personal data, our successors will probably discard. The same cannot be said of efforts to make the files accurate. If there is one sure, positive contribution we can make to those who follow us, whatever social forms they choose, it is to make all our reservoirs of information into truer reflections of the world we see.[5]

Alan F. Westin, who directs the Center for Research and Education in American Liberties at Columbia University, published *Privacy and Freedom*[6] in the fall of 1967. In his book Westin expressed concern with the increase in data-collection by honest and honourable men engaged in respectable work. He, too, favoured more specific laws.

Professor Arthur R. Miller of the University of Michigan Law School brought together many of these points in the following statement:

The risks to privacy created by a National Data Center lie not only in the misuse of the system by those who desire to injure others or who can obtain some personal advantage by doing so, but in overcentralization of individualized information and the proliferation of people having capacity to inflict damage through negligence, sloppiness, thoughtlessness, and sheer stupidity. These people are as capable of damaging others by unintentionally rendering a record inaccurate, or losing it, or disseminating its contents to unauthorized people as are people acting out of malice or for personal aggrandizement.

Our success or failure in life ultimately may turn on what other people decide to put in our file and the programmer's ability, or inability, to evaluate, process, and interrelate information ...

The very existence of a National Data Center magnifies the risks to individual privacy by providing a fruitful and central source of information for federal officials and may even encourage their questionable surveillance tactics.

Professor Miller was compelled to conclude that 'control over the proposed Center must be lodged outside the existing administrative channels,' and 'as repugnant as it may sound in an era of expanding government involvements it is necessary to establish a completely independent agency, bureau, or office that can establish policy under legislative guidelines that direct the Center to insure the privacy of all citizens.' In order to preserve 'our traditional freedoms and liberties', Professor Miller advocated the development of a legislative standard to restrict the recording of sensitive personal information, such as medical and psychiatric data, in the absence of an overpowering showing by the government that its preservation 'is essential to some fundamental national policy'.

Writing in *The Atlantic*, Professor Miller stressed the danger of a 'foot-in-the-door'. Computer technology was not standing still, Miller pointed out; plans had to be made for the future capabilities of the system, which would blur the distinction between statistics and intelligence, and result also in link-ups between federal, state, local, and private systems. He continued:

Testimony before Congress concerning the intrusive activities of the Post Office, the Internal Revenue Service, and the Immigration and Naturalization Service gives us cause to balk at delegating authority over the Data Center to any of the agencies that have a stake in the content of data collected by the government. Some federal personnel are already involved in mail-cover operations, electronic bugging, wiretapping, and other invasions of privacy, and undoubtedly they would try to crack the security of any Data

Center that maintains information on an individual basis. Thus it would be folly to leave the center in the hands of any agency whose employees are known to engage in antiprivacy activities. Similarly, the center must be kept away from government officials who are likely to become so entranced with operating sophisticated machinery and manipulating large masses of data that they will not respect an individual's right to privacy.[7]

Professor Miller urged an independent agency to police not only any national centre, but also local and private schemes, as well as the inter-agency traffic in personal data.

Riding with the 'white hats' has been author and sociologist Vance Packard, whose 1964 book, *The Naked Society*, has done much to awaken Americans to the dangers they face. He warned the Gallagher subcommittee that a central data bank would depersonalize the American way of life, cause citizens to distrust and become alienated from their government, and through error or omission affect lives unfairly. His 'own hunch is that Big Brother, if he ever comes to these United States, may turn out to be not a greedy power-seeker, but rather a relentless bureaucrat obsessed with efficiency'.[8]

Mr Packard did not confine his book or his testimony to the subject of federal operations. He cited the exchange of information within government agencies and between credit bureaux and the many other private organizations.

The newspapers, of course, have been in the thick of the controversy, printing powerful exposés of various threats to privacy. Magazines and journals have invited the proponents and opponents to write articles. On 1 August, 1966, a *New York Times* headline reported, 'Computer Plan For Personal Dossiers In Santa Clara Stirs Fears of Invasion of Privacy'. On 9 August, it was concerned 'To Preserve Privacy'. Talking of the proposed national data centre, one of its editorials stated:

Understandably, this idea has brought vigorous protest, in which we join. Aside from the opportunities for blackmail and from the likelihood that the record of any single past transgression might damage one for life, this proposed device would approach the effective end of privacy. Those Government officials who insist that the all-knowing computer could be provided with safeguards against unauthorized access are no doubt of the same breed as their brethren who 'guaranteed' that last November's Northeast electric blackout could never occur.

The *Wall Street Journal* quoted Congressman Gallagher on the privacy issue (20 December, 1966): 'We cannot be certain that such dossiers would always be used by benevolent people for benevolent purposes.'

The Professionals

Wearing hats to match their suits of organizational grey are the computer men. With their deeper knowledge of the potential of these machines, they are aware of their power both for good and for evil.

Professor John McCarthy of Stanford, writing in *Scientific American*,[9] recognized that in many minds the computer is the ultimate threat. He proposed a bill of rights governing collection, access, and disclosure of personal data, recognizing that to establish it, with technological advances so rapid, would require new thinking on the source and nature of freedom.

The editor of the journal *Computers and Automation*[10] suggested another defence: 'the deliberate propagation by individuals of mistakes, wrong numbers, wrong spellings, aliases, and so forth, on the principle, "garbage in, garbage out".' There spoke the voice of experience! The same journal later took to task the members of the Washington chapter of the Association for Computing Machinery for regarding the national data centre as inevitable; as they described a meeting: 'Laughter filled the room when invasion of privacy, misuse of information by officials, and erroneous damage to individual reputations were discussed. The consensus was that the National Data Center is already feasible and that processing techniques exist now.'[11] *Computers and Automation* spoke with Representative Gallagher, who thought that adequate safeguards were impossible and that the system's users and customers were too obscurely defined. Both, he thought, were essential before a system could be considered inevitable.

The May 1967 issue of *Computers and Automation* carried a 'Resolution on the National Data Center and Personal Privacy' proposed by the very chapter of the Association for Computing Machinery that had laughed off the matter a few months before:

As members of the computing profession, we are concerned not only with technological progress, but also with the larger question of the role of the computer and the computing profession in the American society. Our professional duty requires that we point out the benefits of purely technological innovation and, with the same high level of competence and objectivity, call attention to actual or potential consequences of technological change which may be incompatible with the democratic society.

The question of a National Data Center is particularly difficult to adjudicate because while the chief potential benefits which motivated the proposal – economies in time and money – can be objectively assessed, the disadvantages that are potentially the most critical can be assessed only subjectively. These disadvantages include the invasion of privacy, the denial of constitutional protection against self-incrimination, and the adverse economic and social consequences of misuse of personal data which could result from storage, manipulation, and transfer of personal dossiers.

The conclusion is inescapable that in a democratic society individual rights take precedence and determine a practical upper bound on the efficiency attainable in government.

Technical safeguards against intentional or inadvertent misuse of personal data by users, data-center supervisors, programmers, operators, and maintenance men must provide for absolute accountability in the handling of personal data and must effectively restrict access and use to that which is consistent with constitutional rights. But technical safeguards are not sufficient. Strong laws must be passed and vigorously enforced to provide both an effective legal basis for accountability in the handling of personal data and procedures for redressing and compensating individual injury.

George Sadowsky, Director of the Brookings Institution, charged the authors of the 'Resolution' with ignoring the Kaysen and Dunn Reports.[12] Zealous bureaucrats seeking potential totalitarian control were not, he claimed, behind the plan. While the proposers were aware of the privacy issue, Sadowsky wrote, those responsible for credit information and local-government systems were unregulated even though they were potentially far more intrusive. He urged his colleagues to review those inadequacies of the present statistical system that prompt the proposal for the national centre.

Dr Emanuel R. Piore, vice-president and chief scientist of IBM, was invited to explain how the machines operate to the Long subcommittee in March 1967. He summarized the workings of computers and discussed logical locks in operating systems and physical locks on terminals. He concluded:

> None of these facts and prospects should obscure the plain truth that in the end, preservation of privacy rests not with machines but with men. The effectiveness of all protective measures, however sophisticated they may become, will still depend upon people: operators, service personnel, supervising officers, and all those who decide what information to put into a computer and how to use it.[13]

In another professional journal, *Computer Digest*, Professor Westin's statement to the Spring Joint Computer Conference (SJCC) was quoted:

> At the moment, American society is barely entering the beginning stage of this debate over data surveillance. We can see that three quite different approaches are already appearing. One position, reflected by the initial views of many newspaper editors, civil-liberties groups, and congressional spokesmen, is to oppose creation of data centers and intelligence systems completely. The need for better statistics for policy analysis or of richer information systems for criminal-justice purposes is seen as inadequate when weighed against the increase in government power and fears of invasion of privacy that such systems might bring.
>
> A second view, reflected in the initial thinking of many executive-agency officials and computer scientists, assumes that traditional administrative and legal safeguards, plus the expected self-restraint of those who would manage such systems, is enough to protect the citizen's privacy. The more reflective spokesmen in this group would add that a large-scale decrease in the kind of personal privacy we have through inefficiency of information-collection may well be on its way out, but that this would be something individuals could

adjust to and would not seriously threaten the operations of a democratic society.

The third position assumes that neither the 'total ban' nor the 'traditional restraints' positions represent desirable alternatives. What is called for is a new legal approach to the processing of personal information by authorities in a free society and a new set of legal, administrative, and system protections to accomplish this objective. The fact is that American society wants both better information-analysis and privacy. Ever since the Constitution was written, our efforts to have both order and liberty have succeeded because we found ways to grant authority to government but to tie it down with the clear standards, operating procedures, and review mechanisms that protected individual rights. A free society should not have to choose between more rational use of authority and personal privacy, if our talents for democratic government are brought to bear on the task. The most precious commodity we have now is the few years of lead time before this problem grows beyond our capacity for control. If we act now, and act wisely, we can balance the conflicting demands in the area of data surveillance in this same tradition of democratic, rational solutions.[14]

The Few Years of Lead Time

Professor Westin was speaking in 1967. Numerous articles and several books have since been published and public interest has stimulated parliamentary and congressional inquiries, leading to some legislation in the United States.

As an immediate result of a Commons debate, the British Government set up in May 1970 the Younger Committee on Privacy 'to consider whether legislation is needed to give further protection to the individual citizen and to commercial and industrial interests against intrusions into privacy by private persons and organizations, or by companies and to make recommendations'. The terms of reference excluded the public sector, a severe restriction since the Committee was able to report later that, 'almost all the more credible of [the apprehensions about computerized information] relate to ... use by central or local government, in particular to those in the hands of the police, the Inland Revenue and the health and social services'. The Committee operated with the minimum of publicity and attracted evidence mainly from those, ourselves included, who were trying to stimulate public appreciation of the problems.

The Younger Committee's report was published in July 1972.[15] It was unable to arrive at a single, comprehensive definition of privacy and so could not recommend the establishment of a general right to privacy. Although the conclusions were disappointing, its research was widespread, thorough and met with general approval.

It found that 'the computer problem as it affects privacy in Great Britain is one of apprehensions and fears, and not, so far, one of facts and figures. We cannot conclude that the computer is at present a threat to privacy.' The Committee 'were very conscious of the fact that technology is advancing rapidly' and recommended 'that the Government should legislate to provide itself with machinery for keeping under review the growth in, and techniques of, gathering personal information and processing it with the help of computers'. They went on to propose that this 'independent body with members drawn from both the computer world and outside' should pick up where they themselves had left off and consider 'the need for and practical application of a detailed licensing system which ... we believe at present would be premature'. It should consider also 'whether a useful purpose would be served if each user of a computer appointed a supervisory officer with responsibility for ensuring that the principles we recommend were observed'. The principles are:

1. Information should be regarded as held for a specific purpose and not be used, without appropriate authorisation, for other purposes; and

2. Access to information should be confined to those authorised to have it for the purpose for which it was supplied.

3. The amount of information collected and held should be the minimum necessary for the achievement of the specified purpose.

4. In computerised systems handling information for statistical purposes, adequate provision should be made in their design and programs for separating identities from the rest of the data.

5. There should be arrangements whereby the subject could be told about the information held concerning him.

6. The level of security to be achieved by a system should be specified in advance by the user and should include precautions against the deliberate abuse or misuse of information.

7. A monitoring system should be provided to facilitate the detection of any violation of the security system.

8. In the design of information systems, periods should be specified beyond which the information should not be retained.

9. Data held should be accurate. There should be machinery for the correction of inaccuracy and the updating of information.

10. Care should be taken in coding value judgments.

The Committee proposed that computer users should voluntarily adopt these principles for handling personal information. They recognized that 'a system of control based on the ethical standards and technical skills of those responsible for programming and operating computers would depend on their organisation into a professional association or associations on the pattern of medicine or the law'. But no such professional association yet exists nor could it establish quickly the status of older professions. Because the Committee chose not to recommend legislation, it may be many more years before a British computer professional can turn to the Statute Book for support in opposing a demand from his employer to set up an unethical system.

REFERENCES

In this and subsequent chapters, extensive use has been made of the following reports:

'Special Inquiry on Invasion of Privacy', *Hearings before a subcommittee of the Committee on Government Operations, House of Representatives, 89th Congress, 1st Session. June 2, 3, 4, 7, 23 and September 23, 1965*. US Government Printing Office, Washington, DC, 1966. (Subcommittee Chairman, Cornelius E. Gallagher, New Jersey.)

'The Computer and Invasion of Privacy,' *Hearings before a subcommittee of the Committee on Government Operations, House of*

Representatives, 89th Congress, 2nd Session. July 26, 27, and 28, 1966. US Government Printing Office, Washington, DC, 1966. (Subcommittee Chairman, Cornelius E. Gallagher, New Jersey.)

'Computer Privacy', *Hearings of the Subcommittee on Administrative Practice and Procedure of the Committee of the Judiciary, United States Senate. March 14 and 15, 1967.* US Government Printing Office, Washington, DC, 1967. (Subcommittee Chairman, Edward V. Long, Missouri.)

1. Karl Kaysen, 'Data Banks and Dossiers', *The Public Interest*, Spring 1967. Reprinted in 'Computer Privacy', p. 265.

2. 'Data Banks Overdrawn,' *Electronics*, 19 February 1968, p. 53.

3. Ruggles Report, reprinted in 'The Computer and Invasion of Privacy', p. 202.

4. 'The Computer and Invasion of Privacy', p. 91.

5. Kenneth L. Karst, 'The Files: Legal Controls over the Accuracy and Accessibility of Stored Personal Data', *Law and Contemporary Problems*, Vol. XXXI, No. 2, Spring 1966, pp. 342–76. Reprinted in 'Computer Privacy', p. 199.

6. Alan F. Westin, *Privacy and Freedom*, Atheneum, New York, 1967.

7. Arthur R. Miller, 'The National Data Center and Personal Privacy', *The Atlantic*, 1967, p. 56.

8. Vance Packard, *The Naked Society*, David McKay, New York, 1964, p. 13.

9. *Scientific American*, September 1966, p. 72.

10. *Computers and Automation*, October 1966, p. 7.

11. ibid., March 1967, p. 6.

12. ibid., August 1967, p. 22.

13. 'Computer Privacy', p. 122.

14. *Computer Digest*, Vol. 2, No. 6, June 1967, p. 2.

15. 'Report of the Committee on Privacy', Chairman the Rt Hon. Kenneth Younger, July 1972, HMSO, London, Cmmd 5012.

15

THE NATURE OF THE THREAT
TO PRIVACY

*... Whatsoever I shall see or hear in the course of my profession
in my intercourse with men, if it be what should not be published
abroad, I will never divulge, holding such things to be holy
secrets ... – The Oath of Hippocrates.*

Democracy flourishes on information. That information should
be good; it should be accurate and it should be balanced. For
that matter a non-democratic society needs information too,
though often of a different sort – the government may need
information in order to protect itself. A democratic society,
however, should be sensitive to the needs and welfare of its
citizens, and to look after these it requires information about
its citizens. The more complex the society and its technology,
the greater the need for information becomes. Herein lies the
dilemma.

The problem with 'privacy' is its conflict with other social
values, particularly competent government, a free press, protection against crime, provision of services, collection of taxes,
and the development of community living environments. The
authority providing each of these wants to decide what it
should know about us and when it should be told. We, on the
other hand, resent the intruding official eye. We are curious
about our neighbours, and secretive about ourselves.

To date, democracy has had to make do with rather poor
information – distorted, partial, often non-existent. There has
been a tradeoff between speed and accuracy. However, we
have muddled along in a tolerable manner. In places the ill
effects of poor information and insufficient analysis are evident,
sometimes catastrophically so: it is said that the great stock-
market crash of 1929 could have been lessened if the govern-
men had had good information and taken corrective action.

Most of the time, however, society applies its own corrections, although they may be somewhat late and troublesome.

Today we have at our disposal new machinery of great power. We can collect and analyse thousands of times the information we ever could have before. The conflict between privacy and other social values suddenly becomes acute.

We all have a pretty vague idea of what we mean by 'freedom'. Likewise, our ideas on privacy are partially based on a world which is passing away. If we are to build a computerized society, we must define some of its fundamental concepts in finer detail. The computer, where it has been applied to industry, has often demanded new and greater precision in understanding how a department is to be run. If we are to apply it to society, then here also we should specify more precisely what sort of a world we want to live in. If we do not, we shall find the ability to choose taken away from us – we shall live in an environment wholly created by government agencies and computer experts.

CONSENT

Central to most of the arguments about privacy is the matter of *consent*. Does the individual in question always consciously consent to divulge the information that is wanted? This consent is usually given on assumptions about anonymity and/or confidentiality. Frequently, however, consent is taken for granted by both inquirer and inquiree.

Many 'public' figures, politicians and film stars in particular, not only consent to publicity but thrive on it. The press, however, will frequently extend this implied consent to their relatives and to those who, despite their public duties, prefer seclusion. More important, they will extend it to other times and places. This same extension is the essence of the fear of 'invasion' caused by the computer.

In order to obtain *consent*, inquisitors normally offer immediate rewards, promise *confidentiality*, and assure *anonymity*.

It is essential that we take a closer look at each of these in the context of efficient data-retrieval.

A lot of information is collected from respondents without their volunteering it. Most government forms demand completion and threaten penalties for non-compliance. Vance Packard tells the story[1] of William F. Rickenbacker, who refused to complete the Census Bureau's 'Household Questionnaire', which he described as 'unconscionably long, uncivilly inquisitorial, and absolutely unconstitutional'. He was given a suspended jail sentence of 60 days, fined $100, and spent a great deal more on unsuccessful appeals. Perhaps he should have 'consented' to provide a personal inventory.

Confidentiality

The Bureau of the Census has an excellent record for preserving its records inviolate. Only sworn officers of the Bureau may examine individual reports, and all Bureau publications must be statistical and incapable of pointing to individual respondents. (Under very limited conditions, individuals can obtain details of themselves or near relatives.)

The same cannot be said of other agencies. Representative Cornelius E. Gallagher told the story[2] of a potential Marine officer, well suited to serve his country, whose application was turned down because he had received four parking tickets when he was 17. He next applied to the Navy, was recruited, and then turned down on the grounds that he was a Marine Corps reject. With the relentless record-keeping of the computer, damning facts can be remembered indefinitely and brought to light at any time.

Unless the data-hounds can guarantee that confidences will not be broken, the ordinary citizen will cease to trust the government at all. If people should then start to foul the system with 'garbage' – lies, omissions, and nonsense insertions – the scheme would collapse like the Tower of Babel.

Anonymity

One of the best ways to insure confidentiality is by preserving anonymity. If no identifiers – except those provided for the duration of the survey – are attached to the records, people will bare their souls. This is the method of the Gallup Poll, the Kinsey Report's survey, and all other well-conducted sociological research. It is also the significant attribute of the secret ballot.

The Dunn Report (see Chapter 14, p. 326) stressed the need to remove anonymity to provide valid links between separate records. However, without anonymity, confidentiality is more difficult to preserve. And confidentiality is the normal justification for consent when no short-range recompense is offered. In order to gain the willing response of the citizenry, a national data centre *must* insure confidentiality. Much of Section III of this book will show how this can be achieved.

A large volume of records is collected by government 'authorities' in their dealings with the public, with much of the underlying information provided under mild or severe compulsion. The circumstances may range from an application for a government post, through a draftee's military record, to a convict's record. In each case the individual wishes that the subsequent disclosure of *his* record should be under *his* control. Once such details get into the bank, the risks to the citizen become enormous. Just how enormous, we propose to describe in the next two chapters.

For the present, however, let us remember that our primary safeguard until now has been the high cost of efficient data-collection. Government records have been uncollated, uncentralized, erratic, inaccessible, and difficult to interpret. Hiring professional snoopers comes expensive. Perhaps this is why, according to Representative Gallagher,

Confidential reports on private citizens are often bandied about rather freely, both inside and outside the government ... I found lie-detector exams were freely exchanged among personnel circles in government agencies.[3]

State, Local, and Private Data Centres

Despite the many discouraging forebodings, there is reason to believe that a national data centre can be so hedged by administrative, legal, and technical procedures that it will constitute only a minor threat to individual privacy. At least it would be a minor threat compared with state- and local-government plans, and those of the private sector, which include many schemes like the following:

1. The American Bar Association's plan to establish a data bank for recording disciplinary actions taken against lawyers.

2. The Federal Bureau of Investigation's national crime-information centre.

3. A New Jersey firm's project to market to drug companies computerized data on the buying habits and personal backgrounds of doctors.

4. The Credit Data Corporation's plan to computerize credit information on 70 per cent of the US population.

5. Schemes by the cities of New Haven, Connecticut, and Santa Clara, California, and Alameda County (also in California) to establish computer-stored dossiers on every citizen in their respective areas.

There are currently in operation many varieties of data banks holding dossier information, either comprehensive or partial. Table 15.1 lists some of them.

Table 15.1 Some Types of Data Bank
Holding Personal Information

Category	Types of Data Bank
Police	FBI; security clearances; police-information systems
Regulatory	tax; licensing; vehicles
Planning	property-owners; vehicles; economic data; business information
Welfare	medical; educational; veterans; job openings and unemployment
Financial	credit bureaux; building societies; banks
Market	mailing lists

Organizational	personnel files; membership lists; professional bodies; armed forces; corporation employee dossiers recording intelligence, aptitude, and personality tests, and appraisals and attitudes
Social	computer-dating; marriage bureaus; hobbyist data
Research	medical case-histories; drug usage; psychiatric and mental-health records
Travel	airline reservations with full passenger details; hotel reservations, car rentals
Service	libraries; information-retrieval profiles; insurance records
Qualifications	education records; professional expertise; membership of professional groups; results of IQ and aptitude tests

In each case, sound, cogent reasons have been evinced for the collection and retention of the data in these banks. Each collects its information with the respondent's consent, either expressed or implied. Few of these data banks, however, guarantee confidentiality or anonymity.

It is unlikely that any of these banks applies the necessary techniques for preventing the determined invader from obtaining information if he thought it worth the effort. At present the effort is unjustified, since the dossiers are in most cases incomplete. However, there is little reason to believe that they will stay that way.

There was a time when various railroads, mail carriers, and telephone companies competed with each other. But the economic absurdity of changing trains at corporate boundaries, paying for different stages of a mail run, or being unable to dial up a friend in another phone system became evident very rapidly.

The analogy with information is exact: economic logic demands integration. Is it any wonder, then, that the following should appear in Burroughs' *Clearing House* (December 1966)

NEEDED: A Central Source for Consumer Credit Data

The on-line computer seems particularly suited to accumulate and store credit data from a wide variety of sources. The magnitude of the problem is complicated considering the scope of credit-availability. Personal-loan companies, credit unions, numerous credit-card systems, savings-and-loan associations, banks are all suppliers of credit. Indeed, one of the problems is that these sources of credit must realize that competition cannot be a factor in the mutual exchange of credit information.

To accomplish the task, I suggest that the banking fraternity lead the way in forming a central computerized corporation which would handle credit information for an established region.

If such a central source is created within the United States, it will contain, for virtually every adult American, details concerning not only his income, its source, and where he banks it, but also his home, debts, assets, bill-paying speed, health, insurance, family, and spending habits. For many there will be legal notes and microfilmed reports of investigations too valuable to be lightly thrown away.

Such a system will have many advantages both to those who use it and to those whose dossiers it contains. Because it will be central, complete, and all-embracing, it can be both more accurate and more up to date than present systems. Errors will be picked up more quickly since an applicant who fills out a form, for example, can challenge the rejection of his apparently valid application; once the single record has been corrected, all subsequent applications call up an accurate dossier. Delays would be fewer and shorter, except perhaps for the few unusual individuals with no previous records. Bureau customers, like patrons of department stores or automobile dealers, could have their own programs for evaluating credit risks, so that their replies would be merely 'Yes', 'No', or 'Maybe'.

In the latter case the subject could perhaps call out his own record for discussion with the vendor. However, all that information costs money to collect and it would seem foolish not to use it profitably. By collecting only a little information, mailing lists of immense value to advertisers could be provided ... at a price.

Direct-mail promotion material is already inundating the US Post Office. Services that provide lists of names and addresses can select potential buyers for books, boats, or bust-developers on the basis of an individual's buying history, social habits, or financial status. By attaching to each name not only an address and dossier, but also codes indicating subscriptions and purchases, address labels of likely purchasers can be automatically printed – at a rate of thousands per minute – directly onto the literature to be mailed. Representative Gallagher stated, regarding a similar service:

This scientific-dating business that is going on is all well intentioned, where boys and girls write into a data center and give a lot of information as to the size of their moles and all sorts of other things. That was a lot of fun at the beginning, but now people are running off and selling tape catering to the kinds of feelings that they might have.[4]

Businesses do not keep records only of their customers, of course. They are intensely interested in employees, potential employees, and past employees. They do not confine themselves to the kind of data that can be told to them by an interviewee directly. They also record aptitude- and psychological-test results, although there is more than a little doubt concerning businesses' ability to use them when they have them. Because the US federal government is so overwhelmingly in the public eye, its employment terms have been most questioned: 'Many federal workers,' Gallagher reported, 'are subjected to extensive tests on their sex life, family situations, religion, personal habits, childhood, and other matters'.[5] All this material is not only collected, but recorded and on occasion made available to other agencies. The chairman of the US Civil Service Commission said:

For proper decisions in these areas we must have integrated information systems. This will require the use of information across departmental boundaries. It is here that current efforts to standardize symbols and codes will pay dividends. Direct tape-to-tape feeding of data from one department to another may become common. These systems will mesh well with developing plans for an executive-level staffing program which will be designed to locate the best possible man for any given top-level assignment, no matter where in government he may be serving.

The computer's ability to search its perfect memory and pick out records of individuals with specific characteristics has been applied in the search for candidates for Presidential appointments. A computerized file containing the names and employment data of some 25,000 persons, all considered likely prospects for federal appointive positions, is searched electronically. The talent bank, with its automated retrieval system, broadens the field of consideration for the President in critical decisions of leadership selection.[6]

Note that these are dossiers containing information about people that even the subjects themselves did not know. There might be justification for holding such extensive data on known criminals, or suspected criminals, or suspected spies, or potential criminals, or potential spies, or people who associate with known criminals or work on secret projects for the government, or for a government subcontractor or ... It is difficult to draw a line. The New York State Identification and Intelligence System keeps

detailed criminal, social, and *modus operandi* data on each subject. ... *It is not likely* [emphasis supplied] that data such as wiretap information, names of confidential informants, or unverified tips and rumors would be included.... [Data on motor vehicles] will be readily available through what is called an interface between the two systems.[7]

We leave it to the reader to imagine the records that he and his fellow citizens have contributed to the files of the hundreds of data banks in Britain or America. Then let him ponder the economic arguments that point with unerring logic not merely to the consolidation of financial, medical, insurance, licence, welfare, marketing, and organizational records into respective banks, but to the consolidation of *all* such data into *one* bank where it can be put to better use. (It will, of course, be more accurate, up to date, and complete.)

Having achieved that, we could forever rid ourselves of form-filling by coupling this national bank with its federal brother to create a perfect, individualized, computer-based national system.

The Great Data-Bank Robbery

Perhaps some future Perry Mason dialogue will read something like this:

'Okay, Perry. I've identified that girlfriend of Charlie Weiss', the one who's getting his insurance money!'

'How in the world did you get a line on him, Paul?'

'Through the terminal in the Dead-End Office of Poli-State Life. I asked the little blonde to get me a quote for a policy and as she keyed in the data, one of my men, Harry, was around in back listening on the wires. When I saw the rates, I accepted and paid her $100 – you'll see it on your account soon enough – and she filed it away. Then I said, "Did it cover private aviation?" And she said, "No, that's extra." So I got her to quote again and correct the cover – another $28 – and Harry recorded her dragging the record out of the data bank to look it up.'

'Say, wait a minute, Paul. I see how you could get your record out, but how did you get Charlie Weiss's?'

'The operative who was there took his recorder home and decoded everything – none of it was in cipher and all the characters were in USASCII. . . .'

'Is that the standard code for information interchange I was reading about?'

'Uh-huh. It was a dial-up line and the program is very simple. All the girl at the terminal does to look at an item of data in a record is key in the number of the field in the record. So we dialed up the Poli-State computer from a rigged terminal with the right response circuitry, pushed in his policy number . . .'

'Where did you get that?'

'From his bank statement – it's direct-debited. We just kept to the drill the blonde used for my policy, and then tried every number from 1 to 32 when we struck oil. Beneficiary's name and address was typed out. So we said thank you, goodbye, and the computer wrote the time and signed off. Marvellous things, these computers.'

Paul Drake and Perry Mason are known for their ingenuity – or at least for that of Erle Stanley Gardner – and ability to manoeuvre within the shoals of the law. Very few computer-users are in a position to outwit them and their non-fictional ilk. Even fewer have bothered to try.

We believe, and shall attempt to demonstrate in Section III, that data banks *can* be made robber-proof – at a cost. The more ingenious the potential robber, the greater the cost. We also believe, however, that most of today's data banks are grossly insecure – most of them are not protected at all.

Professor Arthur R. Miller told Senator Edward Long's subcommittee on invasion of privacy (see Chapter 14, p. 334) that, according to computer experts at the University of Michigan,

a programmer with less than a month's training can break the more elaborate encoding procedures currently being used in large data banks within five hours. This has been borne out by the fact that at our institution – and this has been tried elsewhere – we occasionally leave a terminal unattended in an unlocked room to see if our students can work their way into the system by breaking the access code. They have never failed us. They have gotten the system even against some rather complex coding.

Most of today's data banks do not use such elaborate encoding. Wastebaskets are frequently filled with thick pads of 'printout' that has either been discarded from tests of programs or replaced by later information. (It is customary at many installations to issue new copies of periodical reports – not just amendments to update versions from previous runs.) Carbon paper on forms that are continuously printed is also continuous and therefore has all the text on it. Also, rolls of paper showing operators exactly what to type in to obtain access to data are usually left in terminals. If the roll is not in the terminal, it is probably in the wastebin. Tapping wires to terminals is also possible – just like telephone wiretapping. The resultant signals require decoding, but this is usually an easy process today. Most messages are in a well-known code, most commonly USASCII; where data sets are used, they are also standard.

The main uses of this type of snooping would be to provide leads to the kind of information available, the method of storage, and the messages that are used to 'access' it; it is doubtful whether stored information can be obtained directly by these means without great persistance. An intruder can probably learn eventually, however, what to key into the terminal to explore the files directly by himself. He may have to obtain a password or operator-identification code.

Getting into a time-shared system is often quite easy. It may require a key (begged, borrowed, stolen, or legitimately acquired), but usually it does not. It may require a password (obtained from the waste bin or simply by watching the operator). Having gained the computer's attention, the infiltrator can browse among those files not specially protected (most are not). He will need a knowledge of what to key in, which may be learned from studying others' printouts or from reading an operating manual left lying around.

A travelling companion of one of the authors once, bored while waiting for a plane in Miami, tried to break into the airline's data bank. The time was 2 a.m. The airline desk was manned by a tired operator and had several computer-terminals. Knowing that a colleague who was flying back to New York would be amused, he made an exercise of obtaining his colleague's personal record. It was not difficult. He asked the operative some questions about his own booking, and the latter used the terminal to print out details. While it was easy to see which keys she pressed, in any case she left the printout in the typewriter (they always do). There was no difficulty in removing it. The characters she used were printed on it. This particular file is protected by a device requiring the operator to sign in and out with an identification card. This operator did not sign out and so left the terminal effectively unlocked all this time. As soon as she was out of sight, the author's associate printed out his colleague's record. It gave his address in New York and indicated how he had paid his fare. He was travelling with a girl. She was going to Chicago. The printout gave her name and telephone number. Anyone trying to elude his wife's hired detective is well advised to travel by an airline that does not use an unprotected computer.

Sometimes files are protected with special coding or programmed 'locks'. As we shall show in Section III, it is fairly easy to protect computer files from all except the most determined professional infiltrator, in which case the data-bank robbery may have to be an inside job. Data-pilferage requires familiarity in great detail with the workings of the system. Most installations are very short on programmers and analysts, and it would usually not be difficult to place a well-trained agent in a job where he could find everything he wanted to know. In some cases this would be worthwhile.

Programmers are renowned for their high job mobility, and could well carry with them not merely memories of, but written notes and programs from, many systems. Consider the following story from *The Times*

BOAC COMPUTER SECRETS MISSING
Inquiry Among Employees
by our Science Reporter

British Overseas Airways Corporation are investigating the circumstances in which valuable technical information is alleged to have been expropriated by employees for outside use.... The technical material involved ... [amounts] to several volumes that have taken many man-years to compile.

Although much of the information is relevant to airline work, the bulk of the detailed programming schemes can be useful for other computer installations which are intended for advanced schemes, such as feeding computer information over telephone lines to a central machine.[8]

Being armed with such material would make probing into the system easy.

Unless restricted access to a system, for the purpose of protecting privacy, is made both a high-level policy and an integral part of system design, innocent disclosure by conscientious personnel anxious to cooperate is always possible. More culpable, but still short of criminal, are those who browse in the files out of curiosity. In some companies, programmers learn most of their colleagues' salaries from perusing the payroll files, despite attempts to maintain secrecy.

One programmer responsible for writing the payroll program of his corporation is reputed to have been fired for failing to document his program so that others could comprehend it should it need subsequent altering. The first time the job was run following his dismissal, the program kept hunting in the tape files and failed to print out anything at all. After a half-hour the personnel manager and the operator called the Data Processing manager, who was normally forbidden to see the payroll, and asked him to find out what was wrong. He 'dumped' the core, writing onto the printer the exact contents of the memory cells in the central processor, and then with details of the program loop recorded from the lights on the console of the computer, began to analyse what had happened. Toward dawn, he discovered that the program always checked that the programmer's own payroll number was on the tape before printing anyone else's!

Programmers can easily conceal, in complex programs, 'trapdoors' through which to 'pass' at a later date. For instance, the checking of a password could be so written that a master password known to the programmer alone will always work. Such a device may once have been essential for program-testing and never rendered inoperative. Also, many programmers like to demonstrate tricks at the terminals of a system they have worked on.

Many installations frequently run test jobs mixed in with all their other work. With on-line data banks, and many jobs going through the computer together, not even the operator can tell if a test run is getting at unauthorized material. A job could very well copy a whole file onto tape for the programmer to take away and recopy at another installation. Nobody need be any the wiser if he brings the original back in time.

Defences against this type of invasion are also possible, as we shall discuss in Chapters 25 and 26. For the present, however, the warning of Paul Baran should be taken to heart:

Assume that not everyone is as honest and trustworthy as ourselves – but *is just as diabolically clever*.[9]

THE COMPUTER KEEPING WATCH

So far, we have been concerned with the threats to privacy stemming from the gathering together of so many facts about an individual, facts made so available by the computer. Machines can equally well be set to monitor particular situations.

The computer is a tireless instrument of surveillance. Furthermore, the computer makes possible a complex form of surveillance. We shall use the word *cyberveillance* to mean surveillance using the techniques of cybernetics. A computer may be made to search constantly for a particular situation; the search may involve complex logic and, when the computer finds the situation, it may take a programmed action – often notifying a human agent.

During testimony before Senator Long's subcommittee on 16 March 1967, Professor Arthur R. Miller made reference to developments in the technology of optical scanning and the possibility of using these devices in connection with mail-cover operations.[10]

Professor Miller said that it was only a matter of how much one was prepared to spend to produce a device capable of reliable highspeed reading of many typefaces. He expressed doubts about developing a device capable of reading handwriting, or of being inserted inside an envelope to read the enclosed letter. There is much talk, however, of sending mail in the future over telecommunication links in machine-readable form. This is already in operation between some points and is the fastest method of sending mail. Transmitted mail of this form can be protected at a cost.

A lot of mischief could be done with today's mail system – even without opening envelopes. A scanner could record the return addresses on all letters sent to a particular addressee in order to determine the 'associates of a known criminal'. As mail delivery becomes more mechanized, such tasks will become easier. The US Post Office's reputation for restraining its curiosity is far from untarnished.

In the future you may decide to *telephone* your friends in the Sing Sing Alumni Association because you suspect the US mail authorities of recording the addresses of your correspondents (for the FBI's national crime-information centre on behalf of the NYSIIS). However, with a telephone-exchange computer capable of many useful services – such as listing your callers in your absence so that you can call them back, or directing calls to the telephone of a friend with whom you are visiting – cyberveillance is just moonlighting. It would be easy to record the numbers of all callers of a particular party or, for that matter, automatically to make recordings of all calls involving that party.

In all detective work or surveillance the process of sound-recording or photographing *everything* involving a given party requires a vast amount of human work in scanning the results. If *computers* could scan the results, then detective or security agencies could record or photograph thousands of times more material than they do now. Sooner or later this is going to be possible. It will probably be desirable to design such machines for national security. The computer can then be set to work scanning a bulk of recorded print or sound, searching for particular words or phrases. As the technology of voice-recognition develops, it will become possible to instruct machines to listen for particular words or phrases and to note the points at which they occur. In a serious criminal investigation, computers might well be used to record any telephone conversation from a particular area in which callers used certain words or phrases. When the 'Boston Strangler' was baffling police, for example, they could have monitored all the telephone booths in a given area in this way if such a technique had been available. A totalitarian government might well use such methods if they were practical. We are not sure whether they will become practical in the next fifteen years. It might take longer, but sooner or later we shall be able to make machines 'listen' for anything we tell them to.

Letters and telephone conversations are not the only transactions between people open to such scrutiny. Cyberveillance techniques could be used to keep watch on telegrams, bank-account postings, credit transactions, licences, share purchases, airline travel, hotel bookings, car rentals, and so on. Some official, somewhere, invariably has a good reason for using such information. The police want to know when such-and-such a suspect makes an airline booking, if so-and-so attempts to enter the country or to check in at a hotel in Buffalo. The following Social Security numbers belong to people who have evaded bill payment: watch out for their next appearance. What bank accounts in the Phoenix area showed unusual activity on the week beginning May 13? The more information about our daily lives that is stored in computers, the more easily such questions can be answered automatically.

Cyberveillance is a logical extension of lessening the 'dividend-and-income gap' by automating the tax system. In the future the IRS may well monitor even more of our financial life; they might even monitor certain individuals totally. Many people may even ask the IRS to do this as a service, so as to avoid the nuisance of record-keeping and obtaining an accountant's advice in filling out tax returns.

The only limit to cyberveillance is cost, and cost is dropping all the time.

HOMOGENIZATION

Data banks constitute subtle vehicles for the homogenizing of opinions, tastes, and ways of life. The operation is far short of harassment, and more akin to socializing. William H. Whyte, Jr, describes in *The Organization Man* how the suburban 'social ethic' forces changes in the behaviour of suburbanites until they conform to the group norm of a desirable 'outgoing life'.[11] There has been an amazing degree of homogenization within suburban social groups in the United States in response to pressures for conformity. Managers are familiar with the tendency of workers to maximize their superiors' *measure* of their merit, rather than maximize the merit itself. If output is equated with attendance, then employees will put in the hours, without necessarily working efficiently. If wearing a white shirt is considered a virtue, then almost everyone will wear a white shirt.

In like manner, the data bank will provide measures for decision-making. If we are unable to control the measure itself, we shall so arrange our lives that we cause the best measure to be recorded. We already devote much effort to arranging the measures of our income to reflect the minimum legal tax liability.

In an economy based on credit, as in the United States, the maintenance of a good credit rating becomes a matter of great importance in the society. The computers must give a good report when anyone checks our credit rating. When an individual needs to borrow money, acquire a house, obtain goods on credit, or join a club, he will no longer be judged by his family name, his accent, or what school he went to, but by his computerized credit rating. If he has a good rating, he will be considered respectable. This societal pressure has already led to changes in behaviour intended to conform to specified 'norms' – those which enhance one's credit rating.

Such behaviour patterns may be no more objectionable than wearing a morning suit to a wedding or in England changing one's Birmingham accent to that of the BBC. Nevertheless, surveillance leads to conformity, and regimentation to drabness in the quality of our lives. An outsider would say that life in the highly conforming suburbs is a bore. It would be unfortunate if the computer, which is a device capable of promoting immense versatility and individualization, were in fact used in such a way as to stamp wholesale uniformity on our way of life.

INTOLERANCE

In the past, nations and organizations have usually sought to confine the freedom of individual members by offering them a limited set of 'socially desirable' choices. Churches have insisted on the acceptance of definite beliefs, democracies have outlawed certain political opinions, and pressure groups have lobbied successfully for the national application of their parochial tenets. All societies are intolerant to a degree, and yet their improvement ironically depends on those who are the objects of their intolerance, just as change in the law results from a more refined appreciation of justice. A national data bank would provide the intolerant society with a mechanism for institutionalizing its intolerance.

Some of the data stored will be descriptive of the public face each of us reveals to the world: name, origin, profession, education, and physical description. Some data will be less public, yet hardly private, such as age, maiden name, former spouse of a divorcee, job, and hobbies. Certain banks may contain very private information ranging from results of medical examinations to details of income by way of convictions, paternity suits, and alimony. Finally, there may be records of personal details of which even we are unaware, including lie-detector readings and results of aptitude tests and 'multiphasic personality inventories' drawing conclusions from our attitudes on everything from sex to God.

Those in possession of such intimate intelligence will possibly run the gamut from gossip columnists and blackmailers through blacklisters and private eyes to reputable security officials, civil servants, and sociologists zealously in pursuit of the higher good as they see it.

One irony is that those who benefit least from the collected data will often be those about whom most is collected. Criminals will try to avoid data-collection. The wealthy who are to be taxed and the poor who require, in society's interest, welfare and 'guidance', are and will be highly documented. Under some circumstances an affluent society may come to consider poverty as cogent a justification for harassment as a communist society regards wealth.

The way in which a government uses its data ultimately determines the desirability of making it available. Like other of this century's more potent inventions, data banks could be a force for either great good or great evil depending upon how they are used.

Administrative systems generally tend to exaggerate the prejudices of governments, just as representative governments apply positive feedback to current concerns and social preconceptions of the electorate.

Senator Joseph McCarthy would no doubt have raised electronic witch-hunting to a fine art. The Gallagher subcommittee heard testimony concerning the use of personality-testing to detect latent homosexuality in government applicants, thereby reflecting the hiring department's estimate of the nation's intolerance of minorities. Hitler would have used computers to identify Jewish ancestry with more ruthless efficiency than he used his bureaucrats for the same purpose:

> The detailed European census, long in effect even before the advent of the Nazi party, provided a most convenient and efficient tool for Hitler's use when he led the party to control Germany. The census information provided a central data system from which the dictator could draw detailed information on any German citizen, thereby facilitating the power surge of his totalitarian regime.[12]

The superb efficiency of future automation will facilitate the exercise of intolerance – either overtly by governments or as a by-product of other well-meaning social forces. The automation of intolerance is an alarming prospect. We suspect that as the power of the machines increases and the cost of data banks decreases, we shall, whether we like it or not, lose some degree of our present privacy concerning personal information. The forces bringing the data banks into being are greater than the forces seeking to prevent them. If this is so, and if the means to automate intolerance do exist then we believe that the computerized society will have to become a far more *tolerant* society than most societies of the past if true freedom is to survive.

REFERENCES

1. Vance Packard, *The Naked Society*, Penguin, 1966, pp. 237–8.
2. 'The Computer and Invasion of Privacy', p. 33.
3. 'Center for Data on Everybody Recommended', *Washington Post*, 13 June 1966.
4. 'The Computer and Invasion of Privacy', p. 75.
5. 'Special Inquiry on Invasion of Privacy', p. 5.
6. 'The Computer and Invasion of Privacy', p. 37.
7. ibid., p. 163.
8. *The Times*, 26 April, 1968, p. 2.
9. 'Computer Privacy', p. 84.

10. ibid., p. 98.

11. William H. Whyte, Jr, *The Organization Man*, Penguin, 1963, Chapter 26.

12. 'The Computer and Invasion of Privacy', p. 312.

16

MACHINES IN ERROR

Power Failures Snarls Northeast; 800,000 are Caught in Subways Here, Autos Tied Up, City Gropes in Dark – *New York Times headline, 10 November 1965*.

It is extremely rare for well-designed computer hardware to make an error. The logic circuits perform *exactly* the same operation on the data every time they process them. They cannot, like the human brain, deviate now and then, whimsically and irrationally, from their sequence of operations. Circuits do, of course, fail occasionally. Also, it is possible for a strong electrical impulse to change data, although this happens rarely. The machine, however, is designed to check its own operation constantly. The data in the machine are coded with extra ('redundant') 'bits' for error-detection purposes. Using these, the computer will detect any 'bit' of data that is lost or changed in error because of some failure, as soon as the data are moved or manipulated within the machine. A high-speed computer performs millions of checks per second on its own accuracy. Consequently, the same complex calculation could be performed day after day, year after year, without the slightest difference in results.

Nevertheless, the press periodically runs spectacular and often comic stories of computers making monumental errors.

The *Wall Street Journal*, for example, not too long ago ran an article, headlined 'Errant Computer Throws Wholesaler's Business Into Turmoil', about a wholesale grocer who stated in court that his computer 'almost put us out of business . . . When a customer ordered a case of cereal, he might get a whole roomful. And the bill for it might make him think he had bought a shipload.' A customer testified that the computer sent him so many items that he had nowhere to store them. Bills were sent out with wrong prices – for example, one bill charged

$200 for a $13 case of sanitary napkins. 'In some cases,' the grocer claimed, 'it refused to charge at all. The item was shipped, for all intents and purposes, free of charge to the customer.' As for stock-control, warehouse workers were likely to be informed that there were no canned peas, when in fact they were piled up to the ceiling.[1]

Bernard Levin described in the *Daily Mail* how he received 'in the natural course of computerized living' a gas bill 1800 per cent too high. He complained to the Gas Board and an official eventually came round to check, admitting that their computer was at fault. The official was carrying 'some hundreds' of cards, each one referring to a similar error.[2]

The consumer is not always the loser when computers make mistakes: the gas bills of an acquaintance of one of the authors in Canada arrived in the form of credits rather than debits for some months; the computer, it seems, had been working out how much he owed and then *adding* this to his account rather than subtracting it. The authorities, perhaps reluctant to admit the error, permitted him to keep the money in his account. He joyously claimed that it was rather like winning a national lottery – the only trouble was that he could spend the money only on more gas!

The *Daily Telegraph* described how children in Birmingham who had failed their 11-plus examination were nevertheless offered places in the schools they desired to enter by an errant computer. The Birmingham authorities agreed to honour the offer.[3]

Korvette's in New York City advertised a carpet sale with the headline, 'Computer Runs Wild! Huge Overstock of Luxury Carpets'.

Perhaps less beneficial, a computer producing labels for magazine-mailing printed the same address on each, with the result that several truckloads of that issue of the magazine were delivered to the house of one startled subscriber.

SYSTEMS CONTROLS

Almost certainly, none of the errors described above was caused by computer *hardware*; they were caused in some way by the human personnel associated with the machines. They may have been the result of programming errors, operating errors, or of wrong data fed to the machine.

One thing is clear, however: when computer systems do cause errors they can cause spectacular ones – sometimes more spectacular than errors likely to be generated by humans. The unsupervised computer system lacks the *sense of the absurd* possessed by even the humblest clerk.

A group of clerks adding up figures in ledgers, maintaining customer accounts, or updating stock-inventory sheets is going to make *far* more mistakes than a correctly designed computer system tackling the same work. But a clerk will know that he is personally associated with the results he obtains. If he obtains a ridiculous answer, he will definitely notice it. A computer error may cause quite ridiculous results and the computer may still grind on moronically.

Where clerks are updating ledgers, it is necessary to have some control over their possible errors. Schemes have been devised for balancing accounts so that most errors will be found. In a computer system it is also necessary to have such controls. It is possible to devise schemes for detecting an error before it affects the world beyond the computer room. In Section III we shall have more to say about the types of controls required. Let us here merely note that safeguards *can* be devised, and that when a computer system goes berserk – as in the examples above – and plays havoc with the public, it is because the *systems analysts and programmers did not do their jobs well enough.* Either the programs were not sufficiently tested when the system took over, or the systems controls – the design of which is a standard part of professional computer practice – were not implemented as they should have been, or both.

SOURCES OF ERRORS

Figure 16.1 illustrates the possible sources of computer-system errors, discussed below:

1. *Hardware Errors*

As we noted at the start of this chapter, hardware errors are rare in today's machines. Computers contain circuits which detect hardware malfunctions. When a malfunction occurs during a particular operation, the computer may attempt that operation a second time. Often it then works correctly; many errors occur only once and not repetitively. If the machine has a repetitive ('solid') error which it cannot circumvent, it will stop and notify its operator with a bell or flashing red light or other means. On some systems an automatic switchover will transfer the work to an alternative machine. If no alternative machine is available, processing may be held up for an hour or more while an engineer repairs the fault.

The important point is that any error in data caused by a hardware malfunction does not escape undetected. The system may fail to process the data for the time being, but it is designed in such a way that it does not normally put out wrong results.

There is a minute chance that an error will be of such a nature that it can slip through the error-detection mechanism. Computers are designed so as to make this a *very* improbable circumstance. No error-detection system is infallible, although it can be built to be so nearly infallible that it is not worth worrying about. The chances of an error passing undetected can be decreased almost to zero by increasing the elaboration (and hence the cost) of the error-detection circuitry. It is likely that computers in the future will become increasingly error-proof as the cost of logic circuitry drops and computer technology itself increases in reliability.

Probably the most error-prone hardware components of the systems discussed in this book are the telecommunication lines. On the telephone we can sometimes hear stray 'clicks' and 'crackles'. This and other subliminal 'noise' does damage to

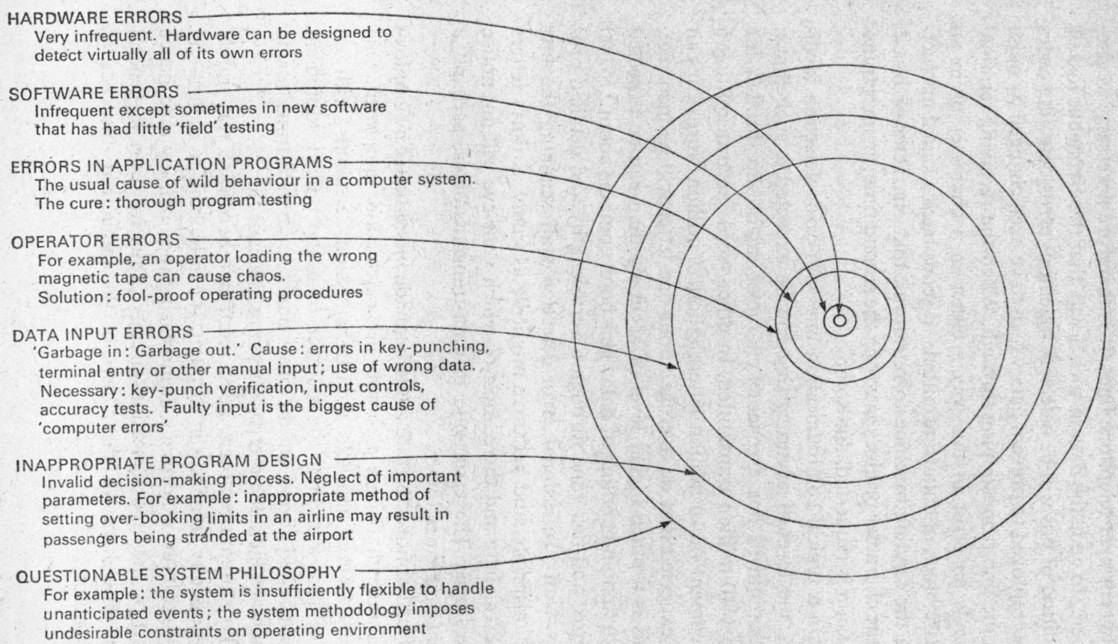

HARDWARE ERRORS
Very infrequent. Hardware can be designed to detect virtually all of its own errors

SOFTWARE ERRORS
Infrequent except sometimes in new software that has had little 'field' testing

ERRORS IN APPLICATION PROGRAMS
The usual cause of wild behaviour in a computer system. The cure: thorough program testing

OPERATOR ERRORS
For example, an operator loading the wrong magnetic tape can cause chaos.
Solution: fool-proof operating procedures

DATA INPUT ERRORS
'Garbage in: Garbage out.' Cause: errors in key-punching, terminal entry or other manual input; use of wrong data. Necessary: key-punch verification, input controls, accuracy tests. Faulty input is the biggest cause of 'computer errors'

INAPPROPRIATE PROGRAM DESIGN
Invalid decision-making process. Neglect of important parameters. For example: inappropriate method of setting over-booking limits in an airline may result in passengers being stranded at the airport

QUESTIONABLE SYSTEM PHILOSOPHY
For example: the system is insufficiently flexible to handle unanticipated events; the system methodology imposes undesirable constraints on operating environment

Fig. 16.1 Sources of computer system errors.

data sent over telephone lines. It is therefore necessary, as elsewhere, to code the data in such a way that the erroneous loss or addition of bits will be detected. Ingenious error-detecting codes are employed to prevent data which have been changed by noise on the line from slipping through. When the receiving terminal detects an error in this way – either in a character or in an entire block of data transmitted – it sends back a signal instructing the sending machine to retransmit that character or block. If the character or block is wrong the second time, again retransmission is requested, and so on.

On a typical communication line with error-detection facilities, the level of accuracy that can be expected is such that a maximum of one erroneous character might slip through for every 10 million transmitted. In other words, in transmitting a document of 10 million letters (about 2 million words), it can be expected that on average about one letter will be incorrect. This is a much higher level of accuracy than can be expected with human handling of data. This book contains about 275,000 words, and has gone through a very lengthy and detailed process with professional error-detection staff scanning the text for misprints, and yet there might be as many as thirty letters in error here and there, most of which your eye will not notice (we hope). The error rate in data-transmission is, perhaps, a thousand times lower.

Furthermore, the rate could be made *much* better by employing somewhat more elaborate error-detection equipment. One manufacturer markets data-transmission terminals with an error rate a million times lower than the typical equipment figures cited above. On the other hand, some data-transmission machines include no error-checking mechanism.

To summarize, then, good computer hardware very rarely permits *any* error to slip through. It can be built to almost any desired level of accuracy in increasing the intricacy (and hence cost) of the error-detection circuits. Thus we must in general look elsewhere for the causes of the publicized misdeeds of computers.

2. Software Errors

Closely associated with computer hardware are the programs for supervision of its various functions generally included under the heading of 'software'. The software is normally provided by the manufacturer and is often independent of the application in question. Programs are designed to organize what is going on in the machine – to coordinate and optimize machine functions under varying loads and job mixes. The software handles input and output, the ordering and scheduling of work, it also deals with error or emergency situations. The software for most computers is becoming very complicated. In addition to the above functions, it provides 'languages' (FORTRAN, COBOL, ALGOL, and others) for the user, and translates what he writes in these languages into the machine's language.

When the software is initially produced it sometimes contains subtle undetected errors which can occasionally cause the machine to function incorrectly. Usually a software malfunction is apparent to the operator: it may stop the machine or cause a job to be rejected, although it normally does not damage the data being processed. After the software has been in use for some months, any 'bugs' in it should have been caught. The software, like the hardware, is generally not the cause of machines making mistakes.

3. Application-Program Errors

A computer does exactly and precisely what its user instructs it to do, by means of his programs. If he has made mistakes in his programs, then the computer will likewise do things wrong. The programs for most applications are written by the particular firm that owns or leases the machine. The complexity of these programs becomes very great for some applications, commonly taking a team of men to produce them. They are often written in a hurry to meet tight deadlines when machines are scheduled to begin operations.

Most programmers make some mistakes. They may fail to

see all of the intricacies in the complex logic they are creating. Running their programs on a computer in order to test them reveals the flaws. They then put their errors right one by one until the program works correctly – giving the right results under all circumstances. Programs written by different members of a team are hooked together and tested in unison until the entire system is performing correctly.

However, the team possibly has not considered every circumstance that might arise in the data, and an eventuality not tested for can cause errors. The story is told of a certain city's officials visiting the house of a lady 107 years old and asking to see her mother. A computer had been programmed in such a way that it could not handle a three-digit age, and thus sliced off the first digit. Using the age '07', the computer determined that the female in question was not going to school and printed out a truant notice! Most program errors that remain after testing are more subtle than this. On complex systems which handle more than one transaction at a time ('multi-programmed systems') – especially real-time ones – the error may involve the relative timing of two events; this type of error can be very difficult to repeat exactly. Tracking down the error, or even knowing when it is going to occur, is then more difficult.

It is likely that the trouble-source in all the stories of errant computers at the beginning of this chapter was one or another error in an 'application program'. When a computer does something *spectacularly* stupid, the cause is usually a programming error.

This sort of chaos can be avoided by insuring that programs are *very* thoroughly tested before they are put to work in the real world. The type of program that might be subject to non-repeatable 'timing' errors, requires a lengthy period of 'saturation-testing'[4] designed to produce the maximum number of potential error conditions. The system should be hammered with all varieties of test data to insure that programs behave properly before they are let loose on the public. Sometimes, unfortunately, this is not done as well as it should be; the computer consequently acquires an undeservedly bad reputation.

4. *Operator Errors*

A story in the *Daily Mail* described a collection of computer errors including a telephone bill on which the charge for the same call had been entered three times.[5] This was probably caused by an operator error.

The operator of a large computer system has a highly responsible job. It is frightening to reflect on how much damage an operator *could* do – files of vital data accidentally erased, financial records mistakenly updated twice with the same data, wrong tapes or programs used, and so on. Computer operators are generally difficult to recruit and quick to leave a position, a situation that enhances the possibility of errors.

It is desirable for systems analysts to devise controls which guard against the sorts of mistakes that operators are likely to make. The computer program can be written so that it will not read or update a file of data until it has checked that it is working on the correct one – each file having a label giving details which the machine can read. Again, the program can total certain fields on the transactions it processes and compare the result with an expected total to insure that no card has been lost. The program can be designed so that it will not run until the operator has carried out certain checking actions.

Installations with well-planned controls of the types described above generally do not suffer from damage caused by operator errors. Where such controls have not been applied, however, results have sometimes been chaotic; moreover, once data are lost or wrongly updated on a system without controls, sorting out the trouble can be difficult.

5. *Data-Input Errors*

If a system has well-designed hardware, software, and operator controls, and the programs have been thoroughly tested, it stands a good chance of keeping its name out of those newspaper columns which delight in reporting computer bloomers. The results it produces, however, are still going to be dependent upon the correctness of the input. Members of the computer industry have an acronym, GIGO: 'garbage in, garbage out'.

Input may reach the computer from punched cards, from cheques printed with magnetic characters, or from a variety of other documents. It may also come from on-line–terminal devices. When punching cards or operating terminals, a worker inevitably makes a small number of mistakes. In the newspaper article mentioned above, a man who was charged £32.6s. for a single telephone call protested, and it was found that the charge should have been 32/6d. This was probably caused by a punched-card operator's error. Similar circumstances often explain items appearing on the wrong bill – the girl probably punched the wrong account number.

The first precaution that is normally taken with keypunching is to arrange to have the punching *verified* by a second girl keying the same data into a verifier machine. If her keying corresponds exactly to the card's, the verifier punches a notch in the card; otherwise a red light flashes. Thus only notched cards are fed into the computer. Where this procedure is followed, errors such as the ones described above are unlikely.

A variety of other types of controls over the accuracy of input are possible, some of which are discussed in Chapter 26. Account numbers, for example, can be made self-checking, by mathematically deriving the last digit or digits from the others; if any digit is incorrectly punched, the computer detects the error, although certain transposition errors could still slip through.

As we have noted, errors are less likely to go undetected on a well-planned computer system than with its clerical predecessor. When Rolls-Royce installed terminals on the shop floor of their factory at Derby to collect shop data, they reported that error rates in shop facts dropped from 2.5 per cent with their manual system to something below 0.5 per cent.[6] The experience of Rolls-Royce has been repeated in factories of many other organizations.

Where a real-time system is in use, the computer can check data – as they are entered – against known facts. When a worker keys in information about the status of the job he is working on, the machine checks that that worker *should* have been on that job at this stage, and that the status report is as expected; if not, the machine may notify a supervisor, who checks what is happening and sees whether an error has been made. In this way much tighter control of shop information has been achieved than was possible before the computer was used.

Nevertheless, we must expect inaccuracies to appear in the input given to a computer, and these will be reflected in output inaccuracies. Good controls, once again, can lessen the problem.

6. *Inappropriate Program Design*

A happily-working bug-free program is still not necessarily devoid of problems. The method it is using may be inappropriate to the situation in question. A program for stock control, for example, commonly decides when to reorder stock and what quantities to order. It uses equations for determining these, employing such parameters for the item in question as stock turnover, delivery time, orders in hand, and sales forecast. However, the equations may have been poorly chosen. They may neglect certain parameters, or produce results which do not minimize cost or which run too high a risk of a stock shortage.

An inappropriate choice of equations may simply fail to maximize profitability; in some cases, however, a poor choice of equations could be more serious. On an international airline, for example, many cancellations of reserved seats occur in the few days before the flight. In many cases seats are sold two or three times because of such cancellations. The airline may therefore allow a limited amount of overbooking in an attempt to fill as many seats as possible. The amount of overbooking permitted decreases as the take-off date nears. If, in this situation, a reservation computer uses inappropriate equations for setting limits on overbooking, the airline runs the risk of stranding passengers at the airport.

This type of problem is somewhat more difficult to solve than the earlier ones, and requires a very elaborate study of the situation in question. Having determined the correct equations – or the right logic – however, the problem is easily put right.

7. *Questionable System Philosophy*

The airline-reservations situation in the preceding section could be put right by modifying a small element of the decision-making process. The system may have been fundamentally sound, but needed 'tuning' or adjusting to the environment it worked with. This is often the case when a system makes elaborate decisions or takes control actions – based for example on probability calculations of some form.

A potentially worse situation is one in which the overall philosophy of the system is questionable. The system may, for example, be insufficiently flexible to handle unanticipated events properly. This has been the case in some systems for factory production-control: the computer laid down a schedule which became difficult to maintain when unforseen circumstances arose, such as a batch of material being abandoned, a foreman deciding to give an apprentice certain types of work not on the schedule, or a panic occurring when everyone rushed to complete a particular customer's order. Shop-floor personnel may consequently begin ignoring the computer's work-schedule and start making their own decisions. When one part of the schedule is ignored, other parts are usually invalidated. The shop personnel thus lose confidence in the computer, and before long the machine's orders are being disobeyed completely.

Here, the entire scheme is at fault, and no amount of tuning would make it sufficiently adaptable. Instead, a scheme is needed which either permits shop foremen a measure of decision-making responsibility, or else deals in turn with each new circumstance on the shop floor as it occurs, rescheduling if necessary. The latter procedure requires real-time collection of data from the shop floor and a computer which re-schedules the work every time its current schedule is overridden by unforeseeable events.

Again, in a system which gathers statistics, produces summaries, or uses statistics to influence events, one would like guarantees that the final figures do indeed represent the facts. This, however, can be more difficult than it may appear at first sight. The method used for collecting or condensing the information will often hide or distort facts. There is a variety of such *methodological distortions*. Data easily collected may be incomplete, or may reflect a bias inherent in the method of collection. Decision-making based upon data so collected reflects this bias.

The system methodology may impose unreasonable constraints upon the type of decisions that can be made or the type of functions that can be performed. Furthermore, where a system is sufficiently complex, it is extremely difficult to change its methodology. An airline-reservations system which handles passenger details, for example, is so complex that the airline will think hard before embarking upon a major change in its operating methods. Production-control systems of the future are probably going to be equally difficult to change. One occasionally hears reports of a top-management meeting being told, 'We can't do that – the computer won't handle it.' At first the idea that a computer can thus dictate what can and cannot be done seems outrageous. However, the strictures of computer methodology will probably force just such compromises in the art of managing machines.

REFERENCES

1. *Wall Street Journal*, 27 March 1968.
2. Bernard Levin's column in the *Daily Mail*, 18 March 1968.
3. 'Computer Mixes Up 11-Plus Results', *Daily Telegraph*, 1 August 1967.
4. James Martin, *Programming Real-Time Computer Systems*, Prentice-Hall, Englewood Cliffs, New Jersey, 1965. (See Section V.)
5. 'Computer Crazy', *Daily Mail*, 11 April 1968.
6. 'Rolls-Royce Speeds the Data', *Financial Times*, 15 June 1967.

17

HARASSMENT

Rex Reed, writer and sometime actor, ordered a bed from a
Manhattan department store. Three months passed. Then came
the long anticipated announcement: the bed will be delivered on
Friday. Reed waited all day. No bed. Having disposed of his other
bed, he slept on the floor. Next day deliverers brought the bed
but could not put it up. No screws. On Monday, men appeared
with the screws. But they could not put in the mattresses. No slats.
'That's not our department.' Reed hired a carpenter to build them;
the department store's slats finally arrived 15 weeks later.
Undaunted, Reed went to the store to buy sheets. Two men came
up and declared: 'You're under arrest.' Why? 'You're using a stolen
credit card. Rex Reed is dead.' Great confusion. Reed flashed all
his identity cards, the detectives apologized – and then tore up his
store charge card. Why? 'Our computer has been told that you are
dead. And we cannot change this.' – *Story in Time Magazine,
23 March 1970.*

It seems certain that in the future we are going to live out our
lives against a background of highly informed and intercom-
municating computers. For the decades ahead the data banks
will continue to grow and multiply at a staggering rate (see
Figs. 1.4 and 1.7). Computer terminals will be everywhere and
gaining access to the appropriate machine for a particular function will be almost as easy as dialling on the telephone today.
The American Congress might slow down the establishment
of a national data centre, but it will not stop the rolling tide
of automation.

Set against this background, our lives will probably be richer.
We shall probably have more leisure time, fewer routine chores,
perhaps fewer forms to fill in, greater affluence, more intellectual pastimes, perhaps less chance of being bored, and more
opportunity for creativity. A man with the intellectual capabilities demanded by the new age will have a greater chance of

achieving something worthwhile. But we shall always be aware of the machines. They will be part of society: their knowledge of us, of our telephone bills and medical histories, our educational attainments and consumer habits, and so on. Probably more of our mail will come from computers than from people – computers trying to sell us things, sending us bills, asking for money for charities, naming ten of our neighbours who take *Readers' Digest* when we do not, notifying us of bank debits for their services, and so on.

To what extent could this constant electronic activity constitute harassment? To what extent will it intrude upon our desire to be left alone? To what extent will undesirable by-products of automation become a nuisance? This depends very much upon what regulators we apply to the systems – we shall discuss these in Section III. In this chapter, let us talk about the potential for harassment.

MISTAKES

First of all, as we saw in the last chapter, computers can represent a decided harassment when they go wrong. Figure 17.1 shows a notice sent by a typical American electric-power company: a punched card mailed by a computer. Several punched-card bills and warnings had been sent before it; however, the computer, probably because of a keypunch girl's error, had sent these to the wrong address, and so none of them had reached their destination. The computer is programmed to send a 'disconnect' notice if a customer's bill is not paid after a given time. However, this notice was also mailed to the wrong address and so no remedial action was taken and the electricity was cut off. The customer's household is without electricity for at least 24 hours, which could be serious. It could certainly be very unpleasant if someone was ill in bed.

The main point here is that the computer is permitted to take an extreme action – the issuing of discontinuation of service – without human intervention. The man who cuts off the electricity then takes his orders blindly from the computer. This perhaps would be all right if the system were infallible, but the fact that the computer can be given the wrong customer address clearly shows that it is not. A simple human mistake, instead of being caught in the common-sense course of manual operations, is here capable of being amplified by the mechanism until it reaches harassing proportions.

A colleague of the authors in Germany once showed them a

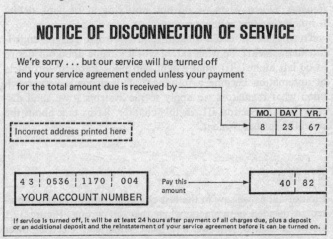

Fig. 17.1 Harassment. The computer, having mailed the two previous electricity bills to the wrong address, now threatens to cut off the electricity regardless of the damage this might cause, for example, if somebody is sick in bed. Note the threat that the electricity will not be turned on again for 24 hours after payment of all charges due plus a deposit. The household may be several days without electricity. Unfortunately, this notice is also mailed to the wrong address.

copy of a bill he had received from a computer requesting the payment of 'zero deutschmark'. He ignored it, but two weeks later the machine sent him a letter reminding him that he had not paid the sum of 'zero deutschmark'. Two weeks after that another and more strongly worded letter arrived. He still took no action other than photocopying the letters and gleefully showing them to his friends. However, the computer persisted and eventually announced that it was referring his failure to pay to company lawyers. So he telephoned the company. They explained to him that there was a minor oversight in the program, assured him that it was being put right forthwith, but requested him to send a cheque for 'zero deutschmark' to simplify the reconciliation. He duly made out a cheque for '0.0 DM.' and mailed it. Two days later the cheque was returned to him from the bank with a polite (non-automated) letter stating that the bank's computer was unable to process such a cheque!

The mailing of the 'zero deutschmark' letters was due to a programming inadequacy soon corrected. Often, however, similar annoyances are inherent in the system and are not put right. Usually the least expensive way to design a system is to design it so that mistakes *can* occasionally happen. Perfection is expensive. The cost of a telephone exchange, computerized or otherwise, is less if one permits it occasionally to miss or misinterpret a dial-pulse. The psychology of the average person dialling is such that he will normally think *he* has misdialled. He will dial again and the next time obtain his connection – hopefully. Similarly, the cost of a computerized invoicing system is less if there is no human intervention. If we design for minimum cost, we often run the risk of permitting the computer to become a nuisance. A public utility usually has a local monopoly and so can take a more casual attitude toward annoyance to customers than can a company sensitive to competition. The public should understand that this type of harassment could be avoided at slightly higher cost – there *is* a choice.

PESTERING

The second potential source of computer annoyance is the excessive or overaggressive use of computers to sell, advertise, beg, or solicit. There is an increasing market for computer tapes containing lists of names and addresses of particular categories of potential customers. These come from a variety of different sources. They will no doubt become much more detailed, more numerous, and much less expensive as the information collected by various types of data banks grows. Equipped with these, the marketing executive's computer can mail a wide variety of different letters and advertisements to possible buyers. In one insidiously successful campaign the machine printed sales letters for a service listing the names of a person's neighbours who were taking that service – automation of keeping up with the Joneses. As yet, few organizations have realized the full potential of computerized marketing; when they do, we may be in for a blitz.

Perhaps pestering with paper is not too bad. One of the authors (who lives in the USA) does not mind throwing away half of his mail. There is a large bin by the mailbox. However, pestering by telephone can be much more annoying. Already some advertisers are using computer listings to telephone possible customers. In some parts of the United States the number of such telephone calls one receives is becoming a definite nuisance. The possibility exists for the future, however, of the machine itself doing the dialling and speaking. A relatively small computer with a voice-answerback unit could call up and speak differently to many people at the same time. It could ask them to respond by using the telephone dial. With a touch-tone keyboard, quite complex responses could be elicited. In the not-too-distant future the machine will be able to interpret certain voice words to a limited extent, and will probably record any that it cannot understand for human checking. This technique could have many beneficial applications if it becomes economical for many marketing organizations to adopt it; however, it could become a plague. The machines have endless capacity and endless patience, and might be programmed to carry on a relentless and highly profitable campaign. How would we control this technique? And how could we differentiate between socially valuable uses of computers on the telephone and mere marketing? Should we permit political parties to solicit campaign support or to poll the views of the electorate with this technique? Should we permit a senator to recruit support on a vital issue by instructing a computer to telephone all his constituents?

CYBERVEILLANCE

Perhaps it will be easy to prevent pestering by machine (although in a society dominated by corporate profits we have doubts). However, there are other forms of harassment of a less obvious nature which we may be induced to accept *for the good of society*.

We are striving with data-processing to run our society more efficiently than it has been run in the past. That is the main purpose of the vast government data banks, econometric models, and so on. The desire to run society efficiently, however, is at some points in conflict with the wishes of individuals in that society to live a life that is free, private, and unharassed.

Perhaps one of our most haunting fears is that of constant supervision. We have used the word 'cyberveillance' to mean constant surveillance using the methods of cybernetics – computers monitoring a situation, processing the data they collect, and taking action when they detect particular circumstances. Once set to work, a computer can keep watch over a situation with infinite patience. There are all manner of situations in our daily lives where cyberveillance might be used to the supposed advantage of society.

Consider the following report in the *Wall Street Journal* concerning the black market for money in South Vietnam:

US soldiers and others who work for the Government or for Government contractors here are paid in MPC (military payment certificates), a scrip convertible by Americans into dollars. MPCs are difficult, if not impossible, for non-American civilians to convert or spend and they have little intrinsic value to blackmarket money-changers. To make a black-market profit with MPCs an American must convert them into regular US currency – or, as generally is the case – into some negotiable dollar instrument like travelers' checks or postal money orders. This can be done legally here, with a dollar of scrip worth a real US dollar. The American then takes his negotiable instrument to a black-market money-changer.

The cash or negotiable instrument is worth about 150 piasters to the dollar on the black market. In official dealings, the dollar is worth 118 piasters.

The Computer Is Watching

The special commission hopes to wipe out the bulk of this illegal activity with a computer scheduled to arrive here shortly. The salary level and MPC bank-account balances of all Americans will be programmed on the computer, and all exchanges of MPCs for negotiable instruments will be fed into the machine. *Any account showing unusual activity will be noted by the computer* [emphasis supplied], and these accounts will receive special audits. Transactions or earnings that can't be documented will be investigated.[1]

This procedure was no doubt highly desirable in South Vietnam, but in civilian life also there are a variety of areas in which it has been suggested that the machines keep watch over our activities. Here it is more difficult to assess what is good for society and what is an ominous step closer to George Orwell's Big Brother. Cyberveillance to spot tax-evasion, on the face of it, sounds like a good idea. Programs in computerized telephone exchanges for tracing malicious anonymous callers (a system currently in use in Sweden) are highly commendable. Computerized surveillance to help crime-detection appears to be potentially valuable. But how far should this be taken? Most citizens would react indignantly to the proposal that a computer monitor their bank account or credit transactions for unusual activity. How would we feel if we knew that a computer were analysing all the telephone numbers we dial as an aid to national security, crime-detection, or for some obscure bureaucratic motive?

So many different facts will be recorded about us with the uses of computers envisioned in Section I of this book, and there are so many diverse reasons for analysing these facts in some form of cyberveillance. Marketing executives want to spot potential customers. Police want to spot potential criminals. Insurance companies want to detect bad risks. Cities want to spot habitual parking offenders. Politicians want to locate potential supporters. Corporations want to spot potential managers. Shops and credit companies want to spot credit risks.

The use of credit is likely to leave a spectacular trail of information for the computers. In a chequeless society, with a great reduction in the use of cash, many of the things we do will be recorded in the form of credit transactions. Buying cigarettes as we leave home, buying petrol and lunch, paying for a hotel room and drinks in the evening, buying presents – all of these will be recorded because we use our credit cards. And someone will have good reasons for analysing it, if he is allowed to. People who want to sell us things particularly want to know of our spending habits and will pay for listings of consumers in differing categories.

Conducting this type of surveillance has been out of the question in the past because of the overwhelming amount of work involved. But in the future the work will be cheap and, once the programs are set up, effortless.

Somewhere in different machines an enormous amount of information about us will be available – financial, educational, medical, business, legal, aptitudes, magazine subscriptions, library usage, and so on. Most of it will be collected without anybody obviously prying into our affairs – a by-product of other mechanisms in society. The extent to which the machines or agencies can intercommunicate to bring the facts together will determine how complete a picture of an individual's existence can be constructed and analysed.

If all the records concerning an individual are associated with a unique 'identifier' – in the United States this would be his Social Security number – then bringing the data together will be easy to accomplish if there is no law to prevent so doing. Even without the unique identifier, it is not difficult for a computer system to build up cross-reference tables linking up with one individual all the numbers with which he is associated. To concentrate on avoiding a unique identifier would be grossly to underestimate computer talents; even today surveillance techniques are no different from those adopted by a manufacturer for watching his product line for items that sell well. Again we find, as we shall continue to stress, that it is in the application of a technique – not in the technique itself – that the danger lies. What we might accept in business and would commend in the fight against crime is elsewhere an impertinent intrusion in our private lives.

FORGIVE AND FORGET

The great fear of many is the capacity of a computer system not merely to remember infallibly, but to trot out, at the most inconvenient times, our past mistakes, mistresses or misdemeanours.

As Professor Charles A. Reich told the Gallagher subcommittee:

> We have already had mention this morning of the great American idea of 'beginning again,' starting anew, getting a second chance, and that is something we would lose by this. We would have a situation in which nobody got a second chance, no matter how young, no matter how foolish, no matter how easily explained the circumstances; we would establish a doctrine of no second chance, no forgiveness.
>
> One life, one chance only. That seems to me very different from the American dream.[2]

Earlier in his testimony that day, Professor Reich showed how this situation could arise and subsequently lead to condemnation without trial:

> Here are people who are not even charged with crime, and yet who may be punished far more severely than the ordinary criminal. Here are people, whose opportunity to have jobs, to earn money, whose reputations and everything else are about to be damaged forever, and they have no trial, no lawyer, no opportunity to find out anything. It seems to me without question a denial of due process of law to send forth bad information about a person in secret that way. It is in this that I see the essence of the evil of the automatic data center. It is in this notion of the petrification: that is, this man is called bad by somebody, hence he is bad forever, and there is nothing he can do about it. There is no remedy in the law.[3]

All the rest of his life, wherever he went, he would have a tin can jangling along behind him. The result is branding with a 'scarlet letter'.

Representative Wright Patman not too long ago 'entered a bill that would restrict release of credit information by banks and credit-rating agencies'.[4] He wanted to protect the consumer who could be 'cut off from credit by false information about which he has no knowledge.' The banks' exchange of information about their customers is, in Patman's view, a 'Big Brother scheme with a vengeance'.

Representative Patman, note, is primarily concerned with the possible effects of false information. Many such errors were chronicled in the last chapter. Here we must concern ourselves with a system working to the limits of its (very considerable) capacity for accuracy.

Consumer-credit agencies are coming to play a vital part in all our lives. We shall increasingly rely on them for help in borrowing cash for large purchases and to meet sudden emergencies. To be cut off by a bad rating contracted many years earlier could seriously affect a person's life in a society accustomed to living on next year's income.

These ratings and the data on which they are based are also being used by investigators assessing job applicants in business and government. A man with a bad credit history is sometimes thought to be a security risk rather than an enthusiastic worker determined to pay his debts and mend his ways. Credit Data Corporation has 25 million records of individuals in New York and California and is currently involved in a legal battle to defend its files from the Internal Revenue Service. As their Eastern Region Vice-President told *US News and World Report*:

We are going to court because of principle. We have told our clients and subscribers that the information (only financial facts about people) is confidential.[5]

We doubt that such integrity is commonplace or that the example is generally emulated.

ERRORS OF OMISSION

Although credit data's restriction of its records to credit information is wholly admirable, it has its adverse consequences. It is extremely difficult to tell when a record is incomplete. We dealt with errors of commission in Chapter 16. To the victim, however, errors of omission can be as disastrous.

Millions of items of 'derogatory' information about people are entered into data banks each year. Very many of these are never examined later to determine whether the matter has been settled to the satisfaction of both parties. Vance Packard cites the instance of a man with three lawsuits recorded against him: one was a scare suit thirty years before over a magazine subscription he had never ordered; the second had been withdrawn after a compromise over a disputed fee; the third case was settled in his favour.[6] These dispositions were not reported.

It costs money to insure that such records are correct and up to date, and the only justification for the expenditure is to make more use of the information – that is, to sell it. There are, of course, economies of scale as well. A single mammoth organization merely records all lawsuit results; two bureaus would both have to select and reject the appropriate proportion of the total.

It appears that we shall have to accept either incomplete and unfair records in many banks, or more complete and fairer records in a unified system. In the latter case we could confront the organization if we disliked its report, see the record, and argue its contents. Errors and omissions, once acknowledged and corrected, would then be removed in one sweep. There would be no need to question every one of many adverse decisions from a multiplicity of sources, none of whom had recorded, say, the withdrawal of an unjustified lawsuit.

As we shall see in Chapter 19, carrying an argument with a data-processing system to a satisfactory conclusion will be beyond the resources of many people. Hiring experts to further our interests, whether lawyers, accountants, or computer consultants, will be expensive.

The adverse consequences of the denial of credit are minor compared with the denial of employment or security clearance due to an adverse 'character reference'. Because of its long memory, a dossier bank could well hold details dating back to one's childhood or college days. If records are kept brief in the interests of economy, there might be no distinction made between the draft-card burner and the gunman. Furthermore, if one were, in an attempt to defend his right of privacy, to prevent an inquirer from obtaining more than a 'Yes' or 'No' answer, he would have no opportunity to challenge the reason for rejection, nor would the human operator be able to exercise a proper discretion.

LIBEL AND BLACKMAIL

If a fellow citizen were to say some of the things about us that many computers repeat to an inquirer, we would soon have him in court. It would be hard, however, to demonstrate malicious intent, and still harder to demonstrate the extent of the damage.

The opportunities for blackmail when dossiers are available cannot be limited. Men in public positions may find themselves particularly vulnerable when the wild oats sown in youth are electronically harvested. People with access to the system may be tempted to find out what is recorded of their superiors. Officials of the party in power may take a look at their rivals in the hope of uncovering dirt. Testifying before the 1966 House subcommittee on the computer's threat to privacy, Director of the New York State Information and Intelligence System (NYSIIS), Robert R. J. Gallati, answered Rep. Gallagher as follows, on being asked about a 'routine police check' of a potential appointee:

The files, like criminal files generally, of course, are confidential. There would be in this particular instance you cite, where a public official would be investigated either by the Civil Service Commission, or, in the case of an appointed official, by the State police, they would utilize our files just precisely, in effect, as they now utilize the central identification files.[7]

If there is sufficient reason to suspect potential office-holders of crimes warranting entry in police files, how much more reason do we have to suspect them of malevolent or idle curiosity?

The zealous official and the enthusiastic bureaucrat represent the most likely danger to the computerized society. Neither the politician nor the civil servant is held in such high esteem by the public or by himself that the traditions of his profession will defend the citizens' interests. Badly protected records, therefore, provide a tool of petty tyranny.

We live in a world of licences. New Yorkers pay $1·5 million annually for the privilege of keeping goats, planting trees, dealing in bait, staging masquerade balls, and enjoying other licensable activities.[8] If the application for any one of these were to be processed with reference to the dossier of the applicant, the clerk would learn much that was irrelevant, and a little that could be more than embarrassing. And yet, in some cases such checking would be highly logical. Should a gun licence be issued to a man with a record of armed robbery or a driver's licence to a man with several convictions for dangerous driving?

Equally appealing to both the bureaucrat and the unwary citizen is the ability of the dossier bank to minimize form-filling. The report in *The Federal Paperwork Jungle* estimated that the corporate and private citizens of the United States filled out 2 billion forms each year, and commended the use of income-tax returns to reduce this burden for some 2 million small businesses.[9] Vance Packard castigates 'the harassment of the citizen by unreasonable forms and regulations'.[10] The US Department of Commerce insisted on the filling in of 16 million forms for its 'Household Questionnaire', and received the help of the courts in punishing a recalcitrant objector to such an inquisition.

With a cashless, chequeless society lubricated by a trail of credit transfers, the inquisitors could couple their computers directly to the banking system and provide a plethora of useful statistics directly from the citizenry's day-to-day transactions. If, however, we cannot protest against the compulsion to fill out impertinent forms by ourselves, how much less can we protest against having the computer do it for us?

In fact, unless we are prepared to abdicate totally to Big Brother, we shall have to keep records ourselves to check on those that are kept about us. The increase in leisure time due to automation will be reduced by our need to monitor the system. Many people with automatic minor-overdraft facilities currently do not even check their bank statements or maintain their stubs accurately. Few subscribers to the telephone networks check the accuracy of their statements; those who do frequently disagree with the bill.

Curiosity is the besetting sin of bureaucracy. Not that the official believes what he sees: men without birth certificates have been told that they do not exist. Nevertheless, there are many questions which are both impertinent, irrelevant, and sometimes unconstitutional. In a country giving equal respect to all faiths or none, any question concerning religion seems improper.

The inquisitive behaviour of the US Post Office is notorious, thanks to the revelations of House and Senate committees, Vance Packard, *Playboy* magazine, and others. A citizen wishing delivery of all that is addressed to him, whatever its nature or origin, may find himself signing an explicit request for communist propaganda. In an excess of zeal, someone might place such requests on record.

Minority tastes can lead easily to stigmatization by odious labels.

Words that categorize people change their meanings from time to time. In nineteenth-century Britain, to call a man a 'democrat' was to insult him; a 'socialist' was a believer in utopia and a 'liberal' could be a prohibitionist. Classification is no easier now, and certainly should not be left to an unsympathetic computer. The young man who, in the bi-polar world of the 1930s chose to support either communists or fascists in the Spanish Civil War would have been sound in neither case if his record in the data bank identified him as a fascist or communist sympathizer.

Stafford Beer has taken a jaundiced view of the ability of an automated library to record objectively the tastes and proclivities of its customers.[11] One can foresee an individual, attracted by a catalogue entry on his domestic terminal, asking for a book later to be classified as subversive. An overzealous official might look down a list of all its readers, compiled by collating each individual's 'profile' (which contains his interests and past reading so that the library could keep him up to date on his 'field').

PREDICTION AND PREORDINATION

Computer-analysis of information stored in data banks will be able to reveal patterns indicating abnormal propensities of individuals or organizations. A systems study of Los Angeles County (which we discussed in Chapter 11) discovered an underlying social pattern that revealed Watts to be one of the greatest potential trouble-spots. Aptitude tests have undertaken similar analyses for spotting the possessors of unusual talents. Psychological and personality tests reveal 'suitability' for certain jobs. Analysis of traffic convictions along with other data on individuals reveals that certain persons are more likely to cause traffic accidents than others.

In business, such 'leading indicators' are eagerly sought after, since they provide forewarnings of things to come, or reveal good marketing opportunities. A manager would be remiss if he did not heed the knowledge he discovered. To what extent should such leading indicators be used in society also? Clearly it would be very valuable to identify an area such as Watts as a potential trouble-spot before the trouble occurs. We want to identify traffic bottlenecks in time to take action before they become more serious. We do not want to appoint a man as Governor if he is a potential psychopath. This is why Aldous Huxley included among the characters in his *Brave New World* a Director of Predestination.

Many organizations now have computerized personnel files which they search to get the right man for a particular job – and to keep the wrong man out. At the moment the criteria for selection are probably biased toward experience, not aptitude.

Many personnel managers would like to have more details on individuals on which to base selection. William H. Whyte, Jr, described a satire he once wrote proposing a 'Universal Card' which would be a passport to organization life, and nothing less than a portable dossier: 'On it would be coded all pertinent information: political leanings, marital relations, credit rating, personality-test scores ... operator's licence and car registration ...'[12] Whyte said that many who took his proposal literally thought it was a splendid idea.

Employment agencies are also beginning to use computers to find jobs, to find applicants to fill particular posts, and to place specialists. The ambitious man of the future may spend much of his time trying to insure that the computers have the right type of ratings about him. The next time there is a promotion opportunity, he wants to be sure that the machine prints his name and dossier among the candidates.

Certain aspects of a man's medical, criminal, or other history might suggest that he is capable of committing a violent crime. After Richard Speck murdered eight nurses in a Chicago apartment in 1966, his psychiatrist, Dr Marvin Ziporyn, stated that Speck had a mental defect which *could be detected* in a small number of other people. Persons having the defect, Dr Ziporyn claimed, were potentially capable of such violence. Again, about one in every 200 men has an abnormal chromosome condition referred to as 'XYY' chromosomes. Each human body cell has 46 chromosomes. Two of these, labelled 'X' and 'Y' because of their respective shapes, determine sexual characteristics. Each cell in a woman's body normally contains two X-chromosomes, while each cell in a man's body normally has one X and one Y. The XYY man has an additional Y-chromosome, which can be detected by analysing his blood. An abnormally high number of XYY men have been admitted to mental institutions or prisons for crimes of violence.

Let us suppose that in a highly informed society of the future, these and similar facts were available to the machines. The computers conclude that there is a much-higher-than-normal chance that a certain individual will commit a crime of violence. To what extent is society entitled to protect itself from this man, who is as yet quite innocent? Should we permit him to take a job as a policeman? If not, what other jobs should we withhold from him? Should we allow him to own a gun, in a country where guns are licensed and controlled? Perhaps we ought to keep such information a secret, in general; but should a prospective wife be informed about it? Should it be left entirely to the individual's discretion whom he tells? If he *does* commit a crime – say, a murder – which a computer predicted as probable, should we sentence him to death?

These are sensational questions, but a whole spectrum of only slightly less sensational ones arise as soon as we assume the existence of a *highly informed* society with the desire to use its information for the benefit of all. If we wish to reduce the probability of accidents, we must restrict those who are demonstrably accident-prone. If we wish society to be as well-managed as possible, we must use information about people in our selection of managers. To maximize efficiency we must extract from our data banks any information they can provide, and put it to use.

At the moment we tend to use analysis to provide *a posteriori* explanations of past events in people's lives. In the future, if we learn to manipulate social data as we have learned to handle commercial and industrial data, we can take action to avoid adverse conditions which are related by the laws of probability to present circumstances.

The concept of social quarantine is distasteful, but it is a very frequent proclivity within groups, whether ethnic, cultural, religious, or national. We find that whole races may be regarded as homogenous in their behaviour – by non-members – and complete social classes condemned by association with the characteristics of a few. The computer provides a means not merely for clothing old prejudices in mathematical guise, but for creating a whole new range of prejudices. After all, a prejudice is nothing more nor less than a decision based on the operation of a model of the future rather than on events as they occur. What is this but 'good management'?

It is a very short step from analysing the readership of a particular magazine, in order to determine the value of buying advertising space, to analysing the occupants of an urban ghetto in order to determine the value of hiring employees from among them. In either case, drawing a correct conclusion is difficult; but in the latter case it could be very easy to justify the unjust.

THE GOALS OF SOCIETY

Non-human systems can be engineered to make their components work in harmony toward a specific desired goal. Human 'systems', be they nations, political parties, or industrial companies, can be led toward common goals, or coerced and threatened into pursuing the goals of others.

Under conditions of external threat most members of human organizations cooperate and concentrate on the achievement of primary objectives. However, under those circumstances that we wish to consider normal, the absence of an overwhelming imperative leaves us free to pursue our personal advantage – that set of goals we individually seek. Simple human overcrowding and the complexity of the products in the marketplace, along with an acceptance of the individual's responsibility for his neighbour following the decline of religious attitudes – 'God ordered their estate; they shall have their reward in heaven' – and the lack of social and racial discrimination of total war have soured the pure milk of laissez-faire. We expect and tolerate regulation and assume that it will be in our general interest. At the same time we require freedom – to pursue with minimum restraint those individual ends we hold most precious.

We pay a price for our freedom. The cost lies in *not optimizing* our total effort in pursuit of a single goal. But that price can be offset against the probability that the single goal is not the one most worth pursuing.

Potentially, we have at our disposal, in modern management science, techniques that can be applied to a whole nation in the same way they are applied to a single firm. In deciding how we shall put them to use we must decide where to draw the line between the freedom of the citizen and the efficiency of the employee.

It is essential to the freedom we cherish that we be able to live our lives in our own way without fear of oppressive supervision. The myth of the 'recording angel' kept millions in subjugation and serfdom for centuries. One needs little imagination to realize the potential if the myth became reality and all our peccadilloes were trotted out on our application for admission to lesser portals than the Pearly Gates. The admissions clerk is unlikely to be as beneficent as St Peter, particularly if we are in dispute with him and he has discovered something interesting from our dossiers.

The cause of privacy has been greatly helped in the past by inefficiency. For many years it has been technically feasible to collect most of the information for proposed national data centre dossiers by manual methods. But it would have cost so much and taken so long that, lacking powerful justification, no one bothered to initiate the project.

The computer will, if it has not already, make data-collection, data-retrieval, and data-presentation quick and cheap. Civil servants will be expected, no less than managers, to make use of information available to them to improve their decision-making. We shall insist that they do not assign an upstairs flat to a cripple rehoused following slum-clearance; we will demand that they 'look it up'.

For the first time in man's history he has the tools to use information with superb efficiency. In doing so he will be forced to ask himself more precise questions than ever before concerning the meaning of freedom and just what constitutes harassment. A delicate and complex compromise will be required. Whatever the answers to these questions, the machines can be programmed accordingly and an appropriate system of laws and controls built. Answers will no doubt differ in different societies, for societies assess differently such values as the authority of government, the efficiency with which controls should be implemented, financial efficiency, and the privacy of the individual.

It will probably be many years before we have answered the

above questions and agreed upon the necessary laws and controls. But meanwhile the technology races on. The data are being collected. Science once again is so far ahead of the lawmakers and sociologists that an interim period of confusion and heated argument seems inevitable.

REFERENCES

1. *Wall Street Journal*, 12 December 1967.
2. 'The Computer and Invasion of Privacy', p. 30.
3. ibid., p. 28.
4. Quoted in *US News and World Reports*, 8 April 1968, p. 83.
5. ibid, p. 85.
6. Vance Packard, *The Naked Society*, Penguin, 1964, pp. 157–8.
7. 'The Computer and Invasion of Privacy', p. 151.
8. *The Naked Society*, p. 223.
9. 'The Federal Paperwork Jungle', *Hearings before the Subcommittee on Census and Government Statistics, House of Representatives Committee on Post Office and Civil Service, 88th Congress, 2nd Session*, May and June 1964. Government Printing Office, Washington, DC.
10. *The Naked Society*, Chapter 15, passim.
11. *The Computer Bulletin*, March 1968, p. 356.
12. William H. Whyte, Jr, *The Organization Man*, Penguin, 1963, p. 164.

18

CRIME AND SABOTAGE

In earlier chapters we talked about the police using computers. We had better redress the balance by talking about criminals using them too. We feel sure that this will be big news in the years ahead.

The following article suggests that they are learning!

The Whiz-Kid Graduates of Sing Sing
The electronic computer may play a significant role in the rehabilitation of released convicts by providing them with jobs. The programme also will help solve industry's programmer shortage.

New York's Sing Sing Prison recently graduated its first class in computer programming, 12 convicts. Two of them were immediately paroled to take jobs as programmers in industry. Another class of 20 inmates started the course the next day.

The Connecticut prison at Danbury has inaugurated a similar school, and the Iowa State prison system is getting ready to do likewise. The Federal prison system has a more advanced computer progamme at Leavenworth, using an IBM 1130 computer to give special training to prisoners with scientific and mathematical skills.[1]

Some computer installations have a staggeringly high cash turnover – some large banks, for example. The London clearing houses turn over many millions of pounds sterling each day. A well-thought-out computer crime might operate over several weeks – even months – netting a proportion of the cash and covering its tracks, making the books appear to balance. Mr J. J. Wasserman, who headed a task force to devise methods of auditing computers used by the Bell System, was quoted in the *Wall Street Journal* as predicting that

within a few years someone will uncover a computerized embezzlement that will make even the $150-million salad-oil swindle seem puny.[2]

Mr Sheldon Danzinger, who runs a New York management-consulting firm, claimed in the same article that he 'could steal a company blind in three months and leave its books looking balanced'. He said that it was a simple matter, given the requisite know-how, to program a computer to fleece a company and fool the auditors.

In a well-designed system handling financial transactions there are many checks and balances which normally prevent loss. In a bank, for example, a 'balancing run' is normally performed at the end of the day to insure that the total of the account balance is in agreement with the totals of cash deposited and cash withdrawn for that day. Any disagreement is investigated immediately. Our criminal programmer would have to understand these checking procedures fully and insure that, after his program had removed cash, the system still appeared to balance. He might do this by adding a large sum to his own account while removing the total equivalent sum in small amounts from many other accounts. Alternatively, and probably more safely, he could corrupt the balancing program itself by adding a few instructions to make it add money to his account and deduct the identical amount from the balance total. To be additionally safe he might make the program first check that the system does indeed balance. If it does not balance, the crime is set aside, because the files might be investigated that night.

If the system appears to balance and the computer prints out the right totals, the bank manager and his employees will go home with that blissful feeling that everything is correct – another day's work completed with all the figures in agreement. They will not know that the computer is fooling them.

Pulling off a big job may require the complicity of several programmers. In some computer crimes that have been discovered the data-processing manager has been the culprit.

The *Wall Street Journal* reported two crimes as follows:

The manager in charge of back-office operations at Walston & Co., a New York brokerage firm, electronically siphoned $250,000 out of the company between 1951 and 1959. By the time the theft was uncovered, the man had become a vice-president.

He programmed Walston's computer to transfer money from a company account to two customers' accounts – his and his wife's.

The computer was further programmed to show the money had gone to purchase stock for the two accounts. Then he sold the stock supposedly purchased, pocketed the cash, and transferred some more.

When a Walston official sensed something was amiss, an examination of the two accounts revealed major irregularities. But the company couldn't figure out the embezzler's system. Because he hadn't stolen any money from customers' accounts, 'what he did was absolutely undetectable without internal auditing.' says William D. Fleming, Walston president. 'Before it happened no one dreamed such a thing was possible, and if he hadn't explained how he did it, we probably still wouldn't know.'

The thief explained to Walston's incredulous directors that he pulled off the elaborate money swap by going into the office early Sunday mornings to punch new computer cards and feed them into the machines.

'It took someone with absolute knowledge of the computer system to do it,' says Mr Fleming. 'This guy was the boss back there. He set up the system and ran the whole show.'

Walston recovered only a fraction of the stolen money. The firm promptly revamped its computer system, instituting a quarterly internal audit and other safeguards designed to foil embezzlement attempts. The former vice-president served a year in Sing Sing prison and is now a furniture salesman.

Even as the Walston theft was being uncovered, a similar embezzlement was beginning at another New York brokerage firm, Carlisle & Jacquelin. From 1959 to 1963, the firm's data-processing manager got away with $81,120 by instructing a computer to write checks to ficticious persons and send them to his home address. The scheme was uncovered when the Post Office accidentally returned one of the checks to the firm and the clerk who received it became suspicious.

Jacquelin refused to discuss the case. 'We'd have to be crazy to give out all the details now so that anyone who wanted to could do it again,' says Mr Muller.[3]

NO FINGERPRINTS

One of the attractive features of a computer crime is that it is often possible to destroy all traces of how it was committed.

The miscreant would normally insure that there were two sets of programs, one felonious and one legitimate. When the crime was ended the felonious set would erase itself, leaving absolutely no trace. Only the normal, valid programs would remain in the system.

It could be devilishly difficult to track down how such a crime was perpetrated. Felonious programs do not leave fingerprints. To find out exactly what happened when no programs can be traced and the printouts were planned so as to leave no clue might involve little more than guesswork. However, like other schemes in the computer world, the planning would have to be done with immaculate care.

Picking one's location for the crime might make it easier. The small town in England from which one of the authors comes is admirably protected by a small number of benevolent and somewhat portly, unarmed policemen who ride bicycles. The idea of these amiable custodians of the law being confronted with a computer thought to have been discreetly filching part of the payroll is amusing. It is difficult to imagine that the police there today could obtain an investigator who would even know where to start.

STEALING PENNIES

A felonious computer program need not steal large quantities of money. A programmer might derive a steady income on the side by taking only a few cents at a time. Suppose he deducted three cents weekly from the tax deductions of all the other employees. Would anyone notice? With a payroll of 8,000 people he might double his income. In one case a firm, in calculating discounts, ignored the fractions of pennies that this process created, always adjusting the sum to the nearest penny. An ingenious programmer always adjusted it down and kept the fractions of pennies for himself. This went on for a long time before he was caught. One wonders how many such cases are never detected.

Another programmer, irresponsible but perhaps not with

deep criminal intent, gleefully programmed the computer in a bank to ignore cheques that overdrew his own account.

Some systems have a 'mistake tolerance'. Any errors less than a given sum are not investigated. This might seem on open invitation to programmed theft. The *Wall Street Journal* described a mutual-fund dividend-payment system in a large bank, where the operation was so large that the mistake tolerance amounted to several hundred thousand dollars. One man designed and wrote a set of computer programs for it, and claimed that he could have paid at least one-half that sum to his own account in small cheques, without the money being missed![4]

TRUSTING THE MACHINE

One trend we have outlined for the years ahead is that of increasingly complex computer applications. There will be far fewer printouts. The machine will inform management only of circumstances that need attention. Many middle-management functions will be taken over by the machine. As this trend continues over the years, management will begin to rely more and more on what the machine tells them. They will be increasingly unable to argue with its figures or decisions. The man who used to work out those figures has long since found another job. Anyway, considering that there are 400 man-years of logical thinking in the programs, what single individual is capable of arguing with the results? Men will become dependent upon the machine – victims of their own superbly intelligent and intricate creation.

Such circumstances offer perhaps even greater opportunities for embezzlement-by-computer – illicit stock transfer, for example, or perhaps more simply an over-ordering of goods which are then shipped and sold by the programmers!

SABOTAGE

Still more alarming, perhaps, is the possibility of computer sabotage. Some corporations in the not-too-distant future will be almost entirely dependent on their computers. In a bank, for example, everything revolves around computerized record-keeping. Airlines, if they achieve today's data-processing ambitions, will utilize one central computer system not only to handle all of their bookings, but to make most of their operating decisions also. If such a system were to run for a number of years – long enough for corresponding manual skills to dissipate – and then were put out of operation by a neatly executed act of sabotage, it is difficult to see how the organization could continue to function. The act of sabotage no longer has the appearance of overt violence. One knowledgeable and malicious programmer in most installations today could erase the vital records and the means of reconstructing them, if he planned this carefully.

Harold Weiss, in an article in *Datamation*, humorously envisaged a foreign power committing acts of sabotage against the United States economy. The agent in charge of the operation addresses an espionage school as follows:

Research showed the increasing vulnerability of the US economy to a relatively few sophisticated machines. There was a great concentration of values at comparatively few locations. The informational life-blood of all the large companies and government agencies flowed through only a thousand or so key computers. . . .

Each agent received intensive training in computer-programming, among other subjects. It was fairly easy to place intelligent, well-trained people in their initial jobs because of the personnel shortage in the computer field. Once they had two years or so of practical experience, their skills became highly salable and they could move from place to place readily. The computer field is characterized by considerable job-jumping, which made our task much easier. An effective agent could therefore be brought to bear against several computer installations in only a few years.

We felt our way cautiously at first to test the lax security of computer installations and see what we could get away with. Our people did numerous little things to reduce the efficiency of the computer operations they worked at and thereby harm the large organizations by which they were employed. We filched cards from program or data-checks, put cards out of sequence, wiped out key tapes and disk-packs, caused subtle equipment malfunctions, and the like. We were able to change program cards or copy magnetic tapes with changes to create program bugs in production programs. The consternation which little things like these caused must be seen to be believed ...

As we gained experience and saw how permissive things were at many computer centers, we got more ambitious. Localized fires turned out to be a very effective technique for computer sabotage. There are often extensive combustible materials present, noxious smoke often results which hampers fire-fighting, and often all the major programs and files of the installation are physically close together and not in fire-resistant vaults. One such fire can knock out a highly integrated organization with most of its information system on the computer or at least cause it to suffer catastrophic consequences. And you should see the Yankee firemen! Just turn them loose with their axes and hoses and you needn't worry about a computer center for a long time. Some places had electrical shorts for months after a fire. It was very pathetic to see the recovery attempts of some computer groups which thought they had good disaster protection. Off premises they typically kept third-generation files, some obsolescent programs, inadequate transaction data, and little if any system documentation.[5]

This is a high-power fantasy, to be sure, but it has one important message for most data-processing managers – which is that today they *are* vulnerable. The security and protection they enjoy enable them to reconstruct records accidentally lost, but in most cases they would not survive a deliberate and planned act of destruction which destroyed their means of recovery also. Much more elaborate security is needed to protect against this eventuality. Very rarely today does an installation deposit a copy of its program tapes and data dumps in a local bank, for example, or somewhere equally secure.

In a lesser sense, installations are vulnerable to loss of personnel also. Mr Weiss' agents enticed groups of key personnel to change to higher-paying jobs, leaving installations vulnerable. (Often key personnel are groups of relatively low-paid programmers or analysts.)

We found that wholesale personnel raids could seriously weaken an information-systems organization. Some of our people had worked up to managerial levels after a few years and could carry out such tasks directly. Others in senior positions found willing accomplices in some computer managers. I recall a data-processing manager who jocularly offered the advice that one should not steal all the skilled personnel of another computer group. He recommended following conservation practices – one senior man should be left on the victim's staff to train a new set of people to be raided at a later date.

UNIONS

The unions could perhaps, do even more damage. Fortunately, as yet there is no effective union for data-processing personnel in most countries, but these might come about in the future. Even today a nationwide strike of computer employees would cripple the economy. In the future, as more and more spheres of the economy come to depend upon working computers, such a strike would be a disaster of enormous magnitude.

Factories relying on computer production-control would be in a state of chaos. Most banks would be unable to operate for more than a day or so. Credit facilities would be suspended. Where computerized telephone exchanges have been installed the telephone service would be in danger. Many process industries controlled by computer would be inoperative, and many areas of the country would have no electricity. Railroads would lack all coordination. Airlines would be unable to make bookings. Most planes would be unable to land at airports dependent on computer-assisted air-traffic-control. Most business firms would be unable to operate because essential functions would be missing.

The British unions' 'work-to-rule' weapon has now spread

across the Atlantic. With the ingenuity of an old Ealing comedy script, union chiefs know that if every hand obeys his bosses' orders to the letter, the system will all but break down. And one can hardly fire a man for obeying his orders. It is appalling to imagine how much damage a mal-intentioned computer operator might do working strictly by the rule – or almost by the rule – in a typical installation with a large data base, especially during times of non-routine activity such as program-testing. Computer systems *can* be planned to protect themselves from operators of malicious intent, but at the time of this writing they rarely are. (It is worth recalling that computer programs always work to rules which, if imperfect, lead to the situations described in Chapter 16.)

It is an alarming reflection that the more complex and inter-related our society becomes, the more vulnerable it is to such strike action. Our society is becoming a machine of immeasurable complexity; unless such a machine is very cleverly designed, immensely expensive damage can be done to it by a well-thought-out act of sabotage or even by a rare technical accident such as caused the great eastern seaboard blackout of 1965.

REFERENCES

1. *Evening News*, 12 June 1968.
2. *Wall Street Journal*, 5 April 1968.
3. ibid., 5 April 1968.
4. ibid., 5 April 1968.
5. Harold Weiss, 'The Week the Computers Stopped', *Datamation*, April 1967.

19

BEWILDERMENT

I fear none of the existing machines; what I fear is the extraordinary rapidity with which they are becoming something very different to what they are at present. . . . Should not that movement be jealously watched, and checked while we can still check it? – Samuel Butler, 'The Book of the Machines', in Erewhon (1872).

As the computer age gathers momentum the 'man in the street' is very likely going to be bewildered by it all, particularly in the next twenty years because general education about computers and their usage is largely unavailable today – a lack of foresight we shall pay for. Subsequent generations will probably take computers in their stride, and the ubiquitous machines will then be better attuned to working with people. Today, however, there are already signs of bewilderment in the populace.

As Bernard Levin wrote not too long ago in the *Daily Mail*, 'I begin to suspect that a long period of purgatory has got to be gone through' before we reach the computerized Kingdom of Heaven; 'we are presently stuck well into it.' Levin continued:

Bank statements (with the exception of those provided for their customers by Coutts and one or two of the other small banks) are now worked out and presented by computer. The only noticeable effect on the banks' customers has been entirely adverse.

Details of cheques drawn, and credits paid in, are no longer given. I know only that on such a date I paid such a sum to an anonymous set of figures, and on such a date I received a sum from another such disembodied source.

I am unable to check the accuracy of the statement without tediously going through the cancelled cheques, and since I do not in fact do so, I have no idea whether the statements are correct or not.

The result is that the computer, as far as banks are concerned, has brought the customer nothing but inconvenience.[1]

Levin went on to describe errors in gas bills (the public utilities always seem to catch it) and then said :

But we are also getting to the point at which it may soon be impossible to catch computerised errors at all. The Post Office, for instance, has long refused, as a matter of principle, to admit that it can get [telephone] accounts wrong, let alone that it does; and I notice an increase in bills, from commercial sources, of the 'punched-card' type, which virtually make it impossible to query the account.

Katherine Whitehorn in the *Observer*, equally bewildered, recommended that her readers send back the punched-card bills with staples stuck all over them in the hope that they would wreck the machines into which they were fed.

Levin, recruiting his readers' participation in the protest, concluded :

I begin to believe that the computer is not the great god we have been led to believe, but a hollow idol, manipulated by crafty priests.

ISFADPM

The International Society For the Abolition of Data Processing Machines aims to bring the bewildered together to combat 'The Beast of Business', and exhorts its members to 'Fight the creeping computer menace! Learn how to

demagnetize your cheques;
add millions to your computerized bank statement;
get ten tons of broken biscuits delivered to people you don't like;
worry a computer;
confuse a computer;
wreck a computer.'[2]

The book contains more than a hundred anecdotes culled from various sources. Many are nonsense, others are innocuous, some are very funny (particularly if anyone really believed them), but a few are genuinely horrific. Nearly all of these result from programming or data collection errors of the type discussed in the last few chapters.

Matusow, an American who has fled from New York to less computerized Britain, believes 'the electronic industry is ... attacking our environment entirely'. He has recruited 2,000 members in England alone, 200 of them within the data processing industry. In their hands his 'Guerilla Warfare Manual for Striking Back at the Computers' could be harmful.

Part of the reason for such hostility is that computer bank statements and punched-card bills are genuinely more difficult to understand than the traditional forms. This is partially due to poor design and partially to cutting costs to a minimum. Some banks which had no alphabetic wording for items on the statement when they first converted to computer operation have now capitulated and issue attractive statements with a worded description of line items – although cheques are still listed by serial number, not by the payee's name. One reader of the *Times* asked, 'Can someone please tell me why the numbers on cheques are now printed in Chinese?'[3]

The use of *numbers* rather than alphabetic descriptions or letters has increased because of the ease with which they are handled by machines. Bank cheques now show the individual's account number on them. Personnel numbers, identification and Social Security numbers are employed increasingly. Telephone numbers have become all digital; Scotland Yard's long-famous WHItehall 1212 has been replaced by a mere collection of digits: 01 230 1212. In a typical newspaper cartoon commenting on this trend, the operator at a computer centre is introduced to a visitor as '76419352784', and the computer is introduced as 'ALFIE'.

Certainly individuals are acquiring many numbers which they must remember. The following list is typical for the United Kingdom:

National Insurance number:	9 digits
bank-account number:	8 digits
savings-account number:	10 digits
home-telephone number:	10 digits (including area code)

office-telephone number:	14 digits (including area code and extension)
car licence-plate numbers:	14 characters (for two cars)
postal code:	6 characters

These total 71 digits, and if one adds such items as safe-deposit–box number, charge-account numbers, passport number, several credit-card numbers, and so on, it soon exceeds 100. Most individuals remember other people's telephone numbers amounting to more than 100 digits.

In the general hostility toward computers, probably a stronger factor than the difficulty in understanding computer forms is the feeling of depersonalization which computers have engendered. If a bill comes from a computer it is more difficult, it seems, to query it. Friendly staff at the local bank no longer seem accessible. One can easily obtain the answer: 'We have no way of checking this figure; it came out of the computer.'

Again, this is probably a temporary phase. The first groping for automation is a struggle. The machines, because of their initial high cost, are only just big enough. Programmers are scarce, time is short, and such systems analysts as can be obtained are lacking in experience and education. But somehow a computer is installed which manages to churn through the work it is given. Later the system will be refined. A bigger and faster machine for the same price has room for far more elegant programs. Bills and statements can then be made as convenient as possible for the user (see Fig. 19.1). And, particularly important, means can be devised to answer the types of questions that may arise.

DIRECT ACCESS FOR QUERIES

Early computer systems used files of data which were not readily accessible because they were on punched cards or tape. The computer could not search its files at random to answer a

Both of these bank statements were produced by computers:

1. A bank statement with no descriptive detail:

014-0-033200   BALANCE FORWARD SEP26 1967   925.40

CHECKS AND OTHER CHARGES			CREDITS	DATE	BALANCE
51.98			379.36CM	SEP27	1,252.78
18.25	60.00			SEP28	1,174.53
21.78	400.00			OCT 2	752.75
407.92	65.00	60.00		OCT 3	201.83
18.00				OCT 4	720.33
8.00	349.96	16.22	892.68CM		
117.45				OCT 6	602.88
15.41	17.24			OCT16	570.23
13.00				OCT17	557.23
29.30				OCT18	527.93
222.86				OCT20	305.07
30.00	1.90SC	.50MC		OCT25	272.67
NUMBER OF CHECKS 19			BALANCE OCT25 1967		272.67

2. A bank statement with descriptive detail*

MARTIN NORMAN, ESQ.

DATE	DESCRIPTION	DEBITS	CREDITS	BALANCE CREDIT—* DEBIT—DR
6 FEB 69	BROUGHT FORWARD			735 6 2*
7 FEB 69	CASH	15 0 0		720 6 2*
10 FEB 69	BOAC	162 15 9		557 10 5*
10 FEB 69	BOUGHT U.S. DOLLARS	45 19 6		511 10 11*
10 FEB 69	TRAVELLERS CHEQUES	103 4 0		408 6 11*
11 FEB 69	DIV ON 800 LONDON & HALIFAX INVESTMENT TRUST CO LTD ORDY SHS		25 11 9	433 18 8*
13 FEB 69	HANDYMAN SERVICES LTD	14 18 3		419 0 5*
15 FEB 69	SALARY		240 3 8	659 4 1*
17 FEB 69	INT ON £500 AQUILLA ENGINEERING CO DEB STK		13 1 0	672 5 1*
20 FEB 69	CATHERINE'S GALLERIES	26 5 0		646 0 1*
22 FEB 69	OF DEPOSIT ACCOUNT		750 0 0	1396 0 1*
24 FEB 69	CAMERA ACCESSORIES LTD	6 010		1389 19 3*
25 FEB 69	MRS ANNABEL CUSTOMER	50 0 0		1339 19 3*
25 FEB 69	CASH CARD WITHDRAWAL	10 0 0		1329 19 3*
25 FEB 69	SOLD 1000 HERODOTUS LTD SHS		530 1 2	1860 0 5*
28 FEB 69	BOUGHT 500 SURE PROPERTIES LTD PREF SHARES	965 12 4		894 8 1*
3 MAR 69	WESSEX DOMESTIC OIL CO	30 0 0		864 8 1*
4 MAR 69	REPAYMENT OF INCOME TAX 1967/68		115 12 3	980 0 4*
5 MAR 69	CARRIED FORWARD			980 0 4*

* A specimen statement from Messrs Coutts & Co., London

A little extra programming helps to lessen confusion

Fig. 19.1

query on request. It may only have had access to a given record once a week or once a month as a cycle of file updating or invoicing came around. Dealing with inquiries from the public was therefore difficult. The only way it could be done was to assign a clerk the job of looking through past computer listings. The machine was programmed to produce piles of paper for this purpose. Even then, the information needed to deal with the query was sometimes unavailable. Answering queries in this way consumed considerable human effort, and so queries tended to be discouraged, giving rise to the impression that an impersonal, incommunicable machine had taken over.

The hope for the future lies in 'direct-access' storage devices. These are coming increasingly into use for a variety of applications, including those in which the public is likely to query bills or statements. Using them, the computer can read any record at random within a fraction of a second, and so can mobilize information as complete as is desired to answer any query. When you telephone the electric-power company to query your bill, the girl you talk to will have a terminal with a screen on which she will display, in a second or so, any facts or figures that could help you. The electric-power company would also benefit in that it will require a smaller staff to handle public queries.

With a scheme such as this, the computer, instead of being bewilderingly remote from its public, is likely to keep them better informed than could have been possible in the pre-computer age. Suppose that you make an airline reservation in the United States for a round-trip flight to Europe. The airline must keep a record of your booking. Now suppose that your adventures in Europe prompt you to change your plans for connecting and return flights. You telephone the airline in Paris to change the booking. The girl who talks to you surprises you by saying 'Oh yes, Mr Jones, your telephone number is 684-0999 in New York City, isn't it? ... Good ... now I know I am talking to the right person. You are booked from Paris to Rome on Friday, aren't you, and you want to take your pet sheepdog on board. What changes can I make for you?'

How does she do it? She has a screen connected to the distant computer which displays your reservation record. If the airline had no computer, the most it could do would be to send a cable to New York and ask you to call back another day. And then they would not have had *personal* details about you.

Computers *can* be programmed to improve the personal touch rather than dehumanize everything.

CRIME BY BEWILDERMENT

One organization in Sweden has used ingenious advertising to promote a more human image. The Swedish Savings Bank Data Centres Company launched a campaign using five toy robots named after the services provided by the centre: SPIN, SPUT, SPAN, SPIR, and FRED! These 'happy Spadab brothers', as the robots were called went under the slogan 'We make human life easier and leave more time for essentials', and each of the bank's customers was given a model of one of them. The robot gnomes and their services are now well known to and accepted by the Swedish public.

Madison Avenue has prided itself on its ability to create public acceptance of totally unfamiliar and novel products, many of which are as complex and awe-inspiring as any computer terminal. Perhaps it is not too much to hope that the advertising industry will turn its talents to a genuine necessity of modern life.

The volume of bills and statements that are unclear or lacking in full information has certainly increased with the spread of computers. There is one corollary to this that is alarming: most persons, faced with bills which are unintelligible or uncheckable, but nevertheless not far wrong, *pay* them. This happens, we suspect, particularly when bills are produced by a machine and thus carry an imposing stamp of computerized authority – an image reinforced when querying the bill proves to be virtually impossible. In his New York flat, one of the authors, for example, has no way of reading the electricity meter and hence no way of checking his large electricity bills. Furthermore, the non-payment of bills from most organizations now carries the vague threat of a black mark on one's credit rating – which is itself recorded by computer. The payee is in a cleft stick. The line of least potential trouble is to pay.

The public, then, is being conditioned to pay bills which it cannot fully understand. Thus if an organization mails bills which are unintelligible, are prepared by computer, and also are *deliberately* made out slightly too high, the vast majority of people will pay them. What is more, it would be exceedingly difficult to prove that a felony had taken place, especially if the overcharge were made to look like a programming error.

An acquaintance of one of the authors received a series of bills which he suspected to be incorrect. Although he had great difficulty in checking them, he succeeded in showing, over a period of time, that 10 of them were wrong. All of the errors were in favour of the organization which had sent the bills. Given ten random errors, the probability of their *all* being on the high side (or all on the low side) is 0·001 – the odds are 1024 to 1 against such an occurrence. The errors may not have been deliberate, but they certainly looked suspicious.

A different sort of case of felony-by-bewilderment is reported to have occurred in certain banks in the US. If one walks into a branch of the bank to deposit cash or cheques into his current account, he may select a deposit slip from a rack of different slips. The slip has a box in which the customer writes his account number. Some customers, however, have their own deposit slips which are printed with their account number in machine-readable characters. To commit robbery-by-bewilderment, you do the following: open an account using an assumed name – preferably in a bank with many branch offices – obtain a quantity of deposit slips with the account number printed in the appropriate box, and place a number of these in the deposit-slip rack in each branch. Members of the bewildered public come in, pick up your deposit slips, and write in the sum of money they are depositing and their name on the slip. Names, however, are ignored by the computer: only the account number in the box registers. The money is therefore deposited in your account, not theirs. Withdraw the money quickly and disappear.

USE OF TERMINALS

Much of this book has been concerned with terminals linked by communication lines to computers. Today the terminals are used by persons familiar with computing equipment; by clerks, secretaries, or others trained to carry out a particular operation; by children using computer-assisted instruction; and so on. Nevertheless, it seems likely that a large proportion of the public may run into trouble when they begin using terminals.

The number of people who will use computer terminals will steadily increase. Today some people use them in their work, and a few use them in classrooms, but only those who need them for their professional work are likely to have them in the home. In the future, however, the domestic-terminal market will probably expand into other fields and the average person, even if he does not have a terminal in his home, will be confronted with them elsewhere. The German railways, for example, considered installing at stations teleprinters which the public could use to obtain train information. If you wanted to travel from Cologne to Nuremberg, for example, you would type in this 'city-pair' and the computer would then determine the best train connections for you. In applying for a job or visiting a medical clinic, a person may be granted a preliminary 'interview' as he sits at the screen of a computer terminal. In registering for evening education classes or joining a political club, you may have to enter the required details about yourself directly into a terminal. Arriving in a new city and desiring a hotel room, you may find a terminal at the airport with which you can interrogate a city file of room availability.

The quantity of information we have to supply when filling out forms of one type or another is probably going to become greater with greater use of computer-planning. This information also may be collected on-line at a terminal; this method provides better control of the data's accuracy, and vast amounts of clerical work that would otherwise be involved are avoided. As we commented in Chapter 12 much man-power would be saved if, when we applied for an automobile registration – or any other kind of licence – we did so at a terminal in the appropriate government office (or at home, or at work, if the appropriate computer could be dialled).

The fewer the steps between the originating of new information and its recording in the computer files, the less is the probability of an error. The usual procedure, involving a pre-printed form filled in by hand, coded and then punched into cards and read into the machine, produces up to 10 per cent errors, which are seldom discovered until the form-filler is long since departed. On-line data-collection is *much* more accurate. The more the computer does and the less the human, the higher the degree of accuracy.

MOTOR VEHICLE REGISTRATION

Let us suppose, then, that Winston Smith wants to renew his automobile registration. He walks into his local government office, the Post Office in an English town, sits down at a machine with a typewriter keyboard and screen, and the following dialogue ensues:

SCREEN: To communicate with the computer please press the Attention key.

User presses Attention *key*.

SCREEN: Farnham G.P.O. Set 8 10/10/84 11.03 AM
Which of the following activities do you wish to perform? Enter one digit.
1: Check personal records
2: Obtain licence
3: New-resident registration
4: Check-rating assessment
5: Other

USER: 2 (*presses* Enter *key*).

SCREEN: Which type of licence do you wish to obtain? Enter one digit:
1: Motor-vehicle
2: Boat
3: Radio or television
4: Marriage
5: Firearms

6: Dog
7: Other

USER : 1 (*presses* Enter *key*).

SCREEN : Motor-vehicle licence. Type 'New' or 'Renew'.

USER: RNEW (*presses* Enter *key*). [*Note*: user's spelling mistake]

SCREEN : Entry not accepted. Type 'New' if you wish to obtain a new licence. Type 'Renew' if you wish to renew a previous licence.

USER : RENEW (*presses* Enter *key*).

SCREEN : What was your old licence number? Type 'Don't know' if unknown.

USER: DON'T KNOW (*presses* Enter *key*).

SCREEN : Enter your personal identification number.

USER : SMIW6079 (*presses* Enter *key*).

SCREEN : Are you Winston P. Smith, owner of blue Ford, number XNJ277F? Answer 'Yes' or 'No' please.

User presses Yes *key or types* Yes, *and presses* Enter *key*.

SCREEN: Please check each statement that follows: After each, type 'C' if correct, type 'X' if incorrect.

The computer then reviews all the information on last year's licence, item by item, updating it as necessary; it checks the insurance file, checks the validity of the driver's address, and inquires about the annual test certificates, and calculates the fee. It asks the user whether he wishes to pay by credit-transfer; if so, it transmits a message to his bank computer, which debits his account. It updates the car log book in its files and produces a summary of the licence on a card which can be read automatically by the transmitter in a police patrol car.

Highly efficient, but is the public going to be able to cope with it?

In this case, probably yes. There will be an assistant to show people how to work the machine if they do not know, and to help them if they reach a deadlock in their dialogue with the computer. A few people may reach a state of paralysis when confronted with the terminal; then the assistant will have to complete their entries for them.

SCHOOLCHILD TERMINAL

In Chapter 7 we discussed schoolchildren doing their homework on a simple terminal. In this case they learned how to use it very easily. They were intelligent and at an age when one learns easily. Some older people had more difficulty with it, however. The terminal operates somewhat as follows:

The keys on the terminal are labelled as in Fig. 19.2. Note that each key represents two possible meanings – either the digit for which it is normally used, or an alternate meaning as shown. The alternate meaning of key 7, for example, is 'divide'. If the child wishes to use a key's alternate meaning, he presses that key and then immediately afterward presses the key labelled '*'. The sequence '7*', therefore, signifies the arithmetic operator 'divided by'. To end each operation the child signals '**'.

Thus if he wishes to divide 3 by 14, he keys in 3, 'divided by' (7), 14, and 'end of operation' (**) in that sequence, thus:

$$3 \qquad 7* \qquad 14 \qquad **$$

and the computer speaks back over the telephone with a distinct human voice:

Your answer: Zero . Point . Two . One . Four . Two . Eight .

Five . Seven . One . Four . Two . . .

and so on up to fourteen digits.

He keys 1* to mean 'decimal point' and 0* for 'minus'; thus to subtract 3·8692 from 14·908, she would key:

141*908   0*   31*8692   **

The machine replies:

Your answer: One . One . Point . Zero . Three . Eight . Eight .

If she failed to write this all down and wants it repeated, she presses

$$8* \qquad **.$$

The machine is equipped with temporary and permanent memory locations which she can use. She refers to these with keys 2 and 3, which are labelled with their alternate meanings 'keep' and 'use'. If she presses 2*0, this means: 'Place a number into temporary-storage number 0 for use in the next prob-

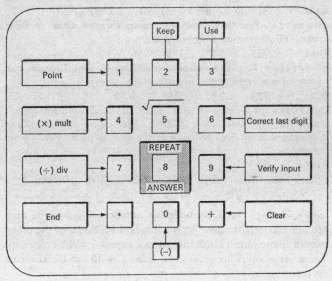

Fig. 19.2 Label to touch-tone telephone keyboard.

lem.' If she presses 2*7, this means: 'Place a number in permanent-storage number 7 for future use.' Similarly, she can use the contents of, say permanent-storage number 9 by keying 3*9.

To calculate pi × 3.78², the sequence would be as follows:

GIRL: **31*78**   **4***   **31*78**   **2*0**   ******

COMPUTER: *Your answer: Holds One . Four . Point . Two . Eight . Eight . Four . Key your function.*

GIRL: **4***   **31*1416**   ******

COMPUTER: *Your answer: Four . Four . Point . Eight . Eight . Eight . Four . Three . Seven . Four . Four.*

If she makes mistakes which the computer can detect, it will inform her, as with other terminal systems, thus:

GIRL :   **31*78**   **4***   **31*78**   **2***   ******

COMPUTER: *A storage location has not been indicated. Please re-enter your problem and specify a storage location after the Use function.*

GIRL :   **31*78**   **4***   **31*78**   **2*97**   ******

COMPUTER: *You have specified a storage location which does not exist. Please re-enter your problem.*

GIRL :   **723**   **7***   **3*9**   *****

COMPUTER : *You have indicated a division by zero. Change your division and re-enter the problem.*

GIRL :   **723**   **7***   **379**   ******

COMPUTER : *Your answer: One . Nine . Zero . Seven . Six . Five . One . Seven . One . Five . Three . Nine . Five . Seven . Your message has not been followed by the End function . Verify your entry.*

GIRL :   **9***   ******

COMPUTER : *Your entry: Seven . Two . Three . Divided by . Three . Seven . Nine.*

Bewildering? To a considerable number of people, if they are not properly trained, it probably will be. Some of the other uses of home terminals discussed in Chapter 7 will be more so. There are many who never even learn how to use the controls on their cameras.

THE NATURE OF THE DIFFICULTIES

The difficulty in using a terminal depends upon three factors. First, the terminal machine itself may be hard to use. This has been true of some elaborately designed machines which exhibit a confusing array of knobs and switches. On the other hand, some terminals are very simple to use. Using some of the machines designed for computer-assisted instruction is literally child's play. Let us then discount the machine itself as a bewilderment factor, assuming that manufacturers learn the lessons to be derived from experience with today's machines.

Second, the application to which a terminal is put will determine the degree of difficulty in using it. The applications discussed above are fairly simple. More difficult dialogues are required by other applications, including some designed for the general public – such as searching for suitable job-openings at an employment-office terminal, planning vacations, or tracking down literature in a library. The applications for which the professional man may use his terminal are in many cases of much greater complexity and often involve programming.

Third, the particular language used to communicate with the machine, or the structure of the dialogue, has a major effect on the relative ease or difficulty of use. In the earlier example of renewing a motor licence, the dialogue is easy for the user. He is told on the screen exactly what to do and the computer is programmed to give him maximum assistance. When he keys in wrong wording, the machine helps him out. It is very important that such applications be programmed with maximum consideration for the user in mind, and that extensive research be conducted on the ways people with no special training react to a particular dialogue structure. If the user becomes bewildered or is embarrassed at the terminal, he is likely to adopt a hostile attitude toward the whole idea of using computers in this way.

Many of the dialogue structures in use today are *much* more difficult for the user than the above illustrations. An airline agent at the screen of a reservation terminal, for example, would be quite bewildered by the system if she had not had extensive training.[4]

With a terminal in the home, the user will eventually be able to dial a wide variety of computers offering different services. It is likely that these will use a host of different dialogue structures and computer 'languages', although some degree of standardization will probably come about. The user will have to go through a sequence for selecting the program he wishes to use.

Using a terminal for problem-solving can involve much greater difficulties than those encountered in the above illustrations. A computer language must be learned, and some skill is needed. A major portion of today's terminal usage is on a high level of complexity. This is typical of the work of engineers, statisticians, architects, and similar professionals performing their basic calculations.

Much of the interesting terminal work, however, is of an even higher level of complexity than this. Writing textile-design programs, testing hypotheses about making money on the stock exchange, simulating a 'system' for the Las Vegas tables – all these could be entertainment for computer hobbyists, but all demand a high degree of facility in using the machine.

Some people seem to be at home with the machines as soon as they make contact with them; others are surprisingly ill at

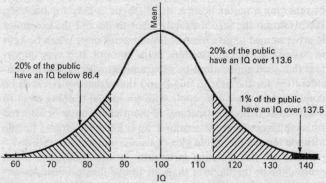

20% of the public have an IQ below 86.4

20% of the public have an IQ over 113.6

1% of the public have an IQ over 137.5

Mean

IQ

Area under a section of the curve equals the fraction of the public having an IQ within that range. (Total area under curve = 1)

Figure 19.3

ease. For some, every program they touch ends in disaster. Dexterity in using computers is a gift, like most gifts, given to men in differing proportions. Even highly trained programmers differ very widely in their capabilities.

Figure 19.3 shows the distribution of intelligence quotients among the general public. The ability to be productive with computers is not the same as IQ. Some people of high intelligence are poor at using machines. However, the distribution of machine capability probably follows a similar *Gaussian* curve: the bottom 20 per cent may never be able to cope with even such simple uses of terminals as the motor licence illustration above; whereas the top 20 per cent will probably be able to learn terminal operations involving programming. Perhaps by creating easier problem-solving languages and teaching people at an early age we can extend the range of peoples' competence. The sophisticated uses of machines, however, will always demand a very high ability, and those at the top of the capability distribution will have a special relation with the machines. Observing the symbiotic relation between a highly proficient user and a computer – communicating perhaps with a screen, keyboard, and light pen – is an extremely powerful and impressive experience. Such men are rare and surely rate a special place in the computerized world ahead.

The gifted will have immensely powerful facilities available to them in the society we visualize, while those who either do not like or do not understand the machines may become increasingly bewildered and hostile. Probably an entirely fresh generation educated in the ways of computers must emerge before the populace feels at home with the new machines.

REFERENCES

1. Bernard Levin, *Daily Mail*, 18 March 1968.

2. Harvey Matusow, *The Beast of Business*, Wolfe Publishing, 1968.

3. *The Times*, 10 July 1968.

4. James Martin, *Design of Real-Time Computer Systems*, Prentice-Hall, Englewood Cliffs, New Jersey, 1967. (See Chapter 8.)

20

POWER

We must form a single unity; one discipline must weld us together; one obedience, one subordination must fill us all, for above us stands the nation. – *Adolf Hitler, speech at Nuremberg, 2 September 1933.*

Organized knowledge can bring power. Computers with their immense capability for organizing knowledge can bring power to individuals, to corporations, to governments, and to the military.

1. *Power for Individuals*

Whereas the people without an affinity for computers may feel bewildered by the ubiquitous terminals, the opposite will be true of those few who get along well with them. The man who knows how to use the machines, and knows how to find out which data banks and programs are accessible, will have a great source of information available to him. The doctor, the engineer, or the factory manager will have his capabilities greatly enhanced. The politician will be able to call upon vast resources of data to support his case. The decision-maker in public office or industry will have knowledge at his fingertips, and knowledge, as Francis Bacon announced, is power.

Power, like wealth, is an attribute which can feed upon itself. The rich become richer. The person with power can use it to acquire more power. Knowledge of how to use the machines will be a key to more knowledge. In the computerized society the gap between the 'haves' and the 'have-nots' will be self-widening.

Two hundred years ago power in society used to reside largely with those who owned *land*. The majority of production came from the land. Ownership of land automatically meant a dominant position in the economy. With the advent of industrialization, land gave way to *capital* as the dominant factor of production, and power now resided with those who owned or controlled capital. Economist John Kenneth Galbraith, in *The New Industrial State*, argued that more recently a shift of power has been taking place from those with capital to those with specialized technical ability:

Power has, in fact, passed to what anyone in search of novelty might be justified in calling a new factor of production. This is the association of men of diverse technical knowledge, experience, or other talent which modern industrial technology and planning require.[1]

Power, Galbraith argued, accompanies whichever factor of production is hardest to replace:

It adheres to the one that has the greatest inelasticity of supply at the margin. This inelasticity may be the result of a natural shortage, or an effective control over supply by some human agency, or both.

In an age of rapidly increasing technical complexity, the natural resource that will be in shortest supply is human talent – talent for management, for complex decision-making, for research, and for bringing the computers into effective use. Galbraith therefore concludes that power resides increasingly with the 'technostructure' – those people who combine to use specialized talent and group decision-making to form the guiding intelligence of corporations. The power to control complex corporations lies not with those who have capital, not with the shareholders, but with the technostructure.

As we shall discuss later, we think that in the computerized society the shortage of talent is going to become extremely acute. As the computer applications discussed in Section I are gradually realized, the control of most functions in society will become more sophisticated and more complex. Society will demand that crucial decisions be made in a much less haphazard manner than today; but this will require an understanding of computer techniques. As the complexity of controls increases, so the talent to deal with it will evidence increasing 'inelasticity of supply', and power will reside more firmly in Galbraith's 'technostructure'. Power for both the individual and the corporation will come, to an increasing extent, from an effective relationship with the machines.

The lawyer with access to and an understanding of legal data banks will be able to serve his clients better. The doctor who can obtain the computer's assistance with diagnosis (and can assess the shortcomings of the machine) will increase his chances of arriving quickly at correct diagnoses and cures. The architect who can use a computer screen and light pen, and make full use of programs developed for him, will command a versatility far ahead of his competitors and will not be bogged down in the details of drawing and calculations. But in each of these cases we shall demand a greater talent than today. These men must understand their own profession, *and* understand the computer programs – and the latter will be very complex.

The high level of talent needed, combined with its scarcity, will again cause an inequality in society, just as ownership of land and capital did in earlier eras (and to a lesser extent still does). The inequality will be of a somewhat different nature because it will not depend so heavily on the position of the family the person is born into. Some persons will be able to acquire the skills in demand through hard work alone. But by and large the ability to acquire power in the world we envisage will be a gift that is scattered at random throughout the populace. Provided that all classes have the same opportunity to be thoroughly educated, it will be independent of colour, race, pedigree, and – to a large extent – of whether father was a tycoon, professor, or dustman.

2. *Power for Business*

In the halcyon days of unfettered enterprise, the pioneers of big business spread across the wide-open spaces of the American continent. Now that the space dimension has been fully exploited, those who seek to grow must master *time*.

The firm that is first in a new market, the financier who most quickly recognizes the leading indicators, and the producer who first develops a new technology will be among those who succeed. In all cases, information is the key, for rarely now does anyone go far by mere luck or enterprise alone.

Big business always seeks the rapidity of response to changing markets that it envies in its more nimble small competitors. However, a random sample of small companies brought together in a group will not show significantly higher growth. In big firms an unsuccessful venture causes decreased profits; comparable failures in small firms put them out of business.

The fashionable response to such diseconomies of scale has been to decentralize, after the highly successful pattern of General Motors and General Electric. Economists studying the 'theory of the firm' have failed to demonstrate that very large firms necessarily have ultimately higher costs than their smaller competitors. Of course, they have shown that a large plant can run out of supplies of men and materials in its neighbourhood and thus incur extra expense in drawing them from further afield. What historically has defeated the multiplant corporation as it grew is inefficient *communication*. Not only are managers ignorant of how to respond to some situations, but they also lose even the incentive to respond.

Just before the Second World War, US Steel commissioned a study which revealed it to be

a big sprawling inert giant ... improperly coordinated ... lack of long-run planning ... antiquated systems of cost-accounting ... inadequate knowledge of the costs or ... profitability ... of items it sold ... inadequate knowledge of markets ... slow in introducing new processes and new products.[2]

Computers will never supply the motivation injected by the new regime at US Steel in the late 30s, but they can provide the vehicle through which motivation is disseminated and directed. Even when everyone is bent on making profits for the company, it is still essential to insure that no one bankrupts another part of the organization in so doing. When a corporation is merely a conglomerate of companies, each making different products independently and neither competing for markets and materials nor using much of each other's output, decentralization is simple. Each part of the organization can be given a target and management suitably encouraged to achieve it.

In a large, highly integrated firm, however, decentralization is simply the admission of defeat, 'Balkanization' a confession of failure. The failure lies in the inability to coordinate efforts among parts of the whole. The drunkard cannot place the cork in his right hand into the empty bottle in his left, because his brain cannot tell his fingers what his blurred vision can hardly see.

The firm must decide what both limbs are to do and give instructions to that effect. Management-information systems are designed to do just that.

One currently popular technique bears a striking resemblance to decentralization – the profit-centre concept. The organization is broken down into a series of small operating units, each of which aims to make a profit on the operations it controls. The managers know both their objectives and how they are measured. A single profit-centre itself may be no more than a single department in one plant – a much finer subdivision than in a decentralized firm. The system can be so tuned that the manager is only held responsible for costs directly under his control. Somebody else will be measured on the costs he does not control and cannot affect. The amount of information needed to run such a scheme is very much higher than with conventional companies. The accounts are in fact kept in multiplicate, and the computer writes them out in the form most appropriate for immediate management needs. The result is a mechanism of control that can direct an organization of any size.

We are moving now into an era in which large corporations whose scale makes them near-monopolies within the geographical limits of single countries may be the most efficient form of organization. Small nations already accept that the world is the market for their industries, and that the industries of the world must provide the competition within their national market. In particular the small nations are under attack, economically, from the United States. To survive, the separate firms within a national industry may combine forces, like those in the

British computer industry. Ten years ago there were many British computer manufacturers; successively they merged, and now there is only one. This one is better able to deal with the American competitors, some of which have set up manufacturing plants in Britain. A similar trend to near-monopolies can be seen in other industries. Economies of scale result from fewer manufacturing plants, less product diversity and duplication, bulk-buying, bigger and better research facilities, and more aggressive and efficient marketing. This is made possible by the removal of geographical barriers with data-transmission, and the improvement in information flow due to data-processing. In the near future the cost of data transmission is going to be largely independent of distance, other than for very short distances. As industrial information systems become more efficient, so the economies of scale are more likely to be more fully realized.

In corporations which manufacture such complex products as computers, aircraft, cars, or telephone exchanges, the need for bigness is apparent. Research, design, and the establishment of production facilities are very expensive, and the more units that are manufactured the more these costs are absorbed. Computers could also justify bigness in industries that have traditionally been small-scale, such as furniture-making and perhaps clothing and shoes. If a customer could walk into a clothing store, choose a fabric and a design, be measured up, and receive the garment within a few days – at the same price as off the peg – he would be less willing to accept the limitations of ill-fitting clothes which require alteration or the two-month delay for hand-tailoring. Carnaby Street styles a few hours old could be available almost immediately. The cloth would be automatically collected from the warehouse and cut by a numerically controlled laser to the customer's exact size. The little man would have more difficulty in competing.

Computerized production, unlike the mass production of earlier days, will enable vast automated plants to turn out products of endless variety. Henry Ford's era, when 'a customer could have any colour of car provided it was black,' will be replaced by an era in which the production machinery responds sensitively and rapidly to changing customer demands and fashions.

Large firms will dominate many industries in the computerized society. Oligopoly is going to make economic sense because it can facilitate production of goods and services at lower prices and often in greater variety than a myriad of competing small firms. In smaller countries today, some industries are regarded as 'natural' monopolies, among them railroads, telephones, mail, electricity, gas and water supply, airlines and broadcasting. In Britain the steel and coal industries are nationalized. In the smaller countries, other such industries and perhaps some manufacturing industries may gradually become to be thought of as 'natural' monopolies, if only to stave off aggressive competition from abroad.

Today companies are taking commanding positions in the economies of other nations, both developed and underdeveloped. Domination of world trade by a few hundred large corporations is probable by the end of this century. Some of these corporations will be instruments of the policies of their home governments; others will defy any *mere* nation. Huge industrial firms must control costs, prices, and their customers' behaviour in the interests of stable markets and commercial success. They must also maximize profits.

When General Motors chose to supply the Japanese market from its US plants rather than its Australian Holden subsidiary, Australia had no other automobile to export to an important market, while it was being inundated with Japanese products.

The British government had to allow a third aluminum smelter to be built in the UK by Alcan. Had it followed its inclination and allowed only two, those of British Aluminum and Rio Tinto Zinc, to be built, Alcan would have set up in Ireland, with which Britain has a free trade treaty through which the Irish smelters' products would have been admitted on the same terms as the British smelters' output.

The world's biggest corporations have sales comparable with the gross national product of a country like Belgium. They can, by choosing where to site plants or reveal profits, severely affect the economies of whole nations. By releasing or concealing technical know-how, or directing the supply of strategic materials they can influence military strength.

If the balance of industrial power swings in favour of bigness, more of the public are going to be 'organization' men, conformist and often unenterprising – their innovations institutionalized into progressive product development. Again we come back to the impact of power on the individual's quality of life. To some, behaving as a mere cog in the technological structure will represent a starvation of spirit in a world of material plenty. In contrast, the benefits conferred by small businesses are not lower prices, but relative freedom for their proprietors and those who choose to associate with them.

On the other hand, the computer industry itself is bringing about many new opportunities for providing knowledge, skill, services, and software. It is not difficult for a small firm to provide new types of programs or better programs than those in existence. The demand for computer consultants is ever growing. Almost everyone involved with computers is in need of education and re-education, and much money is to be made out of short specialized courses. There are innumerable services that the computer can provide, some of which were described in the first section of this book. As the terminals spread, there will always be a market for new services to dial on the telephone. The scope for innovation in these types of markets is endless, but penetrating them may demand more specialized knowledge, more intellectual ability, and probably a higher IQ than setting up in business as a tailor or furniture-maker did in the past. Power, once again, will accompany those on the right-hand side of the bell-shaped curve of computer-aptitude.

3. *The Power of Precedent*

When we describe someone's attributes to a machine we do so by classifying him into a number of groups. We may be interested in a graduate of a particular college, of a particular age, within a specific income bracket, and working for such and such a firm. This kind of information is frequently used by, say, a personnel officer to determine whether to hire a particular individual. He will base his decision on both the similarities and the differences between the present candidate and his predecessors, whose history is known.

Present-day computers generally look for exact matches with precedents. To insure that plenty will be found, systems designers tend to limit the possible classifications. Insurance policies are typical: a few facts suffice to determine a premium, the similarities between one case and another being enough to justify ignoring the differences.

This inclination to categorize everything on the basis of a few key attributes and to ignore all else as basically irrelevant is both the characteristic strength and the limitation of bureaucracy. Public opinion notwithstanding, bureaucracy is a very efficient method of dealing with all matters for which there are classifications and precedents. Because of such efficiency, legislators in both democratic and totalitarian countries delegate great power to bureaucrats who are expected to arrive at decisions with a robot-like indifference to persons.

This is very much in line with our modern concept of equality before the law and a far cry from the distinctions of noble, commoner, and slave of earlier generations. We expect the punishment to fit the crime – for which we have a precedent – rather than the criminal – for whom we do not.

We find such thinking most apparent in court, where the establishment of precedents can be as vital as the establishment of facts. The 'competitive' advantage that the lawyer with access to a computer for searching files of precedents holds over his 'opponent' with no such aid is of major concern to both lawyers and laymen. The possibility that justice may be subject to the length of the lawyer's purse is as abhorrent an idea as that the wealth of the litigant could affect a verdict. There can be no doubt that if information relevant to legal judgments is more readily available to those with more money, justice cannot survive.

But the situation regarding legal precedents is less of a potential danger than the possible introduction into the courtroom of individual histories gleaned from dossier banks. We have already discussed the consequences of such use when the information is wrong (see Chapter 16). Even if it were right, the conclusions drawn from it could well be wrong. In either case, the associated bad publicity would be such that others would try to avoid 'getting involved'.

Most people have a strong propensity to draw invalid conclusions from skimpy evidence. Considering that driving licences are denied to many epileptics – possibly for sound reasons – if we were to find a positive correlation between highway accidents and one-eyed drivers, should all such be denied the right to drive? Such partial correlations will become more frequently apparent as more computers are used for the analysis of statistics about people. Since we can do more through calculations, we shall eventually improve predictive capabilities to such an extent that we can, say, define the characteristics of groups *likely* to fail or succeed in activities all the way from exploring space to knitting socks. We shall need to maintain strict vigilance if we are to prevent having our freedom seriously curtailed merely because, despite our differences from them, similar people have abused their opportunities or have made illuse of them.

4. *The Power of the People in Power*

Political power will be enhanced by improved data-processing, as will personal and industrial power. This will be true in the democratic governments and, frighteningly, also true in totalitarian governments and in the relatively unstable governments of the emerging countries.

Computers can help both in gaining power and in keeping it. Candidates in elections in the US have already used them for performing sensitive, geographically variable analyses of what the voting public wants to hear, constructing their campaigns accordingly. This was first done in a spectacular manner by John F. Kennedy in 1960. Some would say that such use of computers could improve the democratic process by making candidates react more responsively to the wishes of the majority. On the other hand, cynics will say that candidates simply tell each group what they know it wants to hear, and that campaign statements bear little or no relationship to any action a candidate might take once in power. In a Presidential election campaign it is surprising how, travelling across the country, the candidates' arguments vary from section to section. In the last election, it would have done President Nixon great harm if his commercial broadcasts to some of the southern states had been heard in New York and other northern cities. A computer-planned campaign sensitive to the differing wishes of different parts of the electorate works well where nationwide communications are bad. It might be highly successful, for example, in a country like Brazil.

There are other, less innocuous ways of gaining power that could benefit from using computers. There is nothing new about these other methods except the relentless efficiency with which they can now be implemented.

Information is the key to keeping power once it is attained. It must be gathered and summarized into statistics; access to it must be controlled, particularly by governments fearful of democracy; it must be distorted, suppressed, and even invented by tyrants. In the computerized society, government power will be amplified by highly efficient machines for such information processing. Whether the power will be used for good or ill we cannot tell. But any government that reserves to itself all channels of information and all data banks gains absolute control of information and thereby gains absolute power.

The quantity of information currently collected by the United States federal government is enormous. Washington has many 'intelligence' organizations. Agencies such as the Treasury and State Departments have their own intelligence bureaux. Each of the armed services has its own intelligence branches. There are the Federal Bureau of Investigation, the National Security Agency, and America's largest single government agency the Central Intelligence Agency. Reputedly the world's largest computer installations are in the National Security Agency and the CIA. Information is transmitted from all over the world and digested by the machines these agencies use. The handling of such masses of data would have been unthinkable before the computer. The way the data are processed and the conclusions that it is possible to draw from them will no doubt improve as the computer era advances. Today the President can obtain immediate intelligence reports on any critical event anywhere.

During the Arab–Israeli war of 1967 both Washington and Moscow presumably received up-to-the-minute intelligence reports of the situation. President Johnson was in direct communication with the Sixth Fleet in the Mediterranean, and at the same time was using the 'Hot Line' to Moscow. When the Israelis attacked the American intelligence ship *Liberty*, American carrier-based aircraft flew to the scene immediately. The Russians knew instantly of this, so in order to prevent their misinterpreting it President Johnson explained the action over the Hot Line. Never before has information been available to governments in such a fast and effective manner.

Governments will have a tool capable of providing information not only about situations but also, if they choose to use it, about individuals.

Totalitarian governments have always required a network of informers to help maintain control of the political system. By detecting deviations and applying corrective force early and precisely, stability can be maintained. A government can obtain background reports on potential trouble-makers. We have already discussed the potential for harassment if a government has access to the many dossiers on individuals that are likely to exist in a computerized society.

The computer networks, as well as expediting the flow of information a government acquires, may be used to control the information a government puts out.

Confounding an enemy's information is a trick not new in history. Both the Allied and the Axis powers in the Second World War had plans to disrupt each other's economy with forged banknotes which would have destroyed the basis of credit on which the monetary system rested. The equivalence of money and information was stressed in Chapter 4. The American economy would be in trouble if all sources of credit-rating data were simultaneously destroyed. The great stock-market crash of 1929 was amplified by a realization by investors of the unreliability of information about the present state and future prospects of business as reflected in stock-exchange quotations.

Suitable manipulation of the data base on which people depend can lead to loss of confidence and an environment conducive to revolution. Where people have learned to cope with poor information, destroying the information has little consequence. In the nineteenth century, rigging the international trade figures had no impact on trading, but manipulating the information on which share prices were based had a major effect on political as well as commercial empires. Still earlier, the political power of the medieval Catholic Church was destroyed in all those countries in which information was allowed to circulate freely.

Controlling the information needed to detect deviations from a system's goal makes it possible for alternative goals to be achieved. The medieval church used its network of informers and the concept of heresy to prevent people from detecting the absurdity of many of its preachings. The fact that its opponents, when they eventually succeeded, promulgated equivalent nonsense did not matter – the destruction of the integrity of the information in which people had confidence is what lay the basis for revolution.

The prime target of the contemporary *coup d'état* is control of the broadcasting stations, for they are the source of instant information to the general public. The invasion of Czechoslovakia by Warsaw Pact nations in 1968 failed to secure control of Czech radio and television, with the result that the populace questioned the invaders' claimed motives. Other *coups* that failed to control information sources failed completely, for political power rests on control of information even if it is seized initially by military force.

A free citizenry, no less than a free market, depends on the availability of accurate and adequate information. Today the information reaching the public comes largely via the news media. If these stand to gain from major distortion of reports, the country will hardly be 'free'.

In areas served by only one newspaper, the political views

of the publisher are sometimes manifested in the opinion columns of 'his' press. In a democracy, no press 'baron' should be permitted to own all local broadcasting media with exclusive coverage. In some under-developed countries radio sets are given to the people permanently tuned to a station broadcasting propaganda. In many, television is entirely government-controlled. Today it is possible that cable TV in a locality could become controlled by one man.

Political oppression begins with the dictation of opinion, whether by a democratic majority or a theocratical hierarchy. Since opinion is based on information, however gathered, opinion can be controlled by restricting the sources of information. If this restriction proves totally successful, rebels never have cause to appear, for the world appears to be the best of all possible worlds – according to the picture conveyed by the available information. Unfortunately, the rebellions of one generation provide the seeds from which the accepted truths of the next develop. Rebellions are started by minorities.

The defence of rule by democratic majority is not that it is always right, but that it is less often wrong than rule by any one minority. Because the majority is not always right, it is never acceptable for the majority to enshrine its opinions and demand that they be accepted by everyone as absolute truths. Nevertheless, those in power have always done so. Censorship by the limitation of the dissemination of conflicting views has been the traditional approach. All the great tyrants of the past have used this method, from the Pharaohs by way of the Caesars and the Popes to the Commissars. Modern technology from the Gutenberg press to Lord Haw Haw's radio has permitted the widespread propagation of false information in the guise of truth. The computer and the associated electronic means for disseminating information will allow the process to be carried a stage further.

5. Military Power

Computers will enhance military power in a variety of ways. First, they will permit the building of even more terrifying weapons systems. Missile systems already operate without human control once the missile is launched. The state of the art in 'intelligent' robot-control is near the point at which a variety of other delivery mechanisms for nuclear weapons could be built and operate without men. The missile-launching submarine of the future may not need a crew. It can then be designed to operate at great depths for long periods of time, without the conventional problems of providing living space, good air, food, and comfort. Similar robot submarines could be built to police the commercial sea lanes or protect man's new sub-oceanic enterprises. The supersonic bomber flying under enemy radar might find its own way to its target. It could be designed (like the submarine), in evading enemy action, to take far more punishment than a plane with a human crew. A variety of other robot war devices have been proposed, ranging from Wellsian machines that operate on land to a 'Doomsday Machine' that automatically takes massive retaliation if its country is destroyed.

The antiballistic-missile (ABM) issue is probably only the first of a lifetime of such controversies. As soon as any automatic defence system is constructed, a potential enemy must find out how its sensors work and devise a mechanism for fooling them. The United States' Safeguard system might be fooled by an enemy using elaborate decoys. The ABM then has to be made much more complicated if it is to distinguish between the decoys and the missiles carrying warheads. The decoys must then, in turn, be made more convincing. Again, Safeguard might be fooled by a nuclear blast above the earth's atmosphere, creating radar blackouts lasting tens of seconds – long enough for enemy missiles with hydrogen-bomb warheads to slip through. The solution to this problem could be to destroy the missiles closer to the earth – the moment before they land. At altitudes below twenty miles, the radar blackout is not serious. The speed of a missile hurtling toward earth is so great that such a system would require immensely quick action, but it is technically possible. The incoming enemy missile must then be designed to have other means of evasion for the last twenty miles. An atomic explosion twenty miles above the

earth would instantly broil any exposed person, but the retaliatory missile sites could be made safe.

... And so the argument continues endlessly. The machines
for attack stand a good chance of outwitting the machines for
defence; the machines for defence must then be made more
elaborate; and so on. As the research continues, the cost of
superbly intricate systems becomes much lower. Technologists
are usually so impressed by the ingenuity of their latest brain-
child that they want to build and test it. The Semi-automatic
Ground Environment system (SAGE) was obsolete long before
it was completed. (And the cost would have paid for the education of every Negro child in America!) The same is probably
going to be true of our future defensive systems. But still the
race goes on. It may become more complex in the future when
it must be designed to outwit Chinese missiles as well as Soviet,
and the two will use different evasion techniques.

If man should one day tire of technological evolution, it will
probably still race onward because of the struggle for defence.

Spectacular though computerized weapons systems may be,
a more significant contribution that computers can make to
military power would stem from their ability in gathering and
processing intelligence data and using them for decision-
making.

All manner of data that relate to military situations
are currently collected by data-processing methods. Once
in the machines, data can be transmitted almost instantly to
any part of the globe. Information from the battlefield computers is combined with information from intelligence sources,
political sources, surveillance satellites, and from commanders
in other zones to form an overall picture of the situation.
The computers can arrange for appropriate displays
of these data to commanders, the Joint Chiefs of Staff, and
the President and his advisors. In the years ahead, the com-
prehensiveness and the processing power of such information
networks will grow. The ability to display instantly the appropriate information will become more sophisticated and the
computers will give more assistance in command decision-
making.

Such uses of computers already involve a number of implications, all of which will become more significant as the technology improves. The first relates to the speed with which events happen. In the last century, fighting in a foreign country sometimes continued for as long as a week after peace had been declared at home.

Before the Japanese attack on Pearl Harbor in 1941, a warning message was sent by General Marshall in the War Department in Washington to the Commanding General in Honolulu; but because of radio interference and inoperative teletype machines, it failed to arrive until after the attack.

As of this writing, the nations of the non-communist world have four working telecommunications satellites for civilian use, each of which handles many hundreds of channels simultaneously; the United States military has *twenty-four*. This is an indication of the volume of data that they pass around the globe at the speed of light.

The second implication is the interconnectedness of events. Isolated incidents in remote parts of the world that once would have had no effect on events elsewhere now blend into the overall global intelligence picture. What was once a situation solely concerning local authorities may now involve the CIA, the Pentagon, and possibly the President. A border incident or an attack on an American ship would once have been dealt with on the spot by the nearest authority; now it would constitute an instant crisis involving the highest authorities, and the President is flashed a report which links it to intelligence data about the country in question, the state of its relations with the Soviet Union and China, and public opinion at home, and which provides computerized assessments of the possible effects of each of a list of alternative courses of action. There is a large network of ramifying issues: an American military strike is now related to a whole series of problems of which a local commander cannot be aware. Calculations will have indicated probable repercussions in various parts of the world, how hostile countries will react, what support friendly countries will give, and so on.

The third implication of such a fast and well-integrated information flow is that it involves high-level commanders in decisions that formerly would have been made at some lower level: the buck is passed upward at great speed. Often the President is involved in what were previously field decisions. In the Vietnam war the press reported how President Johnson became involved in such command decisions; for example, he personally reviewed each day's bombing targets. Field commanders, it is sometimes reported, resent such interference from on high, but the technology now permits bringing the best brains together on critical situations. To an increasing degree, command will be exercised from a war room in the Pentagon and involve persons and computers in different locations connected via terminal screens.

The pressure on the Commander-in-Chief will be immense. His health may eventually be monitored continuously by computer, as with astronauts now, and often he will have to be sustained, like recent Presidents, with tranquilizers, sedatives, energizers, and other drugs. Such medical systems will become much more elaborate in the future.

Smaller countries, too, will have their war rooms, perhaps commanded by leaders less likely to react with caution. It is alarming to think of some future Latin American or African dictator – perhaps a sick man – sitting in front of a computer display screen that will be essential to the prestige of a small state.

Fourth, with increasing frequency military and political situations are being played out in a simulated fashion on the computers. 'War games' that help to indicate the effects of alternate courses of action are becoming ever more elaborate. As in other areas of computerized life, use of these models tightens the discipline of decision-making. Those involved are forced to answer questions they may never have considered before. They are forced to be precise, to define what they mean exactly, and to quantify what before was vague. In one sense this is good, especially as the military traditionally has a penchant for vague thinking and impulsive decision-making. On the other hand, it also means that increasingly decisions will be made with relentless speed and with less human flexibility.

Actions indicated by simulated war games do not necessarily correspond exactly to real situations, because the computer cannot take individual human reactions into account with sufficient sensitivity. In the 1962 Cuban missile crisis, President Kennedy's assessment of Premier Khrushchev's reactions was all-important. Robert Kennedy states in his book on the subject:

During the crisis, President Kennedy spent more time trying to determine the effect of a particular course of action on Khrushchev or the Russians than on any other phase of what he was doing. What guided all his deliberations was an effort not to disgrace Khrushchev, not to humiliate the Soviet Union, not to have them feel they would have to escalate their response.[3]

The danger with warfare conducted on the basis of information that can move with electronic speed is that there is little time for such vital deliberations. In this regard Robert Kennedy wrote:

If we had had to make the decision in twenty-four hours, I believe the course we would have taken would have been quite different and filled with far greater risks. The fact that we were able to talk, debate, argue, disagree, and then debate some more was essential in choosing our ultimate course.

Such time is hardly likely to be available in some of the computer-assisted crises of the future.

A terrifying situation emerges at points in the computer war games when the probability of losing becomes so high that the only strategy which will minimize losses is a massive pre-emptive strike. The strike must destroy as nearly completely as possible the retaliatory power of the enemy within the shortest possible time. The 1967 Arab–Israeli war provides an example: the sudden and devastating Israeli strike, designed to take the enemy by surprise as much as possible, and based on superb intelligence information, was the optimal strategy. To what extent does this set a pattern for future wars in which intelligence-gathering is fully computerized and the strike capability may be nuclear? The probability calculations of the war games lead only too often to the pre-emptive strike.

In the past the diplomat had a vital role to play in preventing wars. Conventional diplomacy moves slowly, however, and the diplomat is out of his depth in the world of systems analysis. Time and again in cocktail-party conversation one can hear him say, 'I don't know the first thing about computers – wretched things.' He has been trained in different arts. Worse than this, the events in computerized crises move too quickly for him to act. In the electronic and nuclear war situations of the future there will be no time for the art of diplomacy: the tense exchange on the Hot Line will replace it.

Doctor H. Wheeler of the Santa Barbara Center for the Study of Democratic Institutions summed up the effect of computers as follows:

In the next two decades we must expect that the most significant contribution of computers to warfare will be not the enhancement of weapons performance but the revolution in the handling of intelligence information. As we can already see, this revolution makes every incident a crisis and pushes every decision to the highest levels of responsibility: it puts political leaders in charge of military affairs but makes them more likely to use force. The time for decision in response to enemy action has shrunk from days or hours to seconds, and will become in a sense negative when future intentions of the enemy can be *predicted* with reliability. Then the computer will be all-important, and men will have to decide whether to believe what it says.[4]

REFERENCES

1. John Kenneth Galbraith, *The New Industrial State*, Penguin, 1969.

2. Testimony of George Stocking, an economist, during *Hearings before Subcommittee on Study of Monopoly Power*, House of Representatives, 81st Congress, 2nd Session, 1950, Serial No. 14, part 4A, *Steel*, pp. 966–967. Government Printing Office, Washington, DC.

3. Robert F. Kennedy, *Thirteen Days*, W. W. Norton, New York, 1969.

4. H. Wheeler, chapter entitled 'The Strategic Calculators' in *Unless Peace Comes: A Scientific Study of New Weapons*, The Viking Press, New York, 1968.

21

CHANGES IN EMPLOYMENT PATTERNS

When this circuit learns your job, what are you going to do? –
Poster in New York City buses.

Perhaps the most publicized aspect of automation, which is
also the aspect likely at present to give rise to the strongest
feelings among working men, is the question of job displace-
ment. Without doubt, vast changes are going to be needed in
the way men are employed. However, we suspect that most of
the popular accounts tend to overestimate the degree to which
unemployment will be created, and indeed, the degree to which
computers will, in the next two decades at least, increase our
leisure time.

In 1956 a book which described the original 'Leo' computer's
ability to calculate the wages of an employee in two seconds
as 'fantastic' predicted a 32-hour week 'in a few years', and
quoted John Diebold as anticipating shortly a 'five-day weekend'.[1] Today, with machines a thousand times faster, we do
not seem much closer to fulfilling these predictions.

The predictions for unemployment have been alarming in
the extreme. Norbert Wiener in 1954 wrote the following:

The automatic machine ... is the precise economic equivalent of
slave labor. Any labor which competes against slave labor must
accept the economic conditions of slave labor. It is perfectly clear
that this will produce an unemployment situation, in comparison
with which ... the depression of the thirties will seem a pleasant
joke. This depression will ruin many industries – possibly even the
industries which have taken advantage of the new potentialities.[2]

Since Wiener's prediction, there has been a steady succession
of equally dire prophecies. So far, at any rate, they have not
come true. Many employees *have* been displaced by computers,
but in general they have quickly found employment elsewhere,
and usually were transferred to another job in the same organ-

ization. Indeed, it has become an oft-repeated joke in the computer industry that since the installation of the giant machine a company employs more people than it did before.

Since the mid 50s, when the above predictions were made, there have been installed in the United States more than 50,000 computers for use in administrative work alone, each capable of doing the work of a large number of men. At the same time, production automation has been progressing rapidly. Yet as we see from Fig. 21.1, the American unemployment rate has not exhibited an upward trend since 1954. Indeed, from 1961

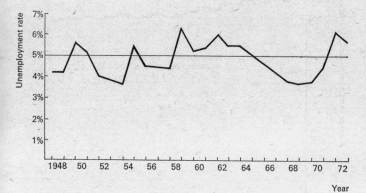

Fig. 21.1 Changes in unemployment rate in the United States since 1948.

to 1965, when large numbers of computers were installed, the unemployment rate fell. Furthermore, as will be seen from Figure 21.2, the number of hours worked per worker per year has fallen only very slightly. In the decade from 1954 to 1964 the fall was equivalent to about one hour per week. If we do not do any better than that in the next decade or so, we shall still be working a 38-hour week on average in 1980.

What has happened? Simply this: the increased productivity made possible by automation has raised the Gross National Product rather than increasing unemployment or decreasing working time. The average annual growth rate of the US Gross National Product between 1953 and 1960 was 3·5 per cent.

This average from 1960 to 1965 was 4·6 per cent. From 1965 to 1970 it has been nearer 3·5 per cent again. The working force has been rewarded with increased affluence rather than with increased leisure time.

But what of the future? The next decade will see *far* more computers installed than the last, and as we can see from the curves in Chapter 1, they will be dramatically more capable.

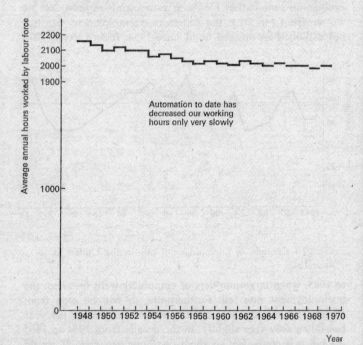

Automation to date has decreased our working hours only very slowly

Fig. 21.2 Decrease in hours worked by the labour force since 1948. Source of figures, US Labour Bureau.

The voices of alarm are still to be heard. In 1964 a report of a Sub-committee on Employment and Manpower of the United States Senate Committee on Labor and Public Welfare included the following paragraph in its summary:

Speculation concerning future developments has led some to fearful conclusions. Several witnesses, among them manufacturers and installers of automated devices and computers, believe that the economy is on the verge of a breakthrough in automation and computerization which would make future reliance on present trends misleading. Among these witnesses were those who held that cybernation would inevitably make substantial portions of the American labor force unnecessary and hence unemployable. Not only blue-collar workers, but white-collar workers and even middle-level management, they thought, would be replaced by electronic and mechanical devices. The result would be a stratified society consisting of a small number of menials performing at low wages functions not worth mechanizing, a few skilled workers and technicians working short hours for substantial incomes, and a scarce supply of high-talented, highly educated policymakers seriously overworked in attempting to solve the problems of civilization. The mass of people would be capable of no productive economic contribution and would be furnished income and activity as a substitute for employment.[3]

This reflected the views, however, of only a small proportion of the witnesses who testified before the committee.

INCREASE IN PRODUCTIVITY

Computers, automation, and improved efficiency in general have the effect of increasing overall productivity per man-hour. This can be measured in terms of the Gross National Product per man-hour, which as of this writing in the US is $5.20. This figure has been increasing at the average rate of about 3·5 per cent per year since 1947. It seems reasonable to assume that the net effect of automation will be at least to maintain this 3·5 per cent growth rate. Some authorities claim that a higher figure can be achieved, and indeed the growth rate in recent years has been somewhat higher. The consensus of opinion among economists seems to favour a forecast of about 3·5 per cent; engineers and scientists, on the other hand – when they address themselves to this question – commonly predict that the use of computers and other new and powerful technologies

TABLE 21·1
1. Assuming a growth rate in Gross National Product per Man Hour of 3·5 per cent per annum

year	1967	1970	1975	1980	1985	1990	1995	2000
No decrease in hrs. worked								
annual hrs. worked	1,920	1,920	1,920	1,920	1,920	1,920	1,920	1,920
GNP per capita	3,800	4,213	5,003	5,942	7,058	8,383	9,956	11,825
Hrs. worked drop by 0·25 per cent annually (average of past 10 years)								
Annual hrs. worked	1,920	1,905	1,881	1,858	1,835	1,812	1,790	1,767
GNP per capita	3,800	4,181	4,904	5,752	6,747	7,914	9,282	10,887
Hrs. worked drop by 0·5 per cent annually								
Annual hrs. worked	1,920	1,891	1,844	1,798	1,754	1,710	1,668	1,627
GNP per capita	3,800	4,150	4,807	5,568	6,449	7,470	8,652	10,022
Hrs. worked drop by 1·0 per cent annually								
Annual hrs. worked	1,920	1,862	1,771	1,684	1,602	1,523	1,449	1,378
GNP per capita	3,800	4,087	4,617	5,215	5,890	6,652	7,514	8,487
Hrs. worked drop by 2·0 per cent annually								
Annual hrs. worked	1,920	1,807	1,633	1,476	1,334	1,206	1,090	985
GNP per capita	3,800	3,965	4,257	4,570	4,906	5,267	5,654	6,070
Hrs. worked drop by 3·0 per cent annually								
Annual hrs. worked	1,920	1,752	1,504	1,292	1,109	952	818	702
GNP per capita	3,800	3,845	3,921	3,999	4,079	4,160	4,243	4,327

2. Assuming a growth rate in Gross National Product per Man Hour of 4·5 per cent per annum

	year	1967	1970	1975	1980	1985	1990	1995	2000
No decrease in hrs. worked	Annual hrs. worked	1,920	1,920	1,920	1,920	1,920	1,920	1,920	1,920
	GNP per capita	3,800	4,336	5,403	6,734	8,392	10,457	13,032	16,240
Hrs. worked drop by 0·25 per cent annually (average of past 10 years)	Annual hrs. worked	1,920	1,905	1,881	1,858	1,835	1,812	1,790	1,767
	GNP per capita	3,800	4,303	5,296	6,518	8,022	9,872	12,150	14,952
Hrs. worked drop by 0·5 per cent annually	Annual hrs. worked	1,920	1,891	1,844	1,798	1,754	1,710	1,668	1,627
	GNP per capita	3,800	4,271	5,191	6,309	7,667	9,319	11,325	13,764
Hrs. worked drop by 1·0 per cent annually	Annual hrs. worked	1,920	1,862	1,771	1,684	1,602	1,523	1,449	1,378
	GNP per capita	3,800	4,207	4,986	5,909	7,003	8,299	89,35	11,656
Hrs. worked drop by 2·0 per cent annually	Annual hrs. worked	1,920	1,807	1,633	1,476	1,334	1,206	1,090	985
	GNP per capita	3,800	4,081	4,597	5,178	5,833	6,571	7,401	8,337
Hrs. worked drop by 3·0 per cent annually	Annual hrs. worked	1,920	1,752	1,504	1,292	1,109	952	818	702
	GNP per capita	3,800	3,957	4,235	4,532	4,850	5,190	5,554	5,943

should *raise* the growth rate, perhaps to 4·5 per cent. It certainly appears plausible that a 4·5 per cent growth rate could be maintained in the United States if this were a deliberate national objective. In Japan the figure is *very* much higher.

ALTERNATIVE LEVELS OF AFFLUENCE

For a given growth rate expressed as Gross National Product per man-hour, assuming that the unemployment percentage does not rise we can make an approximate estimate of how affluent we are likely to become in the next decade or two, and produce some figures which indicate the trade-off between increased leisure time and increased wealth. These are given in Table 21.1. The top half of the table is computed for the growth rate in Gross National Product per man-hour of 3·5 per cent and the bottom half for 4·5 per cent.

Taking the top row in the table, if we continued to work the same hours as today, the Gross National Product per capita would be double the 1967 figure by the late 1980s – specifically, by 1988. By 1988 the public, on average, would consume about twice as much as they do in 1967. If the 4·5 per cent growth rate were assumed, however, the same thing would happen by 1983. Almost certainly, however, we shall not go on working the same number of hours per year as in 1967; and if our working year declines, the Gross National Product per capita will not grow so fast. Taking the bottom row in the top half of the table, if the amount of time we work drops by 3 per cent per year, then we shall work only half as long as today by 1990 but we shall not be much richer. Again, almost certainly this will not happen. Reality lies somewhere between the top and bottom rows.

We invite the reader to choose: of these various alternative futures in Table 21.1, which row would he select? The authors posed this question to a small political group in New York, and the majority chose the row second from the bottom: 4·5 per cent growth rate and working time decreasing by 2 per cent annually. This means that by 1980 we would be consuming about 36 per cent more than in 1967 and be working 1,476 hours per year. Working a 5-day week, 8 hours per day, would leave 13 weeks of vacation and 10 legal holidays. A 5-day week with 7 hours work per day would leave time for an 8-week vacation. Alternatively, people might work a 4-day week, 8 hours work per day, and take off 8 legal holidays, leaving a 4-week vacation.

By 1990, continuing the same growth, we would be consuming approximately 73 per cent more and could contemplate a 20-week vacation if we worked a 5-day week, 8 hours per day; a 16-week vacation if we worked a 5-day week, 7 hours per day; a 12-week vacation if we worked a 4-day week, 8 hours per day; and a 7-week vacation on a 4-day week, 7 hours per day. We would not yet have dropped to a 3-day week, however.

Probably this is not the way it will work out in practice. Advertising agencies will become increasingly skilled at making people buy products they would not otherwise want, and there will be much more emphasis on increased consumption than on increased leisure. Also, a massive proportion of the nation's research-and-development capabilities will probably be spent on space-exploration and defence activities, rather than on increased productivity, and the productivity growth figure of 3·5 per cent might be the one nearer to actual achievement. All in all, then, the third row of figures in Table 21.1 – even if it is not the one we would choose – is perhaps the most plausible, giving an increase in consumption of about 46 per cent by 1980 and a working year of about 1,798 hours – 5 weeks vacation with an 8-hour working day and 5-day week or, perhaps more likely, a 3-week vacation with a working day of about $7\frac{1}{2}$ hours.

Unfortunately, it seems naive to assume that a country has the freedom to select so definitely which of the alternative futures it will build. The vastly expensive mechanisms of society operate with so great a momentum that an enormous force is needed to deflect them from their course. Society is a vast machine immeasurably more complex than any computer system, and to a large extent we are in the grip of that machine. We cannot make a simple choice between the rows in Table

21.1, but if the alternatives are understood there is hope that the course of society's institutions will be slowly directed into those paths most desirable in a changing world.

Sooner or later, we are going to reach the time when people work, on the average, a 4-day week and take off two or three months of vacation time each year. This *could* come about in the next two decades, although we pessimistically expect the mechanisms of our consumption-oriented society will prevent it being that soon. The use of the consequently increased spare time may be a difficulty itself to begin with, but after a generation or so such amounts of leisure time will seem man's natural right. Men will look back upon today's 40-hour week – and vacations so short that people barely have time to adjust to the change in pace and climate – as part of a grim past: the same attitude that *we* hold toward the nineteenth century's 80-hour working week and forced labour of children in the mines.

Homo ludens will find much with which to fill his life. Profit-making recreation agencies will supply him with spectator and participant sports. The universities will widen their age range for students' enrolment, while those students who cannot attend full time will take courses from terminals at home. The Protestant ethic will have outlived its usefulness, with ambition and competitiveness receiving the opprobrium attracted today by sloth and indolence.

DIFFERENCES IN EMPLOYABILITY

Before we reach any such utopia, however, there is a transitional problem that may become severe in the extreme. The authors do not seriously imagine that later in the 1970s they will be working even as few as 40 hours per week. This view seems to be shared by a variety of their colleagues involved in technology – not only those in high-pressure corporations but also many in government, in universities, in space projects, and so on. This view also seems to be shared by social workers the authors meet, doctors, United Nations delegates ... even the local parson. These people have one factor in common: they foresee themselves to be in very great demand in an economy which will have achieved a major degree of computerization. On the other hand, the majority of people in the US are presently in jobs that will be *less* needed.

Table 21.1 averages the growth rates over all of society's job categories. Although this gives a useful indication of the overall trade-off between leisure time and consumption, we must look more closely at the different types of employment. The massive use of computers does not and will not affect all jobs equally; and while the average productivity growth rate may be 3·5 per cent, there will be some jobs for which it can be expected to be very much higher than this, and still others for which it will be lower. For routine clerical work, for example, this rate will rise enormously; but for lawyers, doctors, sociologists, and the like, it may rise hardly at all.

Average figures of employment and unemployment can be misleading. In 1964, for example, the figures for the United States show 74 million persons in the labour force and 3·9 million unemployed (5·27 per cent). However, during 1964, 14·1 million experienced some unemployment (19·0 per cent). The unemployment by occupational category ranged from near zero to 10·6 per cent.

While many – perhaps the majority – of today's areas of unemployment are going to become automated, there are other spheres in which one can foresee only a shortage of qualified people. The shortage will probably be nowhere greater than in the computer field itself. If we are to take advantage of the technology whose future growth is projected by the soaring curves in Chapter 1, the consumption of personnel by the data-processing field is going to be enormous. Figure 21.3 shows an estimate of the growth of personnel in this field in the United States to date. The rate at which computers can be put to use in teaching, in law and medicine, in providing government information, and in the many other ways which are now possible will depend largely upon how many people are available for programming and planning, and for building the requisite banks of information.

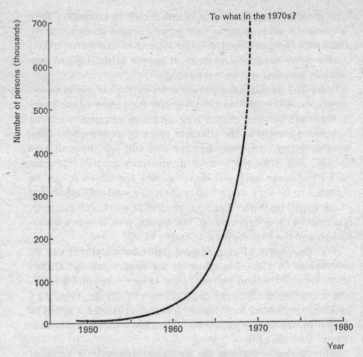

Fig. 21.3 An estimate of the number of people employed in the Information Processing Field in the United States.

Other areas of technology will be very short on personnel also. The United States Department of Labor forecasts an increase in required 'professional, technical, and kindred workers' of 13·2 million during the period from 1964 to 1975; this represents an increase of 54 per cent.[4] During the same period it projects a decrease of 21 per cent in 'farmers and farm managers, laborers, and foremen'. This is a trend which will probably continue well beyond 1975. (Visions of a future computer operator chewing on a straw?!) But the question arises: Will we ever be able to train the man who would have been back home on the farm, or in another declining sphere of employment, to work well in the computerized world? And perhaps in the long run an even more important question: Will he ever be happy there? Some perhaps will. Man is very adaptable. However, one can find many currently in the technical professions, especially older people, who yearn for the world 'far from the madding crowd'. To a large extent the answers to these questions depend upon what new types of work we are going to create, and this in turn depends upon what kind of world we want to construct for the future.

In general, there will be a vast increase in demand for personnel of high IQ, logical ability, and technical aptitude; while at the same time there will be a shortage of the *routine* jobs in which the majority of mankind has been employed so far. There will always be some routine jobs in the foreseeable future which would be either uneconomical or undesirable to mechanize, such as small-shopkeeping, answering letters, bartending, and clearing up the newspapers that blow down the street. In the 1970s there will be many routine industrial jobs not yet reached by mechanization, but their numbers will be decreasing. Many a bank clerk will eventually have to learn computer-programming.

This of course will be, in Norbert Weiner's lasting phrase, a more human use of human beings. Some think that there can be few jobs worse than being a bank clerk. To most people, routine mental work is more monotonous than routine physical work. One of the authors once worked on the factory production line during college vacations. He was doing the work of a fairly simple machine, but his mind was free to roam where it would; he could talk to his neighbours or listen to the piped-in music. However, the person doing a routine clerical job is tied body *and* mind. Eventually, no doubt, we shall evolve a society in which most people have jobs which they can enjoy and find interesting, and those drudge-like jobs that are left will have very short hours.

One difficult problem will be posed by the existence of jobs that are dull but responsible. Automated processes which are prepared to handle all routine divergences from normal operation will still seek the aid of human controllers in the event of impending catastrophe. Such crises would be rare, but their correct solution would require expertise. Like a doctor at sea with a healthy crew, such a man would have to cope with anything that might happen, knowing it is unlikely to. Perhaps he would learn to beat the computer at chess in his idle hours.

THE DECLINING JOB CATEGORIES

During the transitional phase ahead, those sections of the working population in the declining job categories will be faced with three possibilities: first, retraining so that they move to work that *is* in demand, such as jobs created by the computer technology; second, much shorter working hours; and third, unemployment with improved welfare benefits. Probably we shall see an increase in all three of these. The last is clearly undesirable. It is difficult to see how the majority of people can avoid depression and apathy, and feeling rejected by society, if they are unemployed. Most would slip into forms of human degradation, becoming recluses, tramps, alcoholics, and – worse – drug addicts, criminals, or people whose hatred of the society in which they cannot participate takes on forms of violence. There will no doubt be an increase in cults such as the hippies. One aspect of the problem is that it hits the youngest working-age groups severely. A person who can happily opt out of society in his teens or early twenties, drunk with the gay irresponsibility of contagious decadence, will probably not be so happy when he is forty. For Dorian Gray depravity was great, but he had lots of money and a certain talent too. For most hippies middle age will be a nightmare of spreading despair, a communal, cankered hell. Increased unemployment must not be a part of any future we invent.

Most men like to work. A German study found that shorter working time was placed last behind good working conditions, good treatment by superiors, good relationships with colleagues, interesting work, good promotion prospects, and good training facilities. Less than one in eight would have given up work if he were financially independent.[5] Work is a social necessity in the mid-twentieth century. But this may be because of conditioning by society: many aristocrats of the past seemed able to enjoy total idleness without discomfort.

In Britain during 1970 and 1971 unemployment climbed inexorably, to the consternation of the Government and its economic advisers, who were unable to account for the ineffectiveness of traditional stimuli to economic activity. It seems likely that the recession forced many employers to cut back labour which had been underutilized in the past. When prospects improved, instead of rehiring, they relied on the extra productive potential of existing men and machinery, more efficiently deployed, to raise output.

The recession also hit the computer professionals. After the artificial boost provided by changing the country's computer programs over to decimal currency, involving many thousand man-years of work, the bottom dropped out of the market for programming. Several large and many small software houses failed to find any contracts and went out of business; computer-using companies cut back their staff, deferred major projects, and, most significantly, began to get much more work out of fewer people. Decimalization had forced many data-processing departments to bring their management and programming methods up to date, with a consequent large rise in productivity.

In the nineteenth century the 16-hour working day was defended on the grounds that workers would only use extra leisure time to get drunk. We are now less worried about the use of free time; and in the future the need for continual education and retraining to keep pace with technological change will cut into this spare time. Shorter working hours are a basically desirable achievement if we can learn to employ leisure time in a satisfying way.

However, we may face the prospect of more than half the community having a shorter working week while the meritocratic minority work very long hours, some driving hard to maintain the technological acceleration while others sort out the problems it causes. The fraction of the community working long hours would presumably be compensated by high salaries. The balance between the rich and the poor would become different from today, with 5 or 10 per cent of the community becoming very wealthy. That half of society with IQs lower than the median would have little chance of raising their living standard to the higher brackets.

The most desirable alternative, then, is to design the new technology in such a manner that the maximum number of people can participate. The computer industry itself, for example, could design terminals and program the machines so that they do not permanently demand a high IQ of their operators. The computers should be tolerant when their users make mistakes, and tell them in clear English what to do. They should switch to teaching programs when necessary to lead their less-intelligent users gently into the correct ways. This would need much programming and careful (but easy) research. It is not being done in the computer world today except in rare, isolated cases. Today computer-programming and operating is becoming, if anything, more difficult – to some extent needlessly so. There is an industry-wide tendency for people in the know to generate difficulties rather than to simplify: it helps them preserve their professional standing. Efforts to develop new and radically simplified computer languages cut across the steadily engulfing increase in complexity and jargon. In all areas of technology, not just that of the computer, a major and deliberate effort at simplification could increase the range of persons who could participate. The achievement of simplification is in itself usually a complex and expensive process, for it is usually only a superficial simplification. In other words, an elaborate mechanism has been arranged whereby an easy-to-understand human operation such as setting an automatic pilot produces a complex sequence of events. By a conscious and not inexpensive effort, then, a large number of today's clerks, factory hands, and punched-card girls who are going to be displaced by automation can be roped into more technical work.

A good example of what can be done to match work to worker is provided by IBM United Kingdom's scheme for training the blind to write programs. Teaching material was dictated for use on 'Talking-Book' tape recorders or converted to Braille text. Braille-writers were provided along with specially adapted typewriters, the former to provide high-speed output from the computer and the latter for writing programs in COBOL, one of the most widely used of all high-level languages. Although this solves the problems of communication between man and computer, a more intractable difficulty is preparing a 'specification' telling the programmer what is required. Such specifications are usually prepared in typescript by systems analysts following discussion with the user of the computer. The blind programmers have got round the difficulty by tape recording the meetings and making their own specifications in Braille.

CHANGING JOBS

It is inevitable that during the next few decades of transition many persons will have to adopt new kinds of jobs. Blue- and white-collar workers alike will constantly have to be retrained for new types of work. The *acceptance* among the working force of the need for constant retraining and job change is essential. Here again the more intelligent are better off. A man of intelligence often *likes* to change his job: he sees it as a new and stimulating challenge. For the low-IQ portion of society, however, it is a nuisance and sometimes a great source of worry.

Once more, the machines themselves can be used to help in this situation. A computer data bank of job vacancies could be made available by telephone links to all employment offices. The employment office could use the computer to search the files for jobs which best fit the applicant's requirements. The British government's plans for establishing such a system were reported as follows in *The Times*:

The Ministry of Labour is to spend £500,000 on computer facilities for handling Selective Employment Tax calculations. This will be the first stage of what could become a £1 million complex linked to every employment exchange in Britain to provide immediate information about job vacancies ...

Ray Gunter, Minister of Labour, sees it as a further development in shaking off the 'dole queue' image of employment exchanges. With the computer he hopes to match man and job systematically, and provide better labour market information for both employers and workers.

The 970 exchanges in England, Scotland and Wales place 1·5 million ablebodied adults each year and provide other aids to mobility and replacement at a cost of £8,300,000.

Using the computer will not replace individual judgment of Ministry officers, but should make it easier and quicker to match vacancies and applicants, especially where applicants are mobile.[6]

There is many a slip 'twixt plan and implementation of a computer system in British government. So often the government is reluctant to spend the money necessary to make it work. However, this type of scheme should be given every encouragement. It will clearly be necessary in the future we envisage.

RETRAINING

We still have much to learn about the introduction of change, despite a decade and a half of experience installing computers. A job means more to a man than his pay envelope. There are the factors of esteem in his own and his colleagues' eyes, satisfaction in the exercise of his skills, pleasure in his authority and influence over events, and so on. To many a senior administrator, the introduction of computers to his firm has been a greater blow to his pride than to his pocket, and in extreme cases has resulted in sabotage. Someone who owes his job status to experience will only regain it by retraining – *if* he has the basic qualities necessary. If Britain took up baseball and America turned to cricket, few older players, whatever their averages, would adapt with marked success.

Even the desire to make a success of a new job is not enough. Many computer-programming schools have exploited the willingness of people who do not possess the necessary aptitude to pay large sums of money for training that will be wasted. The racket has become serious enough to tarnish the image of other, ethical organizations.

Another major necessity is increased education. Between 1958 and 1965, total annual educational expenditures in the United States rose from $21 billion to $39 billion, and legislation since 1965 indicates that the growth will continue. Unfortunately, the educational expenditure in Europe is not rising as fast, and this can only spell trouble for the future. While the best schools and universities in Europe are second to none, the educational profile of the society as a whole is markedly inferior to the United States', and the gap is widening. Figure 21.4 shows the growth in that proportion of the working population who have received high-school and college education in the US. These curves show the proportion of the *total* working force; thus if we plotted the number completing college and high school, the slopes would be even steeper. The curves approximate straight lines, whereas most curves that relate to technology growth increase exponentially or faster. The demands of technology, therefore, will undoubtedly outstrip the provisions of increased education in the long run.

Furthermore, increased education does not change (at least to any major degree) the IQ profile of society. It increases the number of people who can participate in the high-demand work. However, the low-IQ half of society is not going to be brought into much greater demand.

CREATING NEW TYPES OF JOBS

One solution remains: if we are to automate millions of jobs out of existence, as has already happened in farming and mining and seems certain to happen elsewhere, we must *create* some alternative work for the less capable employees. This is probably one of the most crucial issues in the whole question of unemployment. If we create jobs for this section of the community, they must not be jobs of a useless nature, or work which could more economically be automated. One of the authors remembers seeing a row of workmen in Ankara, the capital of Turkey, crawling across a public lawn cutting the grass with very small shears. On asking about it, he was told

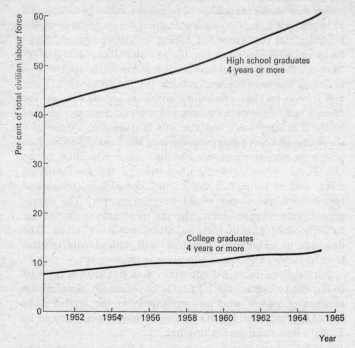

Figure 21.4

that this method prevented them from being unemployed. *This* is *not* the type of work we want to create. Nor do we want to keep Bert on to 'mash the tea because he's a good lad', or any longer have plumbers with mates to 'carry their tools'. With some imagination and government help, a large number of jobs can be created which would make our society much more pleasant to live in.

The United States, for example, is badly in need of public work in its city streets, its slums, its countryside, its parks, and in its hospitals, its libraries, rest homes, and other institutions. In national beautification, it has been estimated that there is full-time work for 1·3 million people today, and in medical institutions and health services for 1·2 million. One could extend this list, and it seems anomalous that so many public needs for unskilled labour exist at the same time as an unemployment rate of 4 million or so. If the government employ otherwise unemployed labour, the unemployment benefits cease and the country has some return for its money.

TO ENRICH THE WORLD

The vast rewards of American industry have been spent mainly on private affluence and little on public amenities. Compare the appearance of such a city as, for example, Geneva with an American city – New York, say, or Los Angeles. In Geneva one finds colourful, perfected flowerbeds, flowing lawns, picturesque steps around the cathedral, a superbly maintained lakefront with yacht moorings and a giant fountain. One can walk – and many people do – for mile after mile in the most beautiful surroundings – without leaving the city. Many gardeners and other workers are constantly busy creating and maintaining an environment that is delightful to live in. In Manhattan, where most people cannot own a car, one might begrudgingly walk around the block for exercise, but hardly for pleasure. Manhattan was once more richly endowed than are most towns with nature's grandeur: the wide-ranging waterfronts, the rock outcrops, the great rolling Hudson and the New Jersey Palisades on the opposite shore could have made it one of the most beautiful cities in the world. In the affluent future that is ours it will be tragic if Americans have to live in megalopolises that they cannot walk in. One of the authors has a neighbour who feels he must have a dog in his flat for protection. He subscribes to a New York dog-walking service and periodically the professional dog-walker arrives in a van. The dog, presumably with many others, is driven off in the van to greener pastures, duly exercised, and returned. Building and maintaining city surroundings like Geneva where it would be a joy to take the dog require the work of many people. These people can be low on the scale of worker-desirability; yet it is pleasant work. The person of low IQ can feel happy and creative planting beautiful flowerbeds or pruning flowering cherry trees, and – more important – feel that he has a worthwhile place in society.

Gardening is a type of *craftsmanship*, which along with many other forms of craftsmanship has become less often practiced in the stampede for technological efficiency. In losing craftsmanship, we have in many spheres lost *quality*. The craftsman has talents of a different sort from those demanded by our mechanized age. More than often he does not have the high logical ability coveted in today's world. He would be lost working at a computer terminal. He works instead with his hands and eyes. His thinking is often visual, perhaps intuitive. Whatever brain processes give him a 'green thumb' or make him skilled with a chisel are not processes which could place him in great demand in today's labour market. The craftsmen who constructed Chartres' stained-glass windows and the Byzantine mosaics might well be unemployed in the computerized world.

In many fields, a return to craft industries would make our living richer. Mass-produced bread and pastries do not compare with those made in small French village bakeries. In a world as affluent as ours, why should we not drink wine out of superbly hand-cut glass, glittering in the glow from a silversmith's candelabra, and eat from fine bone china rather than plastic disposable plates? In the food industry, too, craftsmanship has largely fled. The judges who determine ratings for the famed *Guide Michelin* found no restaurants in Britain or America worthy of Michelin's three-star (highest) or even two-star rating. There were only a handful of one-star quality. Dining in France at La Pyramide or L'Auberge du Père Bise is like existing for a glorious few hours in the previous century. Such perfection cannot be found elsewhere today. And yet one wonders whether such places make much profit. Probably not. It is pride, not profit, that cultivates excellence.

Today, it would be virtually impossible to build cathedrals such as those at Rheims or Canterbury. The requisite craftsmanship does not exist. One can go to a service each Sunday in Oxford or Cambridge in the most magnificent surroundings with a choir and thundering organ as perfect as man can make them. But where in our modern cities?

House servants, too, are something of a past era. Yet in a community where certain people have talents so valuable that we would like them to work as long as possible, does not it make sense to relieve them of their menial tasks? A business executive has a staff to perform all his routine work so that he can concentrate his efforts where they are most needed. In the future we envisage, with the computer terminal in the home, the talented members of our society will spend long hours at home carrying out research and experiments, developing teaching programs, designing buildings, editing film, composing music, writing, and so on. They will probably be gripped by their work and not want to leave it to make the bed and clean the bath, to cook and take their socks to the laundry, or even to write letters, pay bills, and address Christmas cards. Intelligent, capable women will not want to spend all day looking after their children. When we fear unemployment for some, and underemployment for one-half the population, should we not be planning to use the untalented in support of the talented?

By the end of the century, two thirds of all work will comprise administration and service, work which is indifferent to physique: women can do it as well as men. Computer-programming has been a proper feminine occupation since Lady Lovelace programmed Babbage's Difference Engine more than a century ago. As it is a new craft, the sort of barriers to entry inherited from the past in so many older professions were in this case never raised against women.

There have been some exotic forecasts of automation in the home, including robots that can be programmed to do the housework. The following was written by a professor of mechanical engineering at Queen Mary College, London:

The great majority of the housewives will wish to be relieved completely from the routine operations of the home such as scrubbing the floors or the bath or the cooker, or washing the clothes or washing up, or dusting or sweeping, or making beds.

By far the most logical step to allow this variety of human homes and still relieve the housewife of routine, is to provide a robot slave which can be trained to the requirements of a particular home and can be programmed to carry out half a dozen or more standard operations (for example, scrubbing, sweeping and dusting, washing-up, laying tables, making beds), when so switched by the housewife. It will be a machine having no more emotions than a car, but having a memory for instructions and a limited degree of instructed or built-in adaptability according to the positions in which it finds various types of objects. It will operate other more specialized machines, for example, the vacuum cleaner or clothes-washing machine.

There are no problems in the production of such a domestic robot to which we do not have already the glimmering of a solution. It is therefore likely that, with a strong programme of research, such a robot could be produced in ten years. If we assume that it also takes ten years before industry and government are sufficiently interested to find the sum required for such development (which is of the order of $1 million), then we could still have it by 1984.[7]

The authors of this book regard it as most unlikely that any such general-purpose robot will be available in their lifetime, although a variety of special-purpose devices could make housework easier. One might build a self-directing vacuum cleaner which comes out of its cupboard at a given time, snoops around looking for dirt, and then goes back into its cupboard. A self-making bed would not be difficult to design – the more mechanically minded reader might like to try his hand at inventing one. We favour a design in which a blast of air is blown through the bed and the sheets pulled into place, although the owner would have to make sure he was out of bed before switching it on. Whatever gadgetry becomes available, however, it seems that a large amount of manual domestic work will remain if we want an attractive house and garden. A self-motivated vacuum cleaner, however ambitious, is unlikely to be as good as a cleaning lady, and it is difficult to imagine the robot looking after children or taking the dog for a walk.

Part-time household help – nursemaid, cook, gardener, handyman, domestic secretary, and perhaps chauffeur (especially if there is a radio terminal in the car) – would both enable the meritocratic top 10 per cent to produce more and relieve the unemployment problem for the bottom 50 per cent.

There have developed certain emotional overtones about this subject of servants. This is especially so in the United States, where most people consider it beneath their dignity to work in this way. In England once there used to be considerable pride about being in domestic service, and the last traces of this can still be found in rare instances. The 'gentleman's gentleman' and the British army 'batman' were proud, and carried out their jobs with great diligence. One finds remnants of this attitude among college servants who still attend to students' needs in the older universities, and sometimes also among the staff of the traditional European hotels. In parts of Japan and other Eastern countries the service is still a delight. It is conspicuously lacking in even the most expensive American hotels. The servants of the future need to be trained, as every other job needs training. Is it too much to hope that the training could once again cultivate pride in the work? Perhaps we could set up special colleges (the works of P. G. Wodehouse might be required texts). We cannot expect everyone to become a Jeeves, but maybe those who see the course through – for such a course an I.Q. of less than 110 would be mandatory – could be given a bachelor's degree in domestic service!

Most – probably all – of the factors we have discussed in this chapter *can* be controlled by means of a deliberate public effort; it seems highly possible, however, that they will not be. The danger lies in complacency and inertia. The countries of the world are faced with alternatives over which they could have control. Most will opt for the increased riches that can be gained with the new technology. If, however, they do this without dealing powerfully with the employment problems that are a direct consequence, the result will be misery for many millions – even billions – of people. Unemployment, the mass feeling of emptiness and rejection, the sickening purposelessness will in many lead to a hatred of society erupting into civil strife and violence. A conclusion of the United States National

Commission on Technology, Automation, and Economic Progress was:

If unemployment does creep upward in the future it will be the fault of public policy, not the fault of technological change.[8]

REFERENCES

1. E. M. Hugh-Jones, *The Push-Button World*, Basil Blackwell, Oxford, and University of Oklahoma Press, Norman, 1956.

2. Norbert Wiener, *The Human Use of Human Beings*, 2nd edn., Houghton Mifflin Company and Doubleday Anchor Books, New York, 1954.

3. Subcommittee on Employment and Manpower of the Senate Committee on Labor and Public Welfare, *Toward Full Employment: Proposals for a Comprehensive Employment and Manpower Policy in the United States*, US Government Printing Office, Washington, DC, April 1964, p. 20.

4. US Department of Labor, Bureau of Labor Statistics, *America's Industrial and Occupational Manpower Requirements, 1964–75*, US Government Printing Office, Washington, DC, 1967.

5. The study, conducted by the Institute of Empirical Sociology at Saarbrücken, is quoted in Otto Neuloh, 'Automation and Leisure', *Science Journal*, January 1968, p. 79.

6. Richard Wagner, 'Jobs by Computer', *The Times*, 6 November 1967.

7. Professor M. W. Thring, article in the series, 'The World in 1984', *The New Scientist*, 1964.

8. 'Technology and the American Economy', Report of the National Commission on Technology, Automation, and Economic Progress, US Government Printing Office, Washington, DC, February 1966.

Part III: Protective Action

22

HANDLING CHANGE

The major advances in civilization are processes which all but wreck the societies in which they occur. – Alfred North Whitehead.

Our society has largely evolved by a process of chance mutation and adaptation, in the unplanned manner of natural selection. It reacts to major problems only when they starkly present themselves.

Studying the unsuccessful evolutionary byways, however, is a very slow way of learning. Nature wasted a lot of effort on dinosaurs. Man could choose not to do so if he could predict that dinosaurs were not viable. The dinosaurs would be saved a lot of unhappiness too.

The city slum area is a dinosaur of social evolution, an unhappy digression originating in overspecialized reaction to a mutation. Was it necessary? Would we have deliberately planned it, even as an intermediate stage? The dirigible was a dinosaur of air travel. Certain computers have been dinosaurs of their industry. Some of today's mass-transportation systems are dinosaurs still struggling to evolve into viable creatures.

As the pace of technology quickens and its inventions become more and more devastating, so it becomes more dangerous to allow society to evolve by natural selection. If the law of the jungle prevails for the next fifty years and during this time we have the technical innovations that are now being predicted in, say, electronics and biochemistry alone, then society will be thrown into unspeakable turmoil. Educated man might be nature's latest dinosaur, transformed suddenly by a chain reaction in his own cleverness.

In nature, one unviable spur in evolution did not matter too much (except to the creatures involved!) because there were innumerable other paths of development. If the same were true of the development of man's applications of science, then we might be able to weed out the undesirable paths. It is neither true now nor likely to be true in the future, however. The usage of science and the directions of society spread only too rapidly from one country to another. Few can isolate themselves, and in the future, given immensely more powerful communications and more money to spend, fewer still will be able to. We are likely all to go down the same path together. We *must* learn to think about, and to guide, our own evolution.

If we had a large amount of time in which to think about society and to test out the effects of changes that we could introduce, then we would be better able to guide the process. Slow evolution can be controlled evolution. The rate of technical change, however, is becoming too rapid. It is only too apparent that many institutions of today are failing to keep pace with the changes. We are not directing the technology; it, to a major extent, is directing us.

Nothing has been left unscathed by the impact of scientific advance. The beliefs common to the majority in bygone generations are now often the mark of the crank or the eccentric. At one time parents could raise their children as they themselves had been raised and expect them, as a result, to cope successfully with the life they must lead. Now the education and activities of children must be quite different from those of their parents. Even some moral precepts of the past have to be re-examined if society is to survive in today's frenetic environment.

Increased knowledge has revealed the primary role of chance in the evolution of man, where once there was a belief in purpose. But where science has questioned the will of the gods, it has made possible the will of man. Man is at last in a position to guide his own evolution – only to discover that this evolution is so rapid he cannot be certain of controlling it.

There is no shortage of measures of change. There are books on almost any subject with charts and tables accompanied by admiring or lamenting commentary on how fast change has occurred: populations shift and increase; productivity and production expand; transport gets ever swifter and more congested; the statute books thicken; calculations become cheaper and quicker; education and rates of literacy rise; scientific publications multiply even faster than book-publishing; crime increases and its detection techniques change; the strategy and tactics of warfare change beyond the comprehension of earlier generals.

In order to plan for the handling of change, we would like to be able to make predictions about the rate at which it will occur.

Figure 22.1 shows a typical curve of growth. An animal

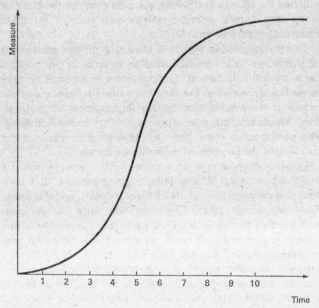

Figure 22.1

population, the profits of an industry, the firepower of a class of weapon, or the speed of a class of computers will tend to follow this development. In the beginning, there is a slow start; the system gathers momentum, exploiting the initial development until 'saturation' sets in. Ultimately, the environment will no longer sustain an increase; predators or competitors appear;

the limit of piecemeal improvements is reached, and the market is fully covered.

The measurement of performance of a particular technology may follow a curve such as that in Fig. 22.1. The increase in the speed of the automobile, for example, has followed such a curve. In many technologies, however, one class of invention replaces an earlier class for performing the same function. When we look at the performance of one class by itself, we find that it follows a curve similar to that in Fig. 22.1; but in studying the change as a whole, we must consider the effect of many such curves superimposed as each class of invention replaces the previous one.

This is illustrated in Fig. 22.2. Here each of four generations of technology or classes of invention replaces its predecessor. Each, by itself, is limited. Developments in engineering take its performance so far, but then the natural restrictions of the technique place a limit on further improvement of performance. Meanwhile, however, a new technology comes into being. The performance curve for the industry as a whole is the envelope of the component performance curves.

This is happening in many industries. The curves in Figs. 1.1 to 1.9 are examples. In computers, one 'generation' has been replacing the previous one about every six years, and this seems likely to continue. In the first generation, price and performance reached the limits imposed by vacuum tubes, delay-line memories, punched-card input and output, and bit-by-bit programming. In the second, transistors broke the 'reliability barrier,' magnetic rings allowed mass-produced memories, magnetic tape made it possible to feed the machine as fast as it could digest data, and symbolic programming languages raised the productivity of programmers. In the third, price dropped another tenfold, 'solid-logic technology' replaced transistors, machine organization raised throughput and allowed attention to be paid to many jobs at once, and higher-level languages brought the machine directly to the user. Communications between men and between machines have greatly improved simultaneously. This will almost certainly continue. We now look forward to monolithic circuitry, large-

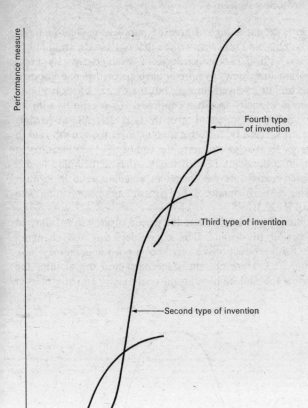

Fig. 22.2 As one type of invention or technology replaces another the growth in performance continues for a sustained period. This has happened in the computer field, for example, as one 'generation' has replaced another.

scale integration, a high degree of parallelism, data-management systems, and other new ideas that change the technology. Similarly with data-storage devices: in high-capacity communication links new inventions have succeeded one another, each technique by itself limited, but the overall effect being an increase in capacity that has continued for a century with an approximately exponential growth rate. In nuclear-particle accelerators, there has also been an exponential growth rate in the energy level achieved, since the end of the 1920s, leading to the gigantic atom-smashers of today. Also contributing to this have been many different classes of machine, each in its turn following a performance-improvement curve something like that in Fig. 22.1.

In other patterns of change the period of growth is followed by a period of decline. The curve depicting the changing market for a product or the spread of a phenomenon may look like Fig. 22.3. Here are the stagecoach and the airship, the daguerreotype and eight-year-old coalminers, and the market

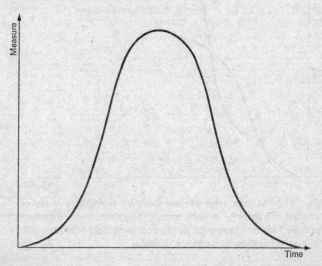

Fig. 22.3

for hula hoops and the unsuccessful Ford Edsel car. This is the picture that describes transients, albeit transients of relatively long duration.

Nearly every system follows one or another of these growth patterns, though not necessarily with quite the artificial smoothness of the curves shown. Locating the position on the curve occupied by any particular system at a given time is difficult, but it is essential to forecasting the future.

The view of the future that ordinarily has been held by the public in this century grossly underestimates the potential for change. We are all, in differing degrees, prisoners of familiarity. This is often true also of the technologist viewing his own discipline and of the industrialist viewing his own product. Both the technologist and industrialist understand in such fine detail how things are done today that they cannot imagine the sweeping changes that will occur tomorrow; the details of these are not yet known. He focuses on the limitations set by the current state of the art. Less than twenty years ago the data-processing industry was predicting, for example, that 'only ten or a dozen very large corporations will ever be able to use the computer with profit'. That view, expressed in 1948, failed to define the limitations in such a way that they could be seen as temporary. Programming ceased to require Ph.Ds, the number of logic circuits could be increased without increasing the probability of breakdown by replacing vacuum tubes with transistors, calculations per dollar could be increased until they so far exceeded man's capability as to justify programming rare decisions as well as routine ones.

The traditional response is based on a limited field of vision, often that of the specialist viewing his specialty. Here is a classic example, quoted by T. Vincent Learson. IBM's President, in an article on 'The Management of Change'. They are the words of the keynote speaker at the annual meeting in New York of the National Association of Carriage Builders earlier in this century:

Eighty-five percent of the horse-drawn–vehicle industry of the country is untouched by the automobile. In proof of the foregoing permit me to say that in 1906–1907, and coincident with an enormous demand for automobiles, the demand for buggies reached the highest tide of its history. The man who predicts the downfall of the automobile is a fool; the man who denies its great necessity and general adoption for many uses is a bigger fool; but the man who predicts the general annihilation of the horse and his vehicle is the greatest fool of all.[1]

Every innovation has short- and long-term constraints on its adoption. Modern marketing in many instances reduces the short-term limitations imposed by consumer ignorance. The longer-term limits may be economic or technological, or they may be social or political.

Each of these limitations has an associated time constant, which defines how quickly the curve can rise. A phenomenon constrained by the rise in national income will have a rate of change of only a few per cent per annum. Another change may be 'price-elastic', becoming nearly universal even more quickly than prices fall. Nylon stockings were a case in point.

When technology has matured, its time constants are long and the associated prices change only slowly. Figure 22.4 shows

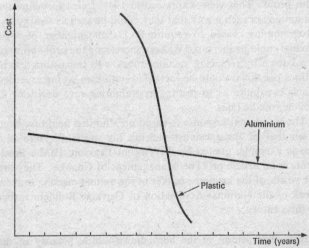

Fig. 22.4

the prices of aluminum and plastic. There are many products in which the one may be substituted for the other. Before the crossover point, price for the aluminum product would have limited market penetration, to a few per cent per annum. The substitute product's arrival is followed by almost immediate market saturation.

As the computer improves in value-for-money, it traces a nearly vertical price curve, like that of plastic in Figure 22.4 It crosses on its way a series of nearly horizontal established-method pricelines. Computer methodology in many fields becomes suddenly cheaper than traditional methods (Fig. 22.5).

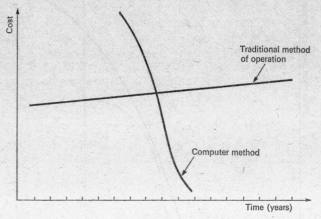

Fig. 22.5

No sooner can one foresee incipient saturation of computers for one application, resulting in a replacement market supplemented with new installations at the rate of increase in national product, than a new market is entered by crossing of another long-standing priceline. In business and industry one application after another falls within the province of the computer. Machines will eventually handle schoolroom drill more cheaply than teachers; using a domestic terminal will be cheaper than driving to the bank, travel agent, or supermarket; data banks will be cheaper than repetitive form-filling. As each area is exploited, the economies of scale force prices ever lower.

One might foresee a terminal in every home, a terminal in every manager's office, and one for every few students at school or college. The price and universality will approach that of the television and telephone. By then the terminal industry will be mature, like the automobile industry today, and the market will be largely saturated with a rate of growth governed

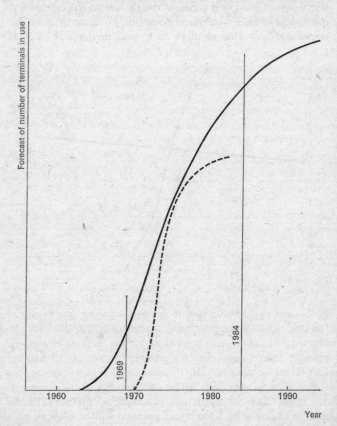

Fig. 22.6

by house-building, replacement, and piecemeal improvement. It will be near the top of the curve in Fig. 22.1.

Figure 22.6 shows a projected curve for computer-terminal usage, indicating the 1969 and 1984 positions on the curve. By 1984 we think that the terminal industry will have matured to the position of the automobile in the late 20s – as an accepted part of life. The dotted logistic shows how sharp a change is required to prevent this growth through new action now. If society wishes to avoid the consequences of computers, it must act with near-revolutionary speed.

The terminal market may reach saturation while the computer industry itself is still far from the top of its curve. The potential for using computers is so complex and all-embracing that it may be centuries before man quenches his thirst for more machine power and greater storage facilities. The fifteen or so years projected in this book are only a tentative beginning.

It is not the market for computers that primarily concerns society. Our concern must be the horizontal lines, characteristic of established ways, whose paths will be crossed – with drastic consequences – by the vertical progress of information technology.

A curve somewhat similar in shape to Fig. 22.6 indicates the change needed in education as we move into the computer era (Fig. 22.7). Unfortunately, education is changing only at a very slow rate. The dotted curve indicates our fear that education may lag disastrously behind what is needed, as indeed it is doing already. We shall discuss this in the next chapter.

Again, the problem with our present law is not only the slowness with which it operates, but also the slowness with which it develops. In few other fields of human endeavour has the rate of increase in fundamental understanding remained constant in the last generation. We still tend to think that man was made for the law, not the law for man. We must frame our laws for the future, not for the past, and design them to cope with what might happen, not merely what has.

The computer itself represents a tool which can help us find out what will happen: we must experiment with the future. Modelling and simulation have been a recurrent theme

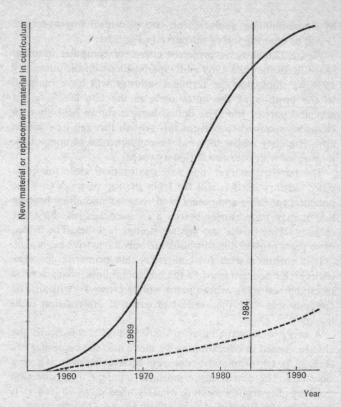

Fig. 22.7

throughout this book. We can simulate the development not only of computers, but of any technological or social advance.

If the buggy-maker had simulated the transport market, he would have related the many factors determining the sales of his product. In so doing, he would have been obliged to quantify his estimates, and with the figures he could then have conducted experiments. If the automobile was not to annihilate the horse and his vehicle, something would have to cause the curve (Fig. 22.1) of value for money in automobiles to reach its plateau before the market for transport was itself saturated. No such limit existed.

As was discussed in Chapter 11, we have already built computer models of urban transportation in cities. These provide a measurement of the congestion to come if we take no steps to introduce improvements and enable us to compare the effects of the possible different courses of action (all expensive). As new inventions and ideas emerge we can use our models to test their potential viability.

Some state and local governments have legislated the prohibition of the automobile in certain areas, with considerable environmental benefit. Others have adopted similar restrictions to regulate and limit their use. It is becoming increasingly desirable to regulate certain other uses of technology. We can rarely prevent 'progress', whatever our doubts about its improving the world; but we should learn how to guide it and restrain its excesses. We now have the ability to predict the excesses without waiting for them to be revealed. We may be unable to determine precisely the date and the extent of the damage that would follow inactivity, but we can at least indicate the range of possibilities.

There will always be unexpected discoveries, but even these we can hypothesize in order to test their consequences. A model can be constructed showing the results to be expected from the availability of a 'morning-after' contraceptive pill and a safe chemical abortifacient. The social consequences depend on the number of people involved, with inevitable repercussions concerning matrimonial and sexual customs and the associated morality. Were such a discovery to be made today, we would be subjected to much pontification about principles and little effort to define the consequences. Ten years from today we would probably be able to count the number of children who would not be born, abortions that would not occur, changes in school and prison populations during the following decades, and divorce-rate reduction. Furthermore, the existence of the data base for such modelling allows us to simulate even unhypothesized events very soon after they occur. Imagine, if you can, the consequences of discovering an enzyme or bacterium which would render seawater fresh enough for irrigation, and you will appreciate the need for a computer's help.

We can now speed up our learning by the generation of artificial experience. We have long been treated to science-fiction scenarios through which men (sometimes of distorted vision) have explored the consequences of future developments. To this sort of speculation we can now add the precision of quantitative techniques.

In an era of change – often discontinuous change – corporations perhaps more than any other institution in society must be alerted to forthcoming changes. Many corporations have withered through lack of foresight or inability to adapt. The President of IBM, T. Vincent Learson, has set forth five principles which he predicts industry management must master if change is not to master management:

1. Give a free hand to those responsible for long-range strategic planning. Make sure that short-range operational planning is dovetailed with those strategic objectives.
2. Spell out a well-defined organizational plan with clear delineation of duties and responsibilities.
3. State precise objectives and establish disciplined measurement systems to check performance.
4. Review programs for the introduction of new products constantly. New products take time. They must be re-examined and, if necessary, their objectives redefined in order to keep in step with the rapidly changing environment.
5. Establish requirements for information. Without information that is refined, summarized, and ready for review, judgments and decisions are only guesswork. Increasingly the information for decision-making will be collected and predigested by massive computer systems.[2]

If we are to manage the evolution of our society, we need a set of principles which broadly parallel these. Long-range planning is now becoming possible using simulation and modelling in an entirely new way. The collection and processing of the requisite information is also becoming possible with the new machines. More difficult is the setting of objectives.

With regard to the more mechanistic aspects of managing

our society, it is not so difficult to agree upon objectives. Reasonable objectives up to 15 years hence in air-traffic control can be precisely determined. Objectives in space exploration can be set, as with the moonshot. Objectives for national defence can be determined in a similar manner, albeit with some debate. The pros and cons of different urban-transportation schemes can be established using computer models, and the desired advantages agreed upon. For those aspects of society in which human values are more deeply involved, however, setting objectives becomes so open to argument as to be difficult – in some cases almost impossible. What value is to be placed on certain aspects of privacy? What value is to be placed on increased leisure time, more beautiful cities, improvements in the arts, or preservation of the countryside? Must a welfare state lead to economic stagnation? How much money should be spent on slum-clearance, noise-control, anti-pollution measures? If better crime-control must mean more police interference and personal dossiers kept by police computers, what is the choice to be?

A society would almost certainly never agree upon a complete set of objectives for its own evolution, but it could certainly establish a more detailed idea of where it is going than that we have today. Its possible evolutionary paths are so multifold that it is probably unreasonable to prepare more than a partial plan. However – and this point is vital – this plan would be worked out with a knowledge of what is expected to happen in the technology in the years and decades ahead. The likely effects of technological changes would be explored with computer models; different courses of action would be simulated; possible future scenarios would be considered and their desirability, or otherwise, reviewed. The chances would be lessened that countries and their institutions might drift into objectionable and often irreversible conditions because of the impact of new developments in biochemistry, the laser, new generations of communication satellites, the ubiquitous computer terminal, or the ever-growing data banks. Just as today the military has early warning in some detail of the implications of possible new weapons, so tomorrow society could have early warnings of the implications of new technology. Undoubtedly we shall at first have difficulty identifying and then quantifying all the significant factors. Eventually, like Sputnik, success will be achieved and social simulators will be taken for granted.

REFERENCES

1. *Columbia Journal of World Business*, Vol. III, No. 1, January–February 1968, p. 59.

2. ibid., p. 61.

23

THE CHANGES NEEDED IN EDUCATION

Today's television child is bewildered when he enters the nineteenth-century environment that characterizes the educational establishment. – *Marshall McLuhan, The Medium is the Massage.*

The most disturbing of all the problems discussed in Section II of this book is the wide variation in people's proficiency with technology. Those well able to cope with computers will be well rewarded; but those who cannot may be nearly unemployable and may go from bewilderment and frustration to outright antagonism.

If this were a technical problem, then eventually a technical solution would be found. We discuss solutions to such problems in the remainder of the book. With foresight and energetic planning, the Purgatory envisioned by some prophets of doom (described in Chapter 19) could, in principle, be avoided. However, we fear that such anticipation will not be shown in practice.

Even the best system engineering and foresighted planning will not change the natural talents of men and women, and thus make everyone's skill valuable to the computerized society.

Let us suppose that the public's ability to cope with the machines is represented by a curve, as in Fig. 19.3 on page 422. It may not be exactly the same curve and, as we commented before, there is not a direct correlation between IQ and one's ability to cope with computers. However, it is likely to be close to a bell-shaped ('normal') distribution, and for the purposes of this discussion that assumption will suffice. The top 1 per cent, then, in a curve as in Fig. 19.3, represents the brilliant practitioners of the new technology. In most professions and fields of work, they will be in high demand. The top 20 per cent will be well off. The shortage of programmers, analysts, operators, CAI writers, and so on is likely to become increasingly acute, and thus will insure them employment. However, the bottom 30 per cent, and perhaps as many as 70 per cent, will be among those for whom many sources of employment will have dried up. They will dislike the ubiquitous terminals and have difficulty in using them in a world in which more and more functions of life are becoming involved with terminals. They will generally be in a lower IQ bracket also, and the automation of low-IQ jobs will affect them seriously.

There are three actions we can take concerning this problem. First – and this is a solution involving the computer men – we can program the machines so that the terminals for purposes affecting the general public are childishly easy to communicate with. Much research is needed in designing terminals and conversation structures so that low-IQ and ill-trained personnel can communicate with them. Except in some work on computer-assisted instruction, little has been done in this area as yet. Many new computers are conspicuously more difficult to operate than their predecessors.

The second action we can take is to create new types of work – in the manner discussed in Chapter 21 – which are beneficial to society but which can be carried out by persons who do not get on with the machines – or those in the lower 50 per cent of our distribution curve. Creating the right type of work could make our world a very much pleasanter place to live in – but few such jobs will be created by private enterprise.

Third – and this is largely the key to our problem – we need to educate as many of those in the existing generation as possible in the skills needed, and to bring up to a new generation which *will* be able to cope with the computers and the changes they bring.

Eventually a fourth solution may emerge allowing the less capable fraction of society not to work at all and to be supported by those who do. However, we doubt whether this will be considered desirable in the next few decades.

There is nothing very difficult about most of the operations we expect the public will carry out at computer terminals. Some of the more elaborate programmes, such as simulating the effect of one's investment policy or working out the shortest routing for delivery with a fleet of trucks, *are* difficult and require substantial training and intelligence. These are for the élite. But the majority of terminal uses that we have foreseen for the 'man in the street' involve him in less difficulty than many operations he considers routine, such as filling in an income-tax form or driving a car on today's complex highways. The trouble is more that of complete unfamiliarity with the terminals and their usage, than of inherent difficulty. A period of relatively straightforward training in this technology would solve many of the problems. Unfortunately, the majority of the public today think that classroom teaching ends on leaving school, and have no desire to take it up again.

Children starting school today are going to start their working life toward the end of the fifteen-year span we have been discussing; some after it. Unless our predictions are wildly incorrect, these people are going to spend almost their entire working lives in a world markedly different from today, in which it will be at least as important to understand and communicate with computers as driving a car is today. The world is now changing so rapidly that teachers should surely ask: What sort of a society will these children live in? How will it differ from today? What changes should we make in their education so that they can make the most of this new world?

EDUCATION FOR LEISURE

A number of factors seem clear. First, there will probably be more leisure time. These children will still be in their thirties at the turn of the next century. If we extend the bottom half of Table 21.1, we obtain for the mid range of the table (hours worked drop by 1.5 per cent annually) a figure of $13,167 for Gross National Product per capita by the year 2010, and 1,002 as hours worked annually. In other words, we could be three-and-a-half times as wealthy as we are today and have 6 months' vacation each year; or alternatively work a 3-day week (8 hours a day) and have 10 weeks' vacation each year. Whatever combination you choose, it seems obvious that today's first-graders are going to have substantial leisure time by middle age. If we do not want them to become television-fed zombies, to become hippies, drop-outs, street-lounging malcontents, or to regress to sitting in the sun like the Latin American peasant, we had better start educating them in how to use their leisure time.

How are we to do this? Some professional educators would perhaps have much to say on this subject. Good, stimulating education in the arts and literature, teaching and practice in handicrafts, gardening, film-making, music, debating, and so on, would seem to have a new importance. A neighbour of one of the authors confided to him recently that since he, the neighbour, left college he had never actually read a book. Most people have difficulty in finding anything better to do with their spare time than family chit-chat, playing golf, watching television or movies, or driving around somewhat aimlessly in the car. America has learned little about how to spend its great affluence effectively.

Historians and other writers seem generally agreed in classifying a few brief periods in the wilderness of man's history as highly 'civilized.' They usually single out the Golden Age of ancient Athens, Italy in the fifteenth and sixteenth centuries, and France from the mid-seventeenth century up to the revolution. On these almost all would agree. A few would wish to add the Tang and Sung dynasties in China. Some, among stout opposition, would add the first and second centuries of the Roman Empire, but that is about all. Contemporary America or England they put very low on the scale of 'civilization'. In attempting to find what so qualified these periods, we note that they all exhibited a rich and splendid use of leisure. The leisured classes were highly informed and paid much attention to music, the visual arts, debate, and intellectual pastimes. They had superbly elegant flirtations and gracious dinner parties. There were few taboo subjects among them in their conversations. Sweet reason prevailed. We could say that their 'education for leisure', however it was obtained, was of a high order.

The leisured class in each of these societies was a minority and it had a slave class or lower class to perform routine work for it. If we are to dream of a richer civilization today, it must be one in which all men can participate if they wish and the machines do the routine work. For the last hundred years or more, civilization has been drowned in the routine toil of our new industry. We have now reached the day when we can emerge from that period.

CHANGE NEEDED IN TEACHERS

The basic need for the coming computer age is to educate the next generation in such a way that they will be happy, constructive, and 'civilized' in a world in which most *routine* operations and *routine* thinking are handed over to the machines. This is a tall order inasmuch as a large proportion of the public today absorbs its long hours in routine. Take away from the clerk or the factory-hand his routine work and he will be at sea. To survive in a world without routine he has to be creative, imaginative, and adaptable; he must welcome innovation rather than shun it. Teachers now spend most of their time teaching *routine*. Much of primary- and secondary-school education consists of drilling into pupils facts and fixed techniques. The emphasis today is mainly on what is necessary to earn a living in a hard-working mechanical world. In the future the world will be less mechanical, less hardworking, and prone to the volatility of rapid change. We are passing out of the black period of the mechanical Industrial Revolution into an age of intelligent computers, ultra-highspeed data-transmission, leisure time, and uncertainty. The pupil today must answer a fixed set of examination questions, leaving him no time to explore. We pump the young mind full of today's set methods rather than draw out its own inner creativity. Many teachers appreciate this but the system as a whole has done little toward implementing an alternative.

Teaching a mind to explore, to debate, to innovate, and to understand diverse viewpoints is immensely more difficult than teaching it fixed methodology and facts. Educating people in the virtues of Athenian civilization, or educating them to appreciate the culture of eighteenth-century France, requires teachers very different from those who could enable them to earn a living in pre-70s Europe or America. The coming computer age may not be Athens, but it would be better to aspire to such virtues than to propagate the leisure ethic of Miami Beach and 'The Lucy Show' in distorted colour. To bring up a generation to survive graciously in a world of greatly increased leisure time and greatly decreased routine, and to lead secure and fulfilled lives in the sliding quicksands of technical innovation ahead, we need schoolteaching different from today's.

Furthermore, the mainstay of today's primary- and secondary-school teachers – teaching routine and drill – will be better conducted by the computers. That the computers can do it is excellent, because it will leave the teachers much more time to concentrate on the above problems, to concentrate on the development of individual minds that can fit into this new world as well as possible. Some teachers are indeed trying to do this today, but they are a small minority. Again, we are placing much greater demands on the teachers if we automate the easy and major part of their job and greatly expand our demands for the difficult part of their job.

Schoolteaching in general for the computer age, we think, is going to be a much more difficult and demanding task. It will require a higher calibre of staff than the majority of teachers today. And it will be a much more stimulating job.

At a time like this, money spent on education is one of the soundest investments in stemming uncertainty concerning the future. To attract the right people into teaching, the authors feel that much higher salaries need to be available. A man with ability who wants to live well usually avoids primary- or secondary-schoolteaching today because the pay is too low. Furthermore, schoolteaching offers very little of the financial incentive to work hard and apply extra effort that occupations in many other fields do: a schoolteacher's salary rises with age and is little influenced by whether he does a first-rate job or not. Consequently, most teachers work far less hard than employees of equivalent talent in a high-incentive corporation.

To remedy this situation, the authors think that both salaries and incentives in teaching should be made comparable with those in industry or other professions. A graduate contemplating a career as a schoolteacher should know that it is possible for him to reach a salary of, say, $30,000 in the United States (say, £7,000 in Britain) – if, and only if, he does a brilliant job and works very hard. Only a *very* minor proportion of his profession would reach this salary, however. His skill and performance should be appraised constantly and he should receive no salary raise that is not a merit raise. We believe that only with such incentives can the teaching profession attract men and women good enough for the difficult changes that are needed.

PRIMARY SCHOOLS

Training to use the machines themselves will take place on many levels. It will probably start at a very early age. In the future when children learn to read, it is likely that a device such as a 'talking typewriter' or computer terminal will be used. It will then seem logical to teach the child to touch-type and to become familiar with some of the responses that may become standard on a typewriter-like computer terminal. Touch-typing is a great asset in communicating with the machines. One sees many computer experts today tapping their communication with the computer with one finger and pausing to search for the next letter. ('I never *can* find the "C".') It might as well be learned at a young age.

Recent research has indicated that much of our learning could and possibly should take place at an earlier age than it does today. Professor O. K. Moore of Yale, who invented the 'talking typewriter' method of teaching reading and spelling, has used it to teach children to read, spell, and touch-type at ages of two and three! Other workers have found that algebra can be taught at four. On the other hand, otherwise normal children in a Teheran orphanage were found to be unable to sit up at age two or walk at four, simply because nobody had bothered to teach them. There has been found to be an optimum age for different basic types of human learning. If one misses learning at this optimum age, it is going to be more difficult later on. In a sense one has learned negatively – learned *not* to do the function in question.

Moore's work is particularly fascinating.[1] He has prepared filmed reports of two- and three-year-old children reading first- and second-grade stories, printing, touch-typing, and taking dictation! In order to do this, one must use an environment that responds quickly to the child's actions. It must be completely non-authoritarian. It must have a means of stimulating a child to prevent him from becoming tired of a procedure. This can be built into the responses of a computer terminal. The child, playing, explores the booth he is placed in. He finds a keyboard. A light touch on a key causes a letter to be typed and the pronunciation of the character comes over a loudspeaker. When the child tires of this procedure the letter is displayed for him to copy. The only key which will work is the one with the character on it. From this simple beginning the stimuli are increased in complexity until the child has learned to read and write. Children spend about half-an-hour per day at play with the machine. Both gifted and retarded children have benefited from the experience. This could be a glimmering of infant-teaching methods that lie ahead.

The policies commission of the United States National Education Association has recommended that school should begin at age four. Perhaps this is the age at which the CAI terminals will first be used, although people with them in their homes will probably experiment on younger children. It is possible that speaking to computers, again over telephone links, will become a common way of communicating with them (it may become the cheapest way); however the vocabulary would consist of a limited set of words, and the user's voice would have to be clear and with no extreme accents. If this comes about, children may learn to speak computer English at a very early age. The computers might be programmed to tell the children stories and to permit them to ask questions. All manner of entertainments would be possible, just as today a child can play at a terminal attempting with a light pen to 'catch' butterflies flitting about the screen. Possibly even children in non-English-speaking countries would learn this limited set of English at a tender age, and the English-speaking could equally learn the rudiments of other tongues.

SECONDARY SCHOOLS

In tomorrow's secondary schools it seems highly desirable that the children should be taught what the machines are, and what they can be made to do. They should be taught roughly how they work, and certainly they should be taught, and have experience in, how to program them. Understanding computer technology is going to be so fundamental to our way of life that it seems much more important to teach it than some of the other subjects taught to us around the age of twelve, such as Euclidian geometry, Latin grammar, and hundreds of unrelated dates in history. There is no reason why fundamental computer-programming should not be taught at the same level as fundamental algebra and geometry. Fundamental programming is not difficult, and children generally love their first encounters with computer terminals. By the age of twelve every child should have attempted to program his first Tic-Tac-Toe game. Bright children should then be allowed to race ahead, as they will certainly want to, and use the machines for more elaborate activities.

It is outrageous that today a major British university should include Latin as a mandatory entrance requirement – even for science students – yet completely ignore computers in its curriculum. Learning to use a computer language surely involves all the use of logic and clear thinking with which some teachers rationalize their medieval addiction to Latin.

UNIVERSITIES

Two major changes are needed in universities to meet the challenge of the computer age. First, there need to be computer-science courses which students can take at both the undergraduate and graduate level – there are excellent examples of these in some universities in the United States. Second, and perhaps more important, there needs to be a widespread acceptance and use of computers in the other faculties.

This area was studied by the 1967 Science Advisory Committee to the President of the United States.[2] One of the major conclusions of their report was that in the near future *'almost all* undergraduates will use computers profitably *if adequate computing facilities are available* [emphasis supplied] ... We believe that undergraduate college education without computing is deficient education, just as undergraduate education without library facilities would be deficient education.' Some undergraduates, the report projected, will be enrolled in curricula in which they could make a 'substantial' use of computers, others in curricula in which computer training would be very useful but computers would only be used in some of their courses, and the remainder in curricula in which the use would be 'casual' – that is, as important background knowledge, and the machines used in one or more courses. The report estimated that 35 per cent of students would fall into the 'substantial-use' category, 40 per cent into the middle category, and 25 per cent into the 'casual-use' category. The breakdown of these categories by study area is shown in Table 23.1.

In 1965, the report estimated, less than 5 per cent of the total college enrolment in the United States had access to computing facilities 'adequate' to their need. The commission considered it practical to supply adequate computing services to nearly all colleges by 1971–2, although the cost of providing this service would be high – by 1971–2 an estimated \$400 million per year, not counting the smaller costs of faculty training and associated research. This is approximately \$60 per student average over all courses. It was considered beyond the capabilities of the colleges themselves to bear all this cost. The report recommended that the federal government share the cost and encourage the expenditure.

It seems unlikely that such a high expenditure will also be made in countries less affluent than the United States. Thus the

United States will probably be much better prepared to take advantage of the computer age than other countries, widening the already serious technology gap. Greater affluence based on more efficient use of technology is a self-amplifying difference. The rich become richer *much* faster than the poor.

In Britain in 1965 a working group under Professor B. H. Flowers of Manchester University recommended the expenditure of £35 million on the acquisition of computers for higher education and research 'to make up lost headway in acquiring computing facilities. *Any attempt to provide less than this computing facility would represent a serious handicap to the progress of research and higher education in this country*' (italics in original).[3] Their forecast was quite correct. Britain's higher education continues to be handicapped by insufficient computing power and the slow implementation of the Flowers proposals for large regional computing centres shared between universities.

A further problem in providing adequate undergraduate computing services is education of the faculty. In some areas this may be a serious problem because of faculty attitude. Many instructors do not *want* to have anything to do with computers. The dons of Oxford and Cambridge in England generally regard themselves as the cream of the world's university staff, but most of them seem almost unaware of the existence of the computer. Any suggestion that computers should be used in their courses would in the main be treated as a barbaric intrusion onto hallowed ground, rather as though one had started to read a comic book at High Table.

In this regard, the report of the Science Advisory Committee recommends the following:

Instruction of no monetary cost to the faculty should be offered in a variety of ways; for example, short courses during the academic year, seminars between regular semesters or quarters, and longer courses during part of the summer. Some of the courses should be so general that a faculty member from *any* discipline can attend and gain something from the discussion. Others should be discipline-oriented and present special techniques that have been advanced to solve particular problems.

TABLE 23.1

Projected Use of Computers in Different Areas of Undergraduate Study

(based on Bachelor's degrees conferred in the US in 1963–64)

Major Area of Study	Number of Students		
	Substantial Use of Computers	Medium Use of Computers	Casual Use of Computers
agriculture	—	—	4,600
architecture	—	600	—
biology	23 000	—	—
business and commerce	28,000	28,000	—
education	—	84,000	28,000
engineering	33,000	—	—
english and journalism	—	—	35,000
fine and applied arts	—	—	16,000
foreign language and literature	—	—	12,000
forestry	—	1,300	—
geography	—	1,200	—
health	1,000	10,500	—
home economics	—	—	5,000
library science	500	—	—
mathematics	9,500	9,000	—
military	2,500	—	—
philosophy	—	—	4,700
physical sciences	17,500	—	—
psychology	7,000	6,500	—
religion	—	—	3,600
social sciences	38,000	38,000	—
others	—	—	12,000
Total (460,000)	160,000	179,100	120,900
Percentage	35	40	25

Computers in Higher Education, the Report of the President's Science Advisory Committee, US Government Printing Office, Washington, DC, February 1967.

The threshold of computer understanding is not a difficult one to cross provided that the faculty member is willing. A two- to six-week course would enable most university teachers to gain an understanding of what is possible, to start doing useful work, and to reach a position in which they could gain further experience and knowledge on their own.

Because computing affects so many different disciplines, a course on the philosophy of computing methods for all (or almost all) students might prove worthwhile. Professor of Philosophy H. W. Johnstone, Jr., of Pennsylvania State University, put it like this:

Scientific method, in the form which is sometimes a required course (or part of one) for all or most undergraduates, is a liberal study. Its purpose is to acquaint the student with the nature of scientific thinking, so that he will see science not as a kind of familiar magic that he takes for granted, but rather as a human achievement. In my view, a similar liberal course ought to be given on computers. The emphasis would be upon the concept of a computer and upon the general methods of using computers. The student who had been exposed to such a course would see the computer as a human achievement rather than as a black box to be taken for granted. He would see how the possibility of using computers to solve problems has revolutionized the ways in which we think about the problems. A person for whom the computer is merely something to be used gains from his contact with it no appreciation of the nature of the contemporary world. Such appreciation presupposes a certain awareness of the nature and method of the computer as such – an awareness that is quite different from the knack of programming. The use of the computer has all at once spread to all aspects of our culture. It is this that struck me as being of primary philosophical relevance. What is relevant is the way the computer has changed the quality of contemporary life – not so much in satisfying our material needs as in causing us to think about ourselves in a new way.[4]

EDUCATION OUT OF SCHOOL

Only a part of our education takes place in a schoolroom or at college. As we move into the computer age we can expect a substantial increase in the other part, which is carried out in industry and by other employers, by professional men training themselves, and by the general public seeking knowledge or skills. There will no doubt be an increase in job-training courses, evening classes, postgraduate refresher classes, leisure-time education groups, and education in the home by television, correspondence, books, records, videotapes, and computer terminals.

In addition to the increasing awareness of education for leisure, this rise in adult education is being brought about by the changes in man's work. Not long ago a man was trained to do a job and then did it, in more or less the same way, for the rest of his life. Now, with so much work being taken over by machines and so many new types of jobs being created, he is quite likely to have to change the nature of his work as many as five times during his lifetime, each change requiring retraining and readjustment.

The professional man is still worse off in most cases, because the knowledge needed for carrying out his work is changing so fast.

HALF-LIFE OF A TRAINING

One might, borrowing a concept from nuclear physics, describe educations in complex disciplines in terms of their *half-life*. Suppose a job needs a certain education to do it well. The half-life of the training is *the time it takes the knowledge to decay to a state in which it can provide only half of the facts or techniques to do the job.*

A medical graduate of a few generations ago could expect that most of his training would be applicable for the rest of his days. Now, however, medical knowledge and techniques are changing so fast that about half of the knowledge a good doctor now needs was not taught ten years ago; thus medical training has a half-life of something like ten years. A physicist's or biochemist's training has a half-life less than this. A training for a computer-systems analyst has a half-life that is perhaps as short as three years. Like Alice, he has to run very hard just to remain where he is. On the other hand a lawyer still has a long half-life, and a historian can put his feet up and relax.

The half-lives in most fields are decreasing. In medicine, for example, it has been estimated that the half-life is likely to drop to five years in the 1970s. This means that a doctor whose original training took five years will need to spend half his time learning new material if his knowledge is to remain current. In fact, less than half his time will be adequate because he acquires the ability to comprehend new facts much faster than when he attended medical school. In effect, he will have been trained to acquire new medical knowledge quickly. The systems analyst similarly comprehends a new machine or technique much more rapidly than the first ones he met. The framed motto that hung on Victorian walls saying 'Knowledge is something that no one can take away from you' needs to be replaced by one saying 'Knowledge decays exponentially.'

Furthermore, the range and depth of knowledge needed in the fast-moving professions is expanding rapidly. The medical practitioner must have access to increasingly large amounts of knowledge. The physicist is faced with an ever-growing body of ideas, techniques, theories, and tools. The design engineer, once happy with a desk calculator, now needs a computer. Submitting work to be handled by the computer operators gives way to using a machine in real-time. The real-time terminal, initially limited and straightforward to use, becomes capable of ever-increasingly diverse and intricate techniques.

Men of forty, in days past, had comfortably mastered their profession. They were by that age experienced, sagely relaxed, and mature in their understanding of the implications of their knowledge. In how many professions in the future will the man of forty be struggling to keep his head above water, conscious of the ever-increasing deluge of complexity, conscious of his dying brain cells, conscious that younger men can use techniques he cannot?

LIFELONG LEARNING

Because of these pressures, and because of the new tools available, education should come to be regarded as something which is continuous throughout life, going on in many different forms and in many different places. It is still regarded by the majority of people as something which occurs during one short period, early in life, in certain fixed buildings. But in the computer era it can go on, in a formal manner if necessary, anywhere a terminal is available on a telephone line; and it needs to go on throughout life.

The educational records that are kept today relate largely, and in most cases entirely, to that education at the beginning of life before one learns to do a job. A person applying for a job now has to submit records only about his school and university education. While this might give a crude indication of his ability in those days, it may be of less relevance now than any unrecorded training from computers that he may have received since. It would be more valuable to let machine-maintained documentation continue with one's post-university learning.

Continuing education, though sometimes directed by a corporation or other employer, is often largely the responsibility of the individual. While education is too important to be left to the educators, it seems inadvisable to leave it to the individual. Some form of counselling or guidance would seem to be called for in a world of continuing adult education. This might be done by a professional counselling organization, but another intriguing possibility is that it will eventually be done by the machines. Several experimental programs have already been written for student counselling.

MACHINE-COUNSELLING

In a Systems Development Corporation experimental counselling project, a fifteen-year-old student sits at a teletype machine and is interviewed by a remote computer. The machine inspects the student's school records and types out his marks in each course for the last term. It asks him if he is having any problems and if so, he types a description of these. These are stored and later printed for a human counsellor.

The machine then asks the student whether he would like to continue the interview or interrupt it for human counsel. Usually the student continues with the machine. His educational and vocational goals are then discussed. Based on accumulated statistics the machine types predictions of his probable examination grades and his chances of success in various chosen activities. The machine assists him to select courses for his remaining years at school. It evaluates his choices and advises him on such matters as appropriate course loads and the relevance of his choices to his possible college major or vocational aspirations. The computer prepares a concise report on the interview for a human counsellor, and retains a complete interview record in case the counsellor wants more information.

There is considerable controversy about the desirability of using a computer to interview a fifteen-year-old in this way. However, for continuing adult education it would seem to be an excellent technique. The computer, which stores the man's educational records, could advise him on what courses, mechanized or otherwise, would be available to him. It could make recommendations based upon a discussion of aspirations and capability. There is probably no alternative way for a substantial proportion of the public to obtain such detailed and continuing advice. The man would be free to disregard any machine suggestions without this being recorded, and in most cases would be mature enough to judge the viability of its comments. He would use it as a tool which might be very valuable in helping him further his education.

Records of citizens' educational attainments are already being included in plans for state- and local-government data banks. These may well contain records of post-university education, and perhaps eventually a summary of counselling where the man in question agrees that he wants this recorded. Such files of information would be useful in searching for the right men to fill certain posts. To help encourage people to continue their ongoing education, the computer might scan an individual's records and mail recommendations to him periodically.

REFERENCES

1. O. K. Moore, *The Automated Responsive Environment*, Yale University Press, New Haven, 1962, and 'Orthographic Symbols and the Preschool Child – A New Approach', *Proceedings of the Third Minnesota Conference on Gifted Children*, University of Minnesota Press, Minneapolis, 1961.

2. *Computers in Higher Education*, Report of the United States President's Science Advisory Committee, US Government Printing Office, Washington, DC, February 1967.

3. *Report of a Joint Working Group on Computers for Research*, HMSO, January 1966.

4. *Computers in Higher Education*, op. cit.

24

THE LAWS THAT ARE NEEDED

There is nothing more difficult to undertake, more uncertain to
succeed, and more dangerous to manage, than to prescribe new
laws. Because he who innovates in that manner has for his enemies
all those who made any advantage by the old laws; and those who
expect to benefit by the new will be but cool and lukewarm in his
defence. – *The Prince, Niccolo Machiavelli, 1513.*

The massive use of computers has potential for great good
in society, but there are also some very serious dangers,
as outlined in Section II. We need laws which will permit
and encourage the good aspects and curtail the bad
aspects.

To foster the good aspects we want our legal structure to
give the maximum encouragement to the inventors and entrepreneurs who will initiate the numerous schemes that this
technology makes possible. There is plenty of scope for the
small man in this. He can invent new and inexpensive terminals.
He can write new software, provide new information services,
set up education programs, and start consultancy operations.
He can construct some of the innumerable data banks that are
now possible. There are many new opportunities to make a
million in this new industry.

Several changes in the law could help this process. For
example, in almost no country is there a clear law protecting
new programs from being copied. One cannot patent a program, and it is generally uncertain to what extent copyright
laws will protect a program that has been recoded but not
copied exactly; they probably will not protect it.

Any law which *unnecessarily* restricts the development of
new products in this fast-changing field is bad. For example,
one of the key factors in the use of computers for the good of
society is the telephone network, and most countries have
inadequate laws governing such use today. They are laws

devised in an era when the telephone network was used only for telephoning.

One significant change in these laws occurred recently in the United States, and is badly needed in other countries. This relates to products which can be connected directly to the public telephone network. The use of an enormous variety of terminal devices is possible for communicating with distant computers. It is very desirable that inventors should be free to design and sell such devices, provided that they do not interfere with the proper operation of the telephone network. We need inexpensive devices that can be used in the home; we need portable terminals that can be used in a public-telephone booth; we need new types of telephone-connected burglar alarms; and innumerable others. American telephone companies, however, at one time prohibited 'foreign attachments' from being connected directly to the dial-telephone lines. You could not connect your electric cooker, tape recorder, light pen, or any other device to the public network except via a relatively expensive 'data set,' sold by the telephone company. This severely limited the design possibilities. Such laws still exist in most countries, and so no-one there invents anything to attach directly to the telephone line. This is bad because the telephone companies are extremely slow in introducing new devices.

In 1968 the prohibition of foreign attachments was ruled, in the United States, to be 'unreasonable, unlawful, and unreasonably discriminatory under the Communications Act of 1934'. So now a designer can attach his own devices to the public lines through a protective connecting arrangement. He still has to use the telephone company's dialling and hook-switch equipment, but this tariff is also under attack. As a result of this change in the law, the electronics magazines soon filled up with advertisements for new and valuable terminal devices.

This type of change in the law is vital for progress. In many countries the telephone network is controlled by a government monopoly with out-of-date laws. Nothing except telephone-company machines can be attached to the public lines, not even with acoustical or inductive coupling (no direct electrical contact). Portable terminals or terminals for the home cannot be marketed because of this law. Many computer personnel think that a technical reason prevents this. That is not so. It is solely a legal reason, and it is doing great harm.

A country with aspirations for technical progress must make sure that archaic laws do not stand in the way. The British Post Office controls all telecommunications and in the past has failed to encourage innovative entrepreneurs. Profits from the telephone service were used to subsidize the mail instead of re-equipping the antiquated telephone exchanges. So the fragile voice network has had to be defended from non-Post Office equipment. With the reorganization of the Post Office as a corporation, some, but not all, of its staff have acquired commercial and forward-thinking attitudes. But there are few who think that it can slough off the past and provide for the data communications needs of the 1980s before the twenty-first century unless its power is eroded.

On the other side of the coin, we need new types of laws to protect us as we hurtle into this era of ever-faster change. One fundamental right is the right to be let alone. To maintain any degree of privacy in the computerized society, we must have new laws.

THE RIGHT TO BE LET ALONE

The history of the law of the right to privacy is a strange one. The need for it was not apparent to eighteenth-century Americans and Englishmen, whose fear was too little contact with their widely dispersed neighbours.

Early cases concerned themselves with 'property', which was deemed to be misappropriated by those who published letters or trade secrets obtained without their original owners' consent. Later, the importunities of the press and of advertisers led to a definition of a right of privacy, and to moves to protect an individual's self-respect. Much ingenuity was needed to translate early legal concepts of property rights, trespass, defamation, and personal injury into a legal principle embodying the right to be let alone.

The foundations were laid by English law, but the need to develop it there was not so strong as in the USA. Not only are the British better protected by their libel law; they are also constrained by protocol and codes of behaviour to respect individuals throughout their public and private lives.

It was lack of such 'taste' that precipitated the article by Warren and Brandeis that was to become the cornerstone of American law on privacy. Samuel D. Warren, who came from a wealthy and socially prominent Boston family, became the subject of the gossip columnists of the *Saturday Evening Gazette* when, following his marriage, he and his wife had started to entertain in the lavish style of the 1880s. His annoyance resulted in an article in the *Harvard Law Review* written in conjunction with his partner and classmate, Louis D. Brandeis:

Political, social, and economic changes entail the recognition of new rights, and the common law, in its eternal youth, grows to meet the demands of society ... The right to life has come to mean the right to enjoy life – the right to be let alone; the right to liberty secures the exercise of extensive civil privileges; and the term property has come to comprise every form of possession – intangible, as well as tangible.[1]

The article went on to describe how the press was 'overstepping in every direction the bounds of propriety and decency. Gossip [had] become a trade'. Discovering no protection against injury to feelings in the law of libel, they continued: 'The common law secures to each individual the right of determining, ordinarily, to what extent his thoughts, sentiments, and emotions shall be communicated to others.' So they took a look at property as it is embodied in the copyright laws, only to conclude that

the protection afforded to thoughts, sentiments, and emotions expressed though the medium of writing or of the arts, so far as it consists in preventing publication, is merely an instance of the enforcement of the more general right of the individual to be let alone ... [If] we are correct in this conclusion, the existing law affords a principle which may be invoked to protect the privacy of the individual from invasion either by the too enterprising press, the photographer, or the possessor of any other modern device for recording or reproducing scenes or sounds.

After an examination of four cases in which the law of contract had been stretched, they decided that:

The principle which protects personal writings and any other productions of the intellect or of the emotions, is the right to privacy, and the law has no new principles to formulate when it extends this protection to the personal appearance, sayings, acts, and to personal relations, domestic or otherwise.

They then set forth certain limitations on the right of privacy:

1. The right to privacy does not prohibit any publication of matter which is of public or general interest ...
2. The right to privacy does not prohibit the communication of any matter, though in its nature private, when the publication is made under circumstances which would render it immune from attack according to the law of slander and libel ...
3. The law would probably not grant any redress for the invasion of privacy by oral publication in the absence of special damage ...
4. The right to privacy ceases upon the publication of the facts by the individual, or with his consent.
5. The truth of the matter does not afford a defence.
6. The absence of 'malice' does not afford a defence.

The Warren–Brandeis article provided the cornerstone of much later judicial opinion. The law is an area in which the States are little United, and only a few went on to legislate the right to privacy in statutes. New York included it in its 1903 Civil Right Law. In other states the legislature or the courts protect their citizens. Others find no 'right to be let alone'.[2]

The federal government, and the states, are of course bound by the Constitution. Since the Constitution is extremely difficult to amend and does not, unlike the common law, 'grow to meet the demands of society,' much ingenuity is required to reinterpret it successfully. Privacy has to be shown to be compatible with liberty and property so that it can then be embodied in state law – that is, justice is an inadequate foundation for privacy, if it cannot be derived from rights embodied in an eighteenth-century document.

According to Karst, the law of tort is presently the prime protection against disclosure of private files.[3] To institute civil action, it would be necessary first to discover the disclosure, then to show it was not explicitly or implicitly consented to, and third to show that damage had been done. There may be a need to show that the First Amendment permits such 'abridging the freedom of speech' of the disclosing party.

Warren and Brandeis, *supra*, considered that the right to privacy ceases on the publication of the matter by, or with the consent of, the subject. It is arguable that consent is implicit in the provision of data to agencies, governmental or private, unless, like the Bureau of the Census, they are government agencies which have undertaken to preserve secrecy. It will be difficult, but essential, to make the dissemination of information without explicit consent of the subject illegal, and liable, therefore, to a criminal, not a civil charge. The civil remedies of damages, restitution, and injunction are too late, and too slow in operation. Criminal sanctions, particularly against officials, would be far more likely to achieve the desired ends.

LEGISLATION REQUIRED

Sooner or later a legal environment must be established relating to computers and data banks. The sooner the better, because the systems designers are not waiting until laws exist.

There have been many proposals for legislation. They vary from amendments to the Constitution to administrative regulations. It is easier to say what is required than to determine how to enact relevant legislation. The general form of some of the laws needed is nevertheless clear.

First, there must be a recognition that modern technology allows one man to gain without another losing. The photograph records and reproduces, and so does the computer, but no extension of the concept of property will protect those who lose no property. Nor are life and liberty threatened thereby. The law must therefore admit of something new rather than seek to devise a stance from within the limitations of the past. Georgia law states that privacy is derived from natural law and is an inalienable right. It would seem appropriate that a democracy could agree to the *right*, without admitting the 'natural' origin, and insist on its preservation and protection on a national scale.

On this fundamental principle would be based laws dealing with the collection, storage, and disclosure of data in government or commercial data banks. The definition of data banks would encompass any organization collecting data for its own use as well as those who propose to make information available to others.

Like automobiles, all private systems would have to meet basic safety regulations. Provided no accident leading to disclosure to outside parties occurs, no restrictions on file contents need be set. Inadequate safeguards, however, could lead to charges of culpable negligence. Those who wish to provide information to customers, subscribers, partners in a joint enterprise, fellow state, local, or federal agencies, or for publication should be more strictly controlled.

For the remainder of this chapter let us consider suggestions for the types of laws that should be considered. Here in Britain an interesting bill is before Parliament at the time of writing, called the Data Surveillance Bill.* Its purpose is to prevent the invasion of privacy through the misuse of computer information. We shall quote some relevant sections from this below. The following, then, are recommendations for laws that should be considered: [4]

* The original draft of this chapter was studied by the group preparing the draft of the Data Surveillance Bill.

1. *A register of data banks should be established.*

This may be a register solely of those data banks which contain information about people. Some authorities have proposed that *all* data banks be registered, but this seems an unnecessary burden. The registrar would have the task of insuring that undue harm did not result from the use the computer systems registered.

The Data Surveillance Bill states that the following information would be registered for each data bank:

a. the name and address of the owner of the data bank;
b. the name and address of the person responsible for its operation;
c. the location of the data bank;
d. such technical specifications relating to the data bank as may be required by the Registrar;
e. the nature of the data stored or to be stored therein;
f. the purpose for which data is stored therein;
g. the class of persons authorized to extract data therefrom.

It goes on to define the main work of the Registrar as follows:

If at any time the Registrar is of the opinion that in the circumstances the information given or sought to be given under paragraphs (f) or (g) [above] might result in the infliction of undue hardship upon any person or persons or be not in the interest of the public generally, he may order such entry to be expunged from or not entered in the register.

The register itself would, of course, be computerized. Perhaps in the future, data-record definitions will be transmitted on-line to the Registrar's computer, and systems designers will be interrogated by it (to avoid filling in and transcribing endless documents).

2. *Facts, not opinions, should be stored.*

One hopes that most organizations providing information will value a reputation for accuracy. To ensure that they do, organizations providing a public service should be restricted to factual data and forbidden to disseminate opinion. For example, a credit report might legitimately say the subject lives on Park Lane, but not that he lives in a 'wealthy neighbourhood'. His wife could be described as 36–24–36, but not as 'beautiful'.

3. *Publicly accessible data banks should hold only data relevant to the purpose in question.*

Certain opinions and evaluations become facts: school grades, court verdicts, and so on. To hold these, relevance must be proved. The tax authorities should not maintain files on non-financial matters irrelevant to tax assessment, such as race, or colour, or prison records.

The Data Surveillance Bill states, with regard to reaching a decision as to what is permissible in a data bank:

The Registrar shall be guided by the principle that only data relevant to the purposes for which the data bank is operated should be stored therein, and that such data should only be disclosed for those same purposes.

4. *All interrogations of data banks concerning the public should be automatically logged.*

As we discuss in the next chapter, a logging tape or file is a desirable part of the system design. This tape can be analyzed to find out what uses are being made of the file and to detect trespassers. With some forms of systems controls, processing the log may reveal attempts at unauthorized entry before they have actually succeeded.

The British Data Surveillance Bill says the following:

The operator of each data bank to which this section applies shall maintain a written record in which shall be recorded the date of each extraction of data therefrom, the identity of the person requesting the data, the nature of the data supplied and the purpose for which is was required.

It seems anachronistic that the bill requires a *written* record. Surely the log should be kept by the computer and become, itself, another interrogatable file. We must use machines to keep watch on the machines.

5. *The public should have the right to inspect records the computers keep about them.*

Probably the best way to protect an individual from misuses of data banks is to show him exactly what the machines have to say about him and to give him the right to protest. He must have the right to have erroneous data corrected, to take issue with unfair and irrelevant data, and to request that his past history be forgotten.

Several schemes have been suggested for enabling the citizens of the computerized society to see what the computers have to say about them.

One is the annual mailing of statements to subjects of data-bank records. Another is the mailing of 'exception reports' providing a printout of adverse decisions – such as refusal to grant credit – when these are based on data-bank records and the individual wishes to question the result. In selecting a scheme we must be careful not to generate an excessive amount of work. Possibly the annual mailing of statements would cause too much protest. The authors of the Data Surveillance Bill thought the following was a reasonable compromise:

Any person about whom information is stored in a data bank shall receive from the operator, not later than two months after his name is first programmed into the data bank, a printout of all the data contained therein which relates to him. Thereafter, he shall be entitled to demand such a printout at any time upon payment of a fee the amount of which shall be determined by the Registrar from time to time; and the operator shall supply such printout within three weeks of such demand.

Every printout supplied in accordance with this section shall be accompanied by a statement giving the following information:

a. the purpose for which the data contained in the printout is to be used ...;
b. the purposes for which the said data has in fact been used since the last printout supplied in accordance with this section;
c. the names and addresses of all recipients of all or part of the said data since the last printout supplied in accordance with this section.

Clearly this cannot apply to police, military, and intelligence data banks. The bill makes the following exceptions, which apply to the above data-logging requirement also:

a. data banks which do not contain personal information relating to identifiable persons;
b. data banks operated by the police;
c. data banks operated by the security services;
d. data banks operated by the armed forces of the Crown.

We must not create more work than we can avoid, and so the mailing of printouts will be done automatically by the computers. Interrogation of one's own record may proceed at an office where suitable display screens are located.

6. *The individual should have the right to take issue with personal data stored about him.*

Under the Data Surveillance Bill the Registrar would handle protests from the public about information contained in the data banks. There must be a tribunal able to determine in detail what the contents should be when the subject and the 'data-banker' are in dispute. One hopes that most data-bankers will value a reputation for accuracy and thus that such disagreements will be few.

The right to protest is expressed in the Data Surveillance Bill as follows:

Any person who has received a printout ... may, after having notified the operator of the data bank of his objection, apply to the Registrar for an order that any or all of the data contained therein be amended or expunged on the ground that it is incorrect, unfair, or out of date in the light of the purposes for which it is stored in the data bank.

The Registrar may, if he grants an order under the foregoing subsection, issue an ancillary order that all or any of the recipients of the said data be notified of the terms of the order.

Thus reacting to the protest will often protect not only the protestor but also other persons about whom data is banked. Such a scheme will do much to keep the computer systems honest and reasonable.

7. *The register of data banks should be open to public inspection.*

The press and other news media play a very important part in keeping a democratic society on the rails. They should certainly be free to criticize and make suggestions about the new world we are building with computers. They, and the public, should have access to the register of data banks (but not, of course, to the personal data stored in them).

The Data Surveillance Bill says:

The register shall be open to inspection by the public, including the press, during normal office hours:

Provided that entries relating to data banks operated by the police, the security services, and the armed forces shall be kept in a separate part of the register which shall not be open to inspection by the public.

8. *Avoiding registration of a relevant data bank should be a criminal offence.*

The bill says:

1. It shall be an offence, punishable on summary conviction by a fine of not more than £500, or a conviction on indictment by a fine of not more than £1,000 or imprisonment for not more than five years or both, for the owner or operator of a data bank to which this Act applies to fail to register it in accordance with this Act.
2. If the operator of a data bank ... (excluding police, military, and intelligence data banks)
   a. fails or refuses to send a printout when under a duty so to do; or
   b. permits data stored in the data bank to be used for purposes other than those stated on the register; or
   c. allows access to the said data to persons other than those entered on the register as having authorized access; or
   d. fails or refuses to comply with a decision of the Registrar; he shall be liable in damages to the person whose personal data are involved and, where such acts or omissions are wilful, shall be liable on summary conviction to a fine of not more than £500 and on conviction

and indictment to a fine of not more than £1,000 or imprisonment for not more than five years or both.

3. A person who aids, abets, counsels, or procures the commission of an offence described in this section or with knowledge of its wrongful acquisition receives, uses, handles, sells, or otherwise disposes of information obtained as a result of the commission of such an offence, shall likewise be guilty of the said offence.

9. *A negligent data-bank operator should be liable for damages.*

The bill says:

An operator of a data bank to which this Act applies who causes or permits inaccurate personal data to be supplied from the data bank as a result of which the person to whom the data refer suffers loss, shall be liable in damages to such person.

10. *A hierarchy of availability categories may be registered for a data bank.*

In many systems data can be classified into various categories according to who can access them. Some categories of persons will be forbidden access to certain categories of data. Some persons may read but not change data. Some may change them but not disclose it. Some data may be disclosed to a limited group under special conditions; the rest may be disclosed with proper safeguards.

What information falls into which category will depend on the organization and its public. A psychiatrist might ask about and record, but never reveal, the sex life of a patient. An insurance company could neither ask for nor record such details, but a market research group might, given suitable safeguards. Inter-agency transfers might be restricted to the highest common factor of records, the 'public' face of an individual, not including his income, marital status, or medical data. Under rare circumstances, however, disclosure might be mandatory to a restricted group of recipients. Typhus carriers and criminals may have to be found quickly, and it would be absurd not to allow one agency to warn another.

Even the courts should be barred from speculative entry into any and every file. If a specific item is sought, a search warrant detailing the requirement would have to be issued. A court would no more be able to call for a speculative look at a whole dossier than to demand a general search of a man's property. The law, then, must specify what data may be elicited, stored, and retrieved by which persons for what purposes.

11. *Aged data should be removed.*

To prevent our past history pursuing us indefinitely, aged data should be removed from certain files as time goes on. California has attempted to 'erase' juvenile-crime records from files of those who have grown older and wiser. Such a scheme is hard to enforce, since other organizations may at a later date distribute the facts gained legitimately before the erasure. The Registrar should have general guidelines for the removal of aged data, and in certain cases should enforce it. The 'removal' of data may occasionally mean not complete deletion but rather transferral to an archival category which has restricted access for all but a few with special needs (for example, research).

12. *Security procedures should be registered.*

To minimize the possibility of unauthorized access, those data banks with sensitive data should register the techniques they have employed to secure them. This will ensure that the systems designers consider building appropriate locks on files which require protection.

SUCCESSIVE TIGHTENING OF THE LAWS

The requirements that we have outlined above no doubt contain loopholes. We believe we can see loopholes in the Data

Surveillance Bill. However, it is a good start. The law is an evolving structure and this bill could provide a foundation on which to build. It is flexible because so much depends upon the office of the Registrar, and this human flexibility is important. As such a set of laws comes into operation, the loopholes we have not foreseen will become apparent and can be sewn up.

The most significant ommission from the Data Surveillance Bill is a mechanism for enforcement. In early 1971, Leslie Huckfield, MP, introduced the Control of Personal Information Bill in the commons. This was based on proposals put forward on behalf of the National Council for Civil Liberties by Joseph Jacob of the London School of Economics. It proposed a 'Databank Tribunal and Inspectorate' which would license operators of data banks, order corrections to contents of databanks, and check the accuracy of police or medical records (which the subject thereof cannot see). The inspectors would have to be drawn from data processing professionals.

It is desirable to make these safeguards as automated as possible; otherwise a massive amount of work could result. If the Registrar, for example, is likely to inspect the interrogation log kept by computer centres, then this log should be in a machine-processable form. The format of the items it keeps should be a format specified by the office of the Registrar. Eventually, it may become a requirement that such logs can be *transmitted* to the Registrar's computer if required. The Registrar's staff can then inspect them on their screens and ask the computer such questions as, 'On what occasions did so-and-so ask for information?' Or, 'How often has this file been used for this purpose?'

The computerized society must eventually frame its laws and safeguards in such a way that computers can police the actions of other computers. Over a period of time appropriate checks and balances will grow up in this fashion, so that we can develop the type of environment we want to live in. Computers can protect us from harassment by other computers.

THE POLICE

Police computers are exempted from some of the provisions listed above. The police argue that this is essential. If you give a suspect access to the police files on him, he can take evasive action and you will never catch him. Many feel that the rising crime rate shown in Fig. 5.1 is more serious than the issue of privacy. This feeling may grow stronger as we climb further up that curve. If the United States does not reduce the slope of that curve, its crime will quadruple in the next ten years.

On the other hand, there has been much misuse of information sources, and other powers, by the police of some countries. There is every reason to doubt their ability to avoid abuse of computer systems.

As Professor Karst says:

We can hardly expect the various police departments around the country to refuse to share investigative leads among themselves. Yet there are serious possibilities of abuse. One recent example was the rather interesting use made of an old criminal file of Mrs Viola Liuzzo, the victim of a murder charged to certain members of the Ku Klux Klan (one of whom was acquitted by a Lowndes County, Alabama, jury). Mrs Liuzzo's record (her 'make sheet,' *not* an investigative file) somehow passed from the Detroit police to the Warren, Michigan, police, to Sheriff Clark of Dall County, Alabama, and even on to Imperial Wizard Shelton of the KKK. The point is that some policemen, like others in all occupations, can be expected to make improper use of information in police files.[5]

There are organizations of policemen which are highly politically oriented. Professor Karst gives an example of one which engages in extreme right-wing activity, and publishes a newsletter which consists in substantial part of lists of names of persons who have been identified by someone as communists or communist sympathizers, often coupled indiscriminately with the names of persons who support the establishment of civilian police-review boards, and so on.

The leaders of this organization are in responsible police positions. Professor Karst claims that they have access to police files. With the deployment of police computers now planned, they might obtain access to police records all over the country.

The unwillingness of the police on occasion to see themselves as the servants of society rather than its masters means that legal rather than ethical controls on their activities are essential.

At the very least, legislation on the Alameda County, California, pattern to separate police and social-service files is essential. Here, although welfare, health, institutions, and probation have been combined into the Social Service Central Index, the Police Information Network is separate as a result of legislative fiat.

Geographical distinctions are losing their significance as communications and mobility increase. For information to be of service, it too must not be bound by artificial boundaries prescribed in an earlier age. But if that information is to be a slave, not a tyrant, logical divisions must be prescribed and enforced by law. The data banks of police, intelligence, and other government departments must be restricted to the specific purpose for which they are intended. The divisions must restrict personal data to those who need to know them and can be trusted with them, and make the possession or divulgence of unwarranted material a criminal offense.

Corporations are obliged to adhere to certain regulations governing their mode of business. Accounts must be rendered in terms set down by law; workshops and offices must meet health and safety standards; contracts are established within a known framework. Such a legal system has grown in answer to needs, sometimes as a result of blatant injustice. We must establish a legal environment for information systems now to avoid injustices we can already predict. Good systems design, like proper accounting, can be legally enforced.

We need well-designed laws, and these must be backed up by tight system controls. We shall describe the latter in the next two chapters. When we have both, we believe that we shall have little to fear from living in the computerized society.

REFERENCES

1. S. D. Warren and L. Brandeis, 'The Right to Privacy', *Harvard Law Review*, Vol. 4, 1890.

2. M. L. Ernst and A. H. Schwartz, *Privacy: The Right to Be Let Alone*, Macmillan, New York, 1962.

3. Kenneth L. Karst, (Professor of Law, University of California, Los Angeles), 'The Files: Legal Controls over the Accuracy and Accessibility of Stored Personal Data', reprinted in *Computer Privacy*, US Government Printing Office, Washington, DC, 1967, p. 175.

4. Data Surveillance Bill, ordered by the House of Commons, 6 May 1969, published by Her Majesty's Stationery Office, London.

5. Kenneth L. Karst, op. cit., p. 195.

25

LOCKS, GUARDS, AND BURGLAR ALARMS

> I believe we in industry must continue to improve existing technological safeguards which limit access to information stored in electronic systems; and we must devise new ones as the needs arise.
> – *Thomas J. Watson, Jr, President of IBM.*

Secure military equipment might be kept in a compound that has several layers of protection. On the outside is a high barbed-wire fence, difficult to penetrate. It is patrolled by guards with rifles and dogs. At the corners are high watchtowers from which searchlight beams sweep across the night. The sentries in the watchtowers have machine guns. Forty yards of thorny undergrowth separate the outer fence from an inner fence, which is electrified. There are tripwires in the undergrowth which activate sirens and lights. Invisible rays also scan the area between the fences. Inside the compound, the bunker with the guarded equipment is locked as securely as a bank vault and is guarded 24 hours a day. To enter the compound, personal identification is needed as well as a password which is changed frequently. Only a few men have keys to the bunker. Security officers inspect the compound at random intervals to be sure that the guards are alert and the defences working. They are informed immediately of any suspicious activity or unusual event.

The information in a computer system can be equally well guarded. In some ways, it is easier to guard because the means of protection are not brute-force measures like dogs and watchtowers but are electronic safeguards. In many ways, however, they are analogous to the methods described above. They can be relentlessly thorough. The sentries will never fail to be alert because they are part of the electronics.

It is important in our discussions of the computerized society to realize that these safeguards *can* be built. Moreover, they are not highly expensive; the cost will vary with the measure of security needed, but a fairly elaborate set of safeguards might add perhaps 5 per cent to the overall cost of the system.

Many of the statements before congressional committees and many of the alarmist writings on the subject seem to betray ignorance of the fact that a computer system can be designed to protect its information with the same unyielding persistence with which it performs other operations. As of this writing, however, most computer systems do not apply these protective measures. This is largely because of lack of appreciation of the problems that we discussed in Chapters 14, 15, and 17. In the case of many computer installations it is because of lack of forethought. (It is staggering to consider that the early police-information systems were designed in such a way that an intruder with a terminal a thousand miles away could – and did – break into the files.) In other systems it is because the owner of the system saw insufficient returns from the expenditure of 5 per cent of the system cost on protecting his clients. Occasionally he had much to gain from selling private information about his clients. An appropriate legal structure is essential in such cases, as well as the capability to build electronic locks.

We anticipate that in the next few years electronic protection of data in computers will become a standard part of the technology. It will be easier, cheaper, and more efficient than today. We cannot urge computer personnel too strongly to pay attention to such safeguards, otherwise they may find that public and congressional outcry will prevent them from building their dream systems. It would be sad indeed if the privacy issue unnecessarily blocked the very real benefits that computers have to offer.

THE DEGREE OF PROTECTION NEEDED

Like a military compound, computer data should be protected with massive safeguards. This depends upon whom we are protecting it from. Protection against casual snoopers is easily achieved. Protection against James Bond or the CIA will need much more ingenious and expensive measures.

We may list six levels of potential protection:

1. *Protection against accidental disclosure of secure information.* This is largely a question of good hardware design and thorough program-testing.

2. *Protection from casual entry by unskilled persons.* It is necessary to prevent clerks, policemen, managers, and government employees from casually browsing in files of sensitive data. The clerk equipped with a terminal should not be able to look up his manager's salary at will. The government official should not be able to look up personal information beyond that intended for his use. A secretary in the personnel department should not be able to read confidential reports on employees. There is no difficulty in 'locking' the files against such prying eyes.

3. *Protection from casual entry by skilled technicians.* As the level of technical knowledge of our supposed marauder increases, we need more elaborate safeguards to keep him out. A man at a terminal with a detailed knowledge of the system might find ingenious ways of making it respond to his questions. Programmers frequently entertain themselves by making a system do something it was not intended to.

4. *Protection against entry by persons who stand to gain financially.* If we assume that the intruder stands to gain financially from restricted information, then we can assume that he might spend money attempting to invade the system, whereas the casual intruder would not. He might infiltrate the programming staff, buy a compatible terminal, use wire-tapping methods, plant bugs, make duplicate keys to the computer room, and so on. This might apply to a private-detective agency or to a hired intruder. We can assume that he will have both technical knowledge and technical equipment. This further raises the cost of defence.

5. *Protection against well-equipped criminals.* We may have assumed in the last category that the intruder was not prepared blatantly to break the law. In this case the laws discussed in the

previous chapter would offer much greater protection than those we have today. Here we assume that a major criminal operation may be underway. It can be an employee embezzling his firm as described in Chapter 18, or it may be the Big Bank Robbery of the computerized society. No masked bandits pull guns on the tellers; instead, a group of graduates change the data on the computer files.

6. *Protection against organizations with massive funds.* Extremely elaborate methods are reported to have been devised by the military and intelligence agencies for obtaining information from other computer systems. These include, for example, the detection of electro-magnetic radiation from computer centres using powerful antennas in a specially equipped truck. Such methods can involve almost unlimited expense, and the defence against them requires an equivalent level of expenditure and capability.

EFFECTIVENESS OF THE SAFEGUARDS

The technical details of computer locks and safeguards is beyond the scope of this book. We shall summarize the *types* of safeguards needed in the remainder of this chapter. A detailed technical assessment of their effectiveness yields the following conclusions. Protection against category 2 above is easy and its cost is barely significant. Protection against categories 3 and 4 needs to be more elaborate, but can be made highly effective. The same is true of category 5, although it is always a mistake to underestimate the ingenuity of our cleverest criminals. Given time, patience, and money, they will probably penetrate very-well-defended systems occasionally, just as they penetrate bank vaults and jewel safes today. If our enemy is in category 6 above, then defence is beyond the scope of the techniques discussed in this chapter.

In other words, the staff and consultants of the quality of an ordinary data-processing installation today should, if they tackle the problem seriously, be able to protect the data from anyone other than the National Security Agency or an organization with similar funds and skills. On rare occasions an ingenious and persistent criminal may slip through the defences. In practice, the degree of protection taken, and hence its cost, will depend upon the assessment of the potential intruders. This, in turn, will depend upon the sensitivity of the data, and their possible value to an outsider. If an outsider is likely to spend money in order to steal the data, then more money must be spent to protect it.

RECOMMENDATIONS FOR SAFEGUARDS

Let us now very briefly recommend types of safety measures that can be used. An installation should not use just one or two of the following. It should use every one that is consistent with its security policy. One measure reinforces the others, and true security is only achieved with a chain of interlocking safeguards. Just as the military compound uses dogs, electric fences, barbed wire, tripwires, searchlight towers, and so on to reinforce each other, so the same is true with the following measures. The specific measures used must necessarily differ from one type of installation to another.

1. *Avoid printouts of sensitive data.*

Many systems for inquiring into files and for changing data on files use visual displays with no printing. This will become, increasingly, standard procedure in the future. In this case there are no printouts of private data to be lost or removed. The computer centre should also ban programs which can print out such data on the system printers.

2. *Identifying the terminal-user.*

On many systems the first step towards making the system secure is to identify the person at the terminal. Until he is positively identified, the computer will not access any sensitive data for him or make any modifications to the files. He is absolutely locked out. There are three types of ways in which a person can be identified:

a. He can be identified by some physical personal characteristic. For example, a device could be used for reading and transmitting his fingerprints or thumb print; or the computer could have a program for recognizing his voice by comparing his speech against a stored voice print. Such schemes are under development, but they are not publicly available at present. Voice-print techniques are now being used experimentally, and if they are successful may come into common use.

b. Next, the individual could be identified by some portable token – for example, a badge, card, or key. He inserts the badge into a terminal badge-reader, or the key into a lock on the terminal.

c. Last, he could be identified by something that he knows or memorizes. He could memorize a password, or answer a pre-arranged set of questions. Techniques of this type require no special hardware. They are the least expensive of the three and can under most circumstances be made reasonably secure. On one system which has achieved a high measure of security, the terminal user must type in his own personal number, and then a unique security code number which is allocated to him and changed frequently. An error in this procedure causes the system to take protective action, which we shall discuss later. The code numbers are never *printed* at the terminals.

3. *Identify the terminal.*

On some systems it is desirable to identify the terminal as well as the user. On other systems the terminal is identified *instead of* the user. The reason for this is most likely that some terminals are located in an area generally regarded as a secure area, whereas other terminals are not. Perhaps only trustworthy personnel are allowed to enter a certain building or a certain room. The terminals in this room are secure, and so the computer as part of its security procedure needs to identify the fact that it is communicating with these particular terminals.

Some terminals have no mechanism enabling the computer to identify them positively. Such a mechanism *has* been included in terminals put on the market recently and is a response by the computer industry to the recognition of the need for security.

4. *Establish user authorization.*

Having established the identity of the user, the computer should next be able to check what he is authorized to do. On many systems he will be authorized to use some of the programs but not all of them. He will be authorized to gain access to certain files and not to others. He may be permitted to read certain files, but not update them or modify them in any way. The computer should, therefore, have a table indicating what each user is authorized to do. The authorization procedure on some systems is very simple, on others highly structured and complex.

In the simplest form, of this procedure the users of the system may be divided up into a small number of categories. The category of each user will determine which programs he is permitted to use. With such a scheme certain programs, or possibly certain files, may be barred from certain categories of users. The category of user will be determined either by a password or security code that he keys in, by a badge or card that he inserts into the terminal, or possibly by the terminal address itself or terminal location. He may, for example, be categorized as a manager, supervisor, clerk, trainee, 'person not a member of the corporation', 'person giving a demonstration', and so on.

Rather than basing the authorization on category of user, it will be desirable on some systems to base it on the individual himself. In this case, the computer may have a table for each user indicating which programs he is permitted to use, or which transaction types he is permitted to enter.

5. *Establish programmed locks on data files.*

When we store sensitive data in filing cabinets, we lock up these cabinets and take careful precautions to ensure that only authorized people can obtain the key. We must do the same when we store data on computer files. The locks, however, are now of an entirely different nature. It is possible to make them at least as secure as the locks on filing cabinets. But instead of having one lock on the entire cabinet, we can have a very elaborately structured set of locks allowing different people to read different parts of the information, and carefully restricting which parts of the data different persons are permitted to modify.

A low-level approach to the problem is to use a password. On some systems the terminal operator creates and maintains his own file. He writes a password on that file, or possibly multiple passwords on different sections of the file; his file cannot then be read by anyone who does not type in the correct password. The terminal-user is free to change the password with which he has locked his files at any time he wishes. Such a scheme is in use today on a variety of different systems and is successful in protecting the files from most persons who would like to pry into them using a terminal. Most of these schemes, however, cannot be regarded as highly secure because an ingenious and persistent intruder might obtain the password. It probably has about the same level of security as the normal locking of a filing cabinet with a key.

The combination of a password system with other precautions can build up a high level of security. One such system stores highly secret information along with unclassified data, both accessible to terminal-users. Only terminals in a secure area are allowed access to the secret information, and then only if the correct document number and password are used.

In another system some users were found casual in their approach, often not bothering to type in a password. In this case the system was re-programmed to make up its own password and lock the file with it.

Sometimes a password permits communication between one

user and another. Just as one person can give the key to his filing drawer to another, so a group of users can share access to a common document or file if each knows the password.

To lock the files in a more secure manner, the password scheme should be linked positively to the identification of the terminal-user and to auditing procedures that we shall mention below. For policing the activities of the terminal-users, the password system should be designed so that the password cannot be obtained by the computer centre operating staff, but can only be accessed or changed by the security officers concerned with the system.

6. *Allow a central authority to control file security.*

Control of a data base can be better maintained if the file is secured by a central authority. This authority sets up *lockwords* on the file. Lockwords present a selective and highly structured scheme for locking the files. They are like combination locks on a safe – locks which require to open them many digits which are different for different records or types of records.

The lockword must itself be linked to the terminal-user and to his authorization to access different types of data. The lockword may be related to

1. an individual user;
2. a category of user;
3. an application program;
4. a terminal or terminal location;
5. any combination of these.

A table is set up indicating just which parts of the file each user, program, or terminal is authorized to read and to write. The computer will not access any records except as indicated by this table.

The lockword can be made as long as is thought necessary. If a combination lock on a safe gives better protection when eight digits are required to open it rather than seven, then it can have eight. A computer lockword could have eighty, although more than eight would probably be unnecessary.

7. *Change the security codes and lockwords periodically.*

The security codes, file lockwords, and other 'combination-lock' sequences should be changed at intervals. They may be changed automatically by the computer, or manually if suspicion arises that an unauthorized person has gained knowledge of them.

8. *Set up a 'police force' for the system.*

To maintain high security in any building or organization, policemen are needed. This is true also of a data-processing system. The system may need physical guards at some terminal locations, and possibly also at the computer centre. It may need a security staff to discourage intruders, make sure that tape or disk stores are locked, attempt to prevent wire-tapping in the building, and so on. But in particular, a secure computer system requires a staff responsible for the security of the data files and for the control of the authorization to use these files.

A security officer may be appointed as the sole person entitled to change the authorization tables or file lockword tables in the system. He will have details of what each individual is authorized to read or change on the files. He will be responsible for issuing passwords or security codes and for insuring that these are used correctly. In a system with terminals in far-distant locations, there may be a security officer in each location. A suitable person with another job, such as an office manager, may be given the responsibility of system security in that location. He will take instructions from the main security officer. The system will send in his listings, which will be described below, of all detected violations of correct procedure that occur in his location.

9. *Make a file-owner responsible for its security.*

In addition to a security officer for each *terminal location*, there may also be one person held responsible for the security of each individual *file*. This man, in addition to his other work, maintains the locks on his particular file and insures that the sensitive data on it are not accessible. This man might be referred to as the 'file-owner', and on some systems he is also responsible for the accuracy of the file. If an individual wants to gain access to the file – or to certain fields on the file – which he does not presently have access to, he must obtain this from the file-owner. On one computer system there may be many different file-owners.

10. *Fragment the files.*

Mechanisms for preventing entry into the files such as lockwords are very flexible. The file can be fragmented into as many parts as desired, all with different lockwords. A person authorized to look at certain records is then locked out of the rest.

It is very desirable to fragment certain files in this way. Where a dossier of information about an individual is kept, the persons needing certain data about him should be locked out of the rest.

11. *Establish 'burglar alarms' on the system.*

Anyone contemplating invasion of the files, either because of curiosity or through malicious intent, should be deterred by the thought that there is a high probability that the system will detect him and immediately inform the appropriate security officers. The psychological deterrent of knowing that the system has effective burglar alarms is great.

It may pay to take action immediately when violations are detected. Probably no action will be taken on the first violation except to log the fact that it has occurred, because most operators can be expected to make single, isolated mistakes.

However, if an operator, having made one mistake, then immediately makes a second, again attempting to enter an invalid code or access an unauthorized file, this may be an indication that he is attempting to do something on the terminal that he should not. The system may then immediately inform the local security officer in the hope that sometimes, at any rate, the culprit will be caught red-handed. The miscreant may even be 'kept talking' by the system until he is caught.

Another approach that has been used is to lock the terminal completely the moment the second violation occurs. The application programs will be written in such a way that no more information will be accepted from that terminal until this condition has been cleared. The only person capable of clearing it should be the security officer for that location.

12. *Establish security audit procedures.*

A log should be kept of all violations of correct procedure in using the system – for example, when a terminal user types in a security code which has not been allocated to him, or attempts to access a file which he has not been authorized to access. Details of these violations will be printed and sent to the security officers. A branch security officer, for example, will receive a listing of all of the violations that have occurred within his branch. A file-owner will be sent details of all unauthorized attempts to read or change records in his file. This log of violations should be analysed to detect any unusual activity. Most violations will be accidental and caused by a genuine mistake on the part of the terminal operator. A sudden departure from the norms in this activity, however, may indicate that some user is tampering with the system, possibly exploring and trying to find a method of gaining unauthorized access. The list of violations may be printed out once a week; on the other hand, it may pay to print it out more frequently on a system containing highly guarded and highly sensitive information. Branch security officers may be sent a list of any violations that occur each night. Then a would-be intruder would have little time to practise.

13. *Maintain very tight security on the authorization records and lockwords.*

It is particularly important to maintain extremely tight security over the authorization records and the file lockwords. If the would-be invader can change these, then most of his problems are solved. No one should be given the authority to read or change these records except the file-owners or the security officers. If any change is made, then the appropriate file-owner or security officer will be sent details of the change the following day. Such changes might be detected on a nightly run by comparing last night's authorization records and file-lock tables with tonight's. If any unauthorized person has managed to make any change in these, it will be detected quickly.

It is recommended that a history be kept which logs all changes made to these security records, indicating who made the change and where it was made.

The codewords and lockwords which the security officers use should be changed very frequently, perhaps every day.

14. *Take measures to prevent wire-tapping.*

With wide use of data-transmission, the possibility exists that wire-tapping will occur. This has been frequently mentioned in public discussion on security in computer systems.

Wire-tapping can be either active or passive. By *passive* we mean that the eavesdropper only listens, and does not originate any data on the communication line. To carry out *active* wire-tapping, he would normally have to have a terminal and model compatible with the type of transmission taking place on the communication line. Even then, the authorization controls and file locks could make it virtually impossible for him to access the precise part of the data base that he is interested in.

There are some particularly ingenious forms of active wire-tapping. For example, 'piggyback' entry into the system refers to the interception of communications in such a way that modifications are substituted for messages that would otherwise occur in the normal man-machine 'conversation.' Often a legitimate user is inactive for a substantial period of time between his sign-on and sign-off, and during this time an active wire-tapper could send messages which would be accepted by the computer. Again, he may be able to cancel the user's name. Such techniques, however, require ingenuity and much technical knowledge, and would be very difficult to implement. We can find no instance in which active wire-tapping has occurred in practice.

Passive wire-tapping, on the other hand, can be conducted with relatively inexpensive equipment. A cheap tape recorder can be used to record the signals passing over a voice line, or sub-voice grade line, in such a way that all of the data can be reconstructed.

A person tapping the flow of data might have to listen for a very long time on most systems before he found any information of particular value to him. The combination of wire-tapping with other techniques may, however, produce results. For example, wire-tapping could be used to obtain the security code that the security officer used to access the system. With a high degree of ingenuity and a cheap tape recorder, passive wire-tapping could help crack the system.

How can we protect ourselves from wire-tapping?

First, we can use 'scramblers'. These are devices frequently seen in Second World War movies for the protection of telephone calls. In the computer age they do not provide a major degree of security. The scrambling method is relatively simple, and it would be possible to translate a recording of what is sent into the correct data signals.

Second, we can use cryptography. This is, in effect, scrambling in a digital rather than an analog fashion. It can be made much more secure than the use of scramblers, and will be discussed below.

Third, we can attempt to make the lines physically difficult to tap. There are, in fact, only a few points at which a telephone line is likely to be tapped. One cannot simply lift the telephone company's manholes in the street and go to work. Even if one did, the vast quantities of cables would make it almost impossibly difficult to tell which was which. The telephone company normally maintains high security over its central offices, and the long-distance links have hundreds of telephone calls scrambled together at a high frequency. It would be impossible without great expense to record and unscramble them. On the other hand, the telephone lines are vulnerable within most public buildings. Here the private-branch exchanges and junction boxes are usually not locked up, and the individual lines are clearly labelled.

The first step to preventing wire-tapping, then, is to make the wires secure in the buildings that contain the computers and terminals. Lock up and secure exchanges and junction boxes. Inspect the wiring periodically. If the wires cannot be tapped in your own building they will probably not be tapped at all.

15. *Take measures to prevent other forms of eavesdropping.*

In addition to passive wire-tapping, a number of other forms of system eavesdropping are possible. A person for example, could attempt to observe security codes that are typed into a terminal or momentarily displayed on a screen. Cameras with telephoto lenses could be used. Carbon copies of printed information and typewriter ribbons should not be left in unlocked waste bins which can easily be rifled. The carbon paper itself is usually thrown away, but is easily readable.

It is possible to tell what some machines are printing, punching, or transmitting by detailed analysis of the sounds they make. This is true of certain terminals, although not of others. This could not be detected by detailed analysis of a recording. Many machines, such as computers, terminals, and occasionally data sets, give out radiation which can be picked up at some distance. It would be difficult to extract information from this, but it is technically possible. Design of computer equipment with protection from this possibility in mind will limit the danger.

16. *Send messages in code.*

Ever since the communication lines were tapped in the American Civil War, cryptography has been considered a means of protecting data transmitted over communication lines. In computer systems handling sensitive data, it could be used both to protect data-transmission and to protect data stored on removable files.

Since the First World War, much has been written about cryptography and code-breaking. The subject, however, has drastically changed its nature with the advent of computers. Where a computer is used to do the coding, an immeasurably more complex form of encoding can be employed. On the other hand, the computer can now be used to do the decoding and to search at high speed through very large numbers of possible transformations. In balance, the sender is generally better off than the potential code-breaker as a result of computers.

One method of encoding is virtually uncrackable. Both the sender and the receiver have a disk pack or other form of storage containing a very large quantity of random characters. These are generated in such a way that it is impossible to predict future characters in the sequence – i.e., no straightforward mathematical method is used. These characters are 'added' to the characters of the message to be sent in fixed-length blocks. At the receiving end the same characters are 'subtracted', and the original message recovered. If this message is intercepted, it cannot be decoded no matter how skilled the interceptor, unless he has access to the disk of random characters being used – and we make sure he cannot obtain this.

In theory, the same disk should not be used more than once if the transmission is to be absolutely uncrackable. In practice, a disk containing, say, 10 million digits could be used safely in most installations for a month or so.

The encoding and decoding in the above example generally needs a computer, although a small special-purpose device could be built to do it. Computer-to-computer transmission can be made highly secure with no extra equipment. Computer-to-terminal transmission cannot similarly be secured without expenditure on special equipment at the terminal or at a unit controlling many terminals in the same building.

17. *Programs should be carefully tested and inspected before use.*

The person attempting to break the data bank may be one of the programmers. Before any programs are allowed to run on a system with 'live' data in its files, they should be very carefully tested to insure that they do what was intended, and they should be inspected to insure that they do not access records to which they have no right. Only programs which have been checked out in this manner should be allowed in a system with sensitive data. They should be stored on its files by a formal procedure. The system should be designed so that it can only use these programs.

18. *Use file-management software which enhances security.*

The software concerned with file-management can be designed to enhance the safety of data on the files. The system should be designed so that a program can *only* access the files via this software. The routines for unlocking the files given suitable passwords or lockwords should be part of this software. This is an important step in protecting the files from the programmers.

19. *Maintain security during program-testing.*

During those periods when the system or its programs are being tested, it is vulnerable to a security branch. This period should be carefully controlled and the personnel involved should be limited. No sensitive data should be used or even left on the file units during testing periods.

20. *Use the system complexity to help safeguard it.*

One factor which helps is that computer systems are becoming exceedingly complex. A large amount of knowledge is needed to break into a reasonably protected data bank. This knowledge itself must be protected. One of the principles of safety should be that we *permit no one man to have all of the knowledge* that he needs in order to break into the system. If you are planning a bank robbery, the first obstacle you meet is that you do not know how the burglar alarm and other security precautions work; neither do you know where the keys are kept. The same must be true of the computer system, and here the complexity factor is on our side. The passwords, authorization procedures, file locks, and audit and alarm techniques raise considerably the level of knowledge that is needed in order to break into the system. Details of how these facilities work must as far as possible be kept secret, and knowledge of them restricted to as small a number of people as possible. Details of the file-addressing, the communication-line organization, and mechanisms of the control program should not be made available to persons with no need for this knowledge. There is a strong case to be made for allowing the computer-user to design his own security procedures instead of using standard offerings from consultants or manufacturers. If he designs his own, their workings can be kept a secret.

If the safeguards are designed so that no one person has sufficient knowledge to be able to break in, there would have to be collusion between different groups of people, or one person would have to go to great lengths to educate himself in other areas and attain system knowledge that he would otherwise not have. The more the knowledge of the system is fragmented, the more the collusion necessary for planning an invasion. This can make it extremely difficult to break in and increase the probability of the culprit being caught. The level of technical knowledge required can be made much greater than that in a major jewel theft.

21. *Maintain tight security over the system operation manuals.*

To prevent a person acquiring the technical knowledge needed to break into the system, the manuals containing this information should be locked up and carefully controlled.

22. *Do not let programmers operate the computer.*

The operating of the computer should be carefully controlled. A programmer should on no occasion be allowed to operate it, as this might give him an opportunity to feed in his own programs and by-pass the system controls.

This is one aspect of fragmenting the technical capability of the system personnel. It should be divided in other ways also. Do not grant one type of person more power than he needs; he needs multiple capabilities to penetrate the defences.

23. *Maintain good physical security*.

In addition to all of the electronic safeguards there should be good physical security – locked cabinets and vaults, burglar-proof tape stores, document-shredders, guards at the computer centre, and so on. Physical access to the computer room should be limited to essential people. Some terminals should be in limited-access areas.

24. *Maintain staff loyalty and competence*.

The loyalty and competence of staff members involved with security is essential. The management should insure that these are maintained.

There are many more technical arguments concerned with computer security than we have space to list here. We hope that the above summary will convince the reader that data in computer systems *can* be locked up. In the years to come, security techniques can become much more rigorous than they are today. Equipment and software will be better designed for security. The powers of the computer can be used in our favour in this matter.

System security can be made virtually uncrackable, as on the systems concerned with our hydrogen bombs and missiles. There will always be clever methods of breaking in, but we believe that at reasonable cost intrusion can be made so difficult that we may regard the data as securely locked for all practical purposes.

26

SYSTEMS CONTROLS THAT
ARE NEEDED

We approach the new (technologies) with the psychological
conditioning and sensory responses of the old. This clash naturally
occurs in transitional periods. – *Marshall McLuhan.*

If we have a sound legal structure and efficient locks and
burglar alarms on sensitive data, as indicated in the previous
two chapters, we shall have laid the foundations for building
abuse-free systems. We can suitably fragment the data and
restrict access to the fragments to those persons who legitimately need them. There will still be data robberies, no doubt,
but like bank robberies they will be the exception rather than
the rule – if we design the safeguards well, they will be rarer
than bank robberies. Above and in addition to these security
measures, we need systems controls.

Data-processing systems interact with their environments by
collecting information from them and disseminating information to them. The environment of all these environments is,
for our purpose, human society, which is itself a system. Its
informational inputs provide inadequate and noisy guides to
the closeness with which it approaches goals it is unable to
define, let alone agree.

As we break the totality of society into smaller parts, the
goals of the more or less organized subsystems thereby disclosed
become better defined. There are groups of nations with objectives laid down in treaties, and single states with five-year plans;
businesses with profit objectives; local governments with projects and budgets; police forces with laws to enforce; colleges
with students to teach; laboratories with prototypes to create;
accounting departments with employees to pay. As we sub-
divide still further, we have computer systems with specified
ability to store and manipulate data and, further, an electronic

circuit with the sole objective of adding numbers or, narrower still, a silicon chip for unerringly converting one electrical pulse into another whenever called on to do so.

At the lower levels in this hierarchy, objectives are absolutely clear. The system controls are precisely attainable. As the level of complexity increases, the goals of the system become more difficult to define, the difficulties of monitoring the system increase, and the very targets against which we measure performance become hard to determine. Nevertheless, it is evident that the purpose of the subsystem is to take the input from its environment and convert it to the form required by the greater system of which it is part. No man is an island; the computer is not even a promontory. Both are closely linked into the society to which they belong.

The problem of defining the objectives for a computer is difficult enough, but defining the limits of the subsystem's enveloping system is more complex still. The environment of the American Internal Revenue Service's computer is not merely the IRS computer department. Does it include the whole of the IRS, part or all of the federal government, all taxpayers, all citizens, just Americans, or foreigners too, or perhaps the whole world?

One thing by now must be apparent. We cannot define goals clearly, unambiguously, and simply once the system we seek to control gets beyond a certain size. What we can do, however, is to set limits which constrain it to certain paths which do not cause damage. The diplomatic 'system' may be instructed by the Chief Executive to further the nation's interests by all means short of war. Here we have an ill-defined goal and a specific constraint. A corporate personnel 'system' could be told to find out everything pertinent to an applicant except his religious, political, and moral opinions. Again, it will be more difficult to determine how close to the goal it gets than whether impertinent questions are asked.

The goals and constraints of American society as a whole are formalized in the Constitution and the Bill of Rights. These hardly give adequate guidance to a society in the position of Faust, theoretically able to choose between untold wonders that science can bring. Further, they are documents that are almost unchangeable in a time of precipitous change. American society itself is a subsystem of international society, which has virtually no goals or constraints of its own and where famine coexists with obesity.

What controls are needed?

When a system fails to transform its input into the desired output, it is in error. The systems designer's problem is to provide a mechanism for detecting errors and correcting them before any harm is done. Mistakes can be as simple as getting a phone number wrong, or as complex as failure by the super-systems. All centre on one simple idea, that of *regulation*. We have to devise some means of regulating all systems, from the simplest to the most complex. A Watt governor regulates the speed of a steam engine. Totalling and balancing operations control the cash handled in a bank. The police and courts control citizens' adherence to law.

A system under control will, after a temporary disturbance, return to its stable condition. As discussed in Chapter 8, deviations from stated goals must be detected, signals fed back to the process, and corrective action taken. On simple systems where the goals can be stated with precision, this mechanism can be designed with precision. On complex systems, design is more difficult. There is a danger that controls will not be applied because of lack of understanding of the goals and constraints, and the inadequately controlled system may in some degree impair the operation of its supersystem.

We can list a number of types of controls that are needed on computer systems:

First, we need controls over the accuracy of the circuit operation. When a circuit adds 2 and 2 and obtains 5, this fact must be detected immediately and corrective action taken.

Second, we need controls over the correct functioning of the systems – in other words, programs must perform as intended. A program to mail invoices will count how many it sends out, add up their values, and check these against the quantities it was expected to produce. If an error is detected, the programming staff is informed ('feedback') and the program checked.

One human faculty not possessed by the computer is a sense of the absurd, but a good programmer always tries to build one into his programs. Most files contain extra information implied by the grouping of items into messages and records. For instance, a multi-flight journey booked on an airline reservation system is checked for location continuity and time continuity. The computer can be programmed to watch for the unusual – not just the absurd – and, unlike humans, it will spot all such cases.

Third, protection against failure is needed. On some systems this is not too important. A three-hour machine failure may simply mean that the operator has to work overtime. On others it could be disastrous. In air traffic control, complete redundant computers process all data simultaneously and the results are compared. Any discrepancy precipitates emergency action. In some systems men are part of the redundancy procedures and the systems engineer has to design a man–machine complex in which men and machines check each other.

Fourth, protection against external intrusion is needed on some systems. As we discussed in the previous chapter, systems can be effectively locked against those who would pry into or change their data.

CONTROLS THAT INVOLVE PEOPLE

The above types of controls are those that are firmly in the province of the engineer or systems analyst. Now we come to the more difficult ones, where administrators, lawyers, and politicians enter the act. To avoid confusing the argument later, let us note that the above controls can be established with great thoroughness.

Once we can lock unauthorized persons out of the data files with the methods described in the previous chapter, then our controls should be concerned with those who *are* authorized. The authorization procedure itself is our focus of attention.

In the terminology of this chapter, a breach of security is the invalid combination, in the system output, of an information input with a person with undesirable characteristics. Our objective is establishing a checking system that prevents the creation of such an erroneous set. The output set we are trying to avoid has three members: information, person, and character. We can seek to eliminate any one of them. We can in principle limit information to a subset of all possible items, limit persons to a subset of all possible people, or we can limit the character of the persons who have access to information. This means in effect limiting information that can be brought together, denying access to it to all but a few, and restricting the uses to which it can be put.

LEGITIMATE INFORMATION

There is no information about people that cannot for some particular purpose be justifiably collected. Once that purpose is served, however, retention of all the details is indefensible unless they are to be used again. If the citizen to whom they pertain provided them for one purpose, they ought not to be used for another without his consent. This was why the Data Surveillance Bill (see Chapter 24) stressed that data in a data bank should be used only for the stated purpose, and why we stressed in Chapter 24 that old data should be destroyed. Lacking such laws, a government agency should provide its own rules.

The Census Bureau clearly provides the proper standards in this respect. Other agencies should observe the same rules regarding disclosure. The Census Act states that 'information furnished' the Census Bureau will be (1) examined only by 'sworn officers and employees' of the Bureau or of the Department of Commerce, its parent organization; (2) used only for 'the statistical purpose for which it is supplied'; and (3) 'com-

piled in such a manner that the data supplied by an individual
... cannot be identified'.[1]

The Census Bureau once refused President Truman's request
for information because he was not a sworn employee. Lesser
officials will have to suffer indignities with understanding, not
anger.

Policing the agencies will be more difficult, especially if the
types of laws discussed in Chapter 24 are not passed and the
only sure method is to limit the information they can collect.
The Inland Revenue should be restricted to financial facts. If
it needs confirmation, say, of a taxpayer's medical history, it
should ask the Department of Health, Education and Welfare
to evaluate his status according to IRS rules and send back an
answer, rather than to provide all the details so that the IRS
can do its own evaluation. To achieve these limited inter-
changes, the various agencies must know each other's file-
content classifications so that they can ask for feasible answers.

STATISTICAL FILES V. DOSSIERS

It is important to separate the concept of statistical files from
the concept of dossiers about individuals. Many of the uses to
which the information in a national data bank will be put
are statistical. In statistical files the identification of the individual should be stripped off. Once that is done the files can be
used much more freely. It is still necessary to make sure that
people cannot be identified by some other means, such as home
location, a high salary, the fact that he has thirteen children
or cats, weighs 300 pounds, or has a combination of factors
that makes him unique. Judgement is required as to what degree
of security is necessary.

In general, not only the name and identifying numbers should
be removed. but all other information not relevant to the
statistical purpose in question.

In the case of the proposed national data centre, arguments
have been put forward for keeping information about an
individual obtained from many different sources in one centre.

Leaving the sets of information in the separate systems where they were collected, however, would be much safer. If each of the separate systems is well guarded and controlled, it would be very difficult for a private eye or official to assemble all information about an individual.

In the interests of individuals' security, the national data centre should be an index to government files, not the repository of their contents. The centre can then be the sole organization with access to the files of more than one department, and can be rigorously policed with elaborate security precautions. It concentrates the risk, since it provides the only means by which a dossier can be built up.

On the national data centre files would be a table of all the various agencies' reference numbers of every individual. Using these numbers, the national data centre could call up each agency and fetch the records that correspond. The regulations of the agency would not allow it to retain such information, however, beyond the operation for which it was required. All applications for material collating the records of two or more departments would have to be processed through the centre and demonstrated not to infringe privacy. The centre would strip identifying data from its output, and insure that intelligence requests did not sneak in disguised as statistical analyses of populations with one member.

The central system would only be able to call other systems; it could not be called from them. Subversion of a single 'peripheral' agency would not, therefore, lead automatically to subversion of all the others. Anyone trying to get at the peripheral data banks would have to break into them individually, or go through the well-defined centre.

To manage this legitimately, an inquirer would have to establish his 'need to know' and satisfy thereby the criterion for access. The centre would be the only body that could authorize the granting of a request for statistical information, and it would set various levels of security classification for different individuals and purposes. No request for information which included identifiers would ever be met.

ENCODING SENSITIVE DATA

In many cases where private data are stored, they may be encoded first and the cipher key only produced on order of the courts or other suitable authority. As we commented in the last chapter, cryptography since the advent of the computer can produce code which is virtually indecipherable. Disks containing many millions of randomly generated characters provide the key for encoding and decoding. Without the relevant disk, the world's greatest code expert is unlikely to be able to read any of the data, because there are no repetitive patterns or algorithms used in its encoding.

In Britain, following reform of the abortion laws, the Medical Association approved a suggestion for such a coding scheme to safeguard the secrecy of the medical records of women receiving abortions. The *Observer* reported as follows:

Mr Diggory, [a consultant gynaecologist] said yesterday: 'The records should be given a code – to be broken only under the direction of a court of law – by the doctor who carries out the abortion before they are sent to the Chief Medical Officer at the Ministry of Health.'

Doctors are still bitterly opposed to the Ministry's insistence that the names of women be submitted with abortion records. The details are then available to a police officer 'not below the rank of superintendent or a person authorized by him' if an offence against the Act is suspected.

There is no dispute that a central record of abortions is necessary for statistical purposes and for bona fide scientific research. But Mr Diggory asked: 'Why should the Chief Medical Officer want to know a woman's maiden name as well as her present name and Health Service number? Of what possible interest is this to statistics or research?' The BMA has insisted throughout on the 'strict confidentiality' of records. It is argued that the new regulations will damage the confidential relationship between doctor and patient. It is also feared that the names may be too easily available to police. This, it is felt, might frighten women into seeking out unscrupulous doctors and back-street abortionists.[2]

It is clearly very important that files such as these be

safeguarded, especially when they are kept with the efficiency and accessibility that the computer brings.

CONTROLS OVER WHAT DATA ARE COLLECTED

The next type of control relates to what data are collected. The working environment in which some data are gathered is simply not conducive to precision.

Information has to be organized into patterns relevant to its ultimate use. This, however, is very seldom the form in which it is most cheaply collected for insertion into files. Take the case of a credit agency that wants to offer its clients relevant case-histories, including adverse information such as litigation. The best way for the agency to obtain this information is to go through all court records, updating its files on each individual accordingly. Because the agency is storing results for future use, the scrutiny associated with an immediately relevant item will be absent. A few errors will get through even conscientious operatives, and *there is no immediate check*.

Some errors of commission will be picked up by programming checks built into the system. Errors of omission are more slippery. A court case may be recorded as pending long after its disposition if the system has no self-purging mechanism which automatically warns of incomplete records due for updating.

In addition to patent errors, large files also accumulate two other kinds of troublesome material – the decaying and the unevaluated. The former is characteristic of biographical details. As Professor Reich was at pains to tell the Gallagher committee (see Chapter 14), 'information gets less reliable, the further away it is from the source'.[3] References and testimonials, to say nothing of college degrees, lose their relevance as their subject grows older. The facts remain facts in the environment of their collection, but they may be inappropriate in the context of the use to which they are put. But, 'every normal human reaction is going to be to give more weight to those things in the file than ... the maker of the file ever meant'.[4] It is this difference between the initial collection and ultimate use that was stressed in the arguments against permitting personality-test scores to influence non-medical decisions like promotions in or appointments to the civil service. There will be justification for *a posteriori* analysis of the relationship between performance and a previous score in order to draw general conclusions, leading to the design of new tests for the *a priori* evaluation of candidates. But the fate of an individual should depend on a re-examination by an appropriate instrument at the time of the decision, not by the translation of an inappropriate finding from another investigation.

The FBI ... also holds untold amounts of raw information – unevaluated data such as rumors and grudge reports about people.[5]

Equally unevaluated, in the legal sense, are the FBI's Uniform Crime Reports:

The Crime Index figure for 'murder and nonnegligent manslaughter,' for example, is, in the words of the report: scored on the basis of police investigation as opposed to any decision of a court, coroner, jury, or other judicial body.[6]

In other words, the facts could be legally wrong. This would not matter much if there were a constant relationship between arrests and convictions, or arrests and crimes, and the statistics were used merely for guiding general policy. However, in the words of an American Civil Liberties Union officer,

the widespread use of incomplete and unexplained arrest records has long concerned the American Civil Liberties Union. We have been deeply troubled by the adverse consequences to an individual flowing from the recording of an arrest not followed by indictment or convictions, where the true nature of the conduct leading to arrest (such as peaceful participation in civil rights or peace marches) is not disclosed. In our correspondence over the past few years with the FBI about the arrest record problem it has been clearly established that too frequently local law enforcement officials report arrests to the FBI but fail to report later disposition of the case. Countless persons against whom charges have been dropped or who have been acquitted must still suffer the harsh consequences of a wrongful taint of criminality when seeking employment or other privileges. These problems are even more grievous in the all-too-common case today of those arrested for the valid exercise of constitutionally protected rights. No reliable procedure exists for differentiating such arrests in present FBI records from arrests made for the normal incidents of criminal conduct.

The union has frequently suggested methods to eliminate employment discrimination based upon the bare record of an arrest, urged better reporting of arrests and ultimate disposition in each case, and pressed for avenues of legal redress for improper use of arrest records. Yet the problem remains, and will be accentuated by the creation of a central pool of information. Such a pool will serve only to multiply the deprivation of the civil liberties of those who are wrongly arrested or arrested and even convicted for merely exercising their rights. Inaccurate and prejudicial data will be made available to a greater number of police officials and through them to still greater numbers of unauthorized persons.[7]

The problem in this case lies in the outer ring of Fig. 16.1; the system's philosophy is questionable, and it may serve the wrong ends.

Representative Gallagher cited a case in which a man had a serious problem because he was charged with a hit-and-run accident in New York City, when in fact the sergeant had made an error and written down the wrong licence number.

When the material collected is very bad, it becomes a problem even to prove it wrong. One cannot merely cable Associated Press and, like Mark Twain, say: 'The report of my death was an exaggeration.' One must be able to confront the source of information. But this may not be an individual; it may be a system.

The design of the system, the way it is put to use, and the nature of information collected will to some extent determine the interpretations that result from it. The relationships between items of information to be processed, and the relative emphasis on different items, may result in distortions in the interpretation of data.

One witness during a Senate hearing stated: 'The programmer could invent a private life.' He went on to suggest that the individual will lose the ability to 'establish a past history different from that jointly provided by the programmer of raw data and the interpreter of processed data'.

The error now is no longer a tangible incorrect field in the record like an incorrect bank balance. It is more subtle – a fault embedded in the nature of the system and the people who use it.

EVOLUTION OF SYSTEMS CONTROLS

Systems controls at the lower levels, then, are definable and can be designed in a fail-safe manner. But lower-level systems are merely components of larger systems. The output of an inner subsystem is part of the input to its surrounding system. There may be many inner systems in an outer. Many messages go into a file; many files go into a data bank; many calculations are performed by a program; and many programs go into a data-processing system. Each level has its feedback loops and checks to guarantee its output. In many instances the checks take into account the relationship between the components which, as a set, provide the inputs.

Eventually we reach a system level at which the components are data banks and data-processors. The lives of people and the fortunes of corporations generate input data. The output may not yet be defined. It will, however, be the input to the supersystem, the society in which we all live. Some of the sub-system outputs or sets of outputs may be damaging to society. How can we identify them and take corrective action? The control mechanism must now consist of people. The detection of undesirable system behaviour must come from the human beings affected and must generate feedback which modifies the system operation.

In the computerized society of the future it is essential that this feedback mechanism be sensitive. We do not know exactly what effects the massive use of computers is going to have. We want to take advantage of them in every way we can. But in allowing this new world to evolve, we must apply flexible and effective systems controls so that it evolves in such a way as to make society a better environment to live in. We must react to complaints in a sensitive manner and take action accordingly. An official such as the British Registrar discussed in Chapter 24 could help in this. An additional approach would be a form of systems 'Ombudsman' – the subject of the next chapter.

REFERENCES

1. Thomas F. Corcoran in 'Computer Privacy', US Government Printing Office, Washington, DC, 1967, p. 163.

2. *Observer*, 21 April 1968.

3. 'The Computer and Invasion of Privacy', US Government Printing Office, Washington, DC, 1966, p. 24.

4. ibid., p. 27.

5. *US News & World Report*, 8 April 1968, p. 84.

6. Andrew McNeil, *The World of Crime*, Hodder & Stoughton, 1968, p. 185.

7. 'The Computer and Invasion of Privacy', p. 182.

27

THE OMBUDSMAN

In human society, as it has existed for thousands of years, a monster has made its appearance – a monster called science. Whether we like it or not, it is breaking up everything in its path. Science calls into question moral concepts, social structures, beliefs. It is obvious that its forward march cannot be impeded, but it is certain that there will be a lot of broken glass along the way. Some are endeavoring to stick the pieces together; those are the conservatives. I think that this is not enough and that what is needed is to re-create moral precepts, to re-create a social framework. The question goes beyond the social order. It concerns also the individual, what he thinks and believes within himself. I think that modern despair, which is sometimes discussed, is an individual more than a social despair. – *French President, Georges Pompidou, Time, 2 March 1970.*

LIMITATIONS OF LEGAL CONTROL

We live in an age of 'big government' and the 'administrative state'. National and local bureaucracies employ millions of people and affect the lives of all. Proposals for a national data bank are meant to make these bureaucracies and their private-sector counterparts more efficient and more humane. A proper set of laws governing their operations will contribute to these ends. Nevertheless, administrative malpractice resulting from incompetence or malevolence will be inevitable.

The ordinary citizen is in a very weak position when faced with a professional and monolithic bureaucracy, because he is inexperienced, ignorant, and poor by contrast to his opponent. Traditionally, his defender against maltreatment has been his parliament or congress.

The courts and the laws they interpret must confine themselves to the correctness of the procedures upon which they are called to adjudicate. They cannot discuss the merits of the procedures themselves. The legislature cannot address individual cases, but must make laws of general applicability. Neither the judiciary nor the legislature can act quickly. Their respective controls – the application of judicial procedures and accountability through party processes – are too slow to monitor the activities of a well-run administrative machine.

There is a social need for quick and efficient administrative decisions, and the price we pay is inherent in the wide range of minor injustices we tolerate. As the bureaucracy grows, however, and its operation increases in efficiency, more and more decisions are made. More and more people are affected, and instances of injustice inevitably multiply.

At the state and local levels little tyrannies of licensing boards, commissions, and regulatory agencies exercise discretionary authority with remarkably little constraint. Occasional scandals reveal complicity in corruption or laxity. More often, policy standards are ill-defined, procedural safeguards conspicuous by their absence, and the operating rules of the agency so well concealed from public scrutiny as to be ineffectual. Balances and checks sometimes fail to provide the legislative and judicial branches with meaningful constraints on the executive.

At the United States federal level, congressmen with their ample offices and aides (by international standards) are almost overwhelmed by 'casework', as their constituents seek rapid redress against the heavy tread of executive operations. Legislation is one of the casualties, and bad laws cause more hard cases in the future.

Much of the helplessness of the citizen stems from his bewilderment with the operations of the officialdom whose actions he feels to be unjust. The cost, in time and money, of determining how the system works may be beyond his means. The law is ponderous and expensive to invoke, uncertain in its outcome, and sometimes powerless. And now, in the interests of efficiency, we are proposing to pile on the Ossa of an already mountainous executive bureaucracy several Pelions in the form of data banks.

The private citizen needs help – *specialist* help – to ensure that the quality of his life will be preserved (even though that may be beyond definition). We propose, therefore, a solution that has been successful in Europe in protecting the citizen from the inevitable shortcomings of bureaucracy: an *Ombudsman* should be appointed in this case, with powers to supervise those operations of all data systems which affect the lives of individual citizens.

HISTORY OF THE INSTITUTION

At the beginning of the eighteenth century, Charles XII of Sweden created the office of *Högsre Ombudsmannen*, or 'Supreme Procurator', to make sure that public servants performed their jobs properly and that laws and regulations were complied with. Despite a change of name (to *Justitiekansler*, or '*JK*') and method of appointment, this officer still gives his major attention to supervising public servants. He answers, however, to the 'King in Council', not to Parliament. The latter therefore appoints its own *Justitie-Ombudsman* under powers granted by the 1809 Constitution, and he is known as the '*JO*'.

Complaints against officials raised by individuals or aired by the press lead to inquiries by the Ombudsman which might result in prosecution. During the inquiry, all material is made available to him. Swedish civil servants are legally liable for their actions and can be condemned, fined, suspended, or obliged to pay damages for any errors. The documents on which their decisions are based are nearly all public: exceptions are made for papers dealing with private lives and state security. The *JO* is in effect a prosecutor who rarely prosecutes, since his reprimand or criticism is usually sufficient. He cannot in fact change any decision of a court or administrative official; he can only bring the official to court if his influence as Parliament's representative and the force of his arguments fail to achieve his ends.

He presents an annual report to Parliament, but he is dependent only on the law. He decides for himself what to investigate and what action to take, and does not take orders from Parliament. He is selected by agreement between parties from among justices of the courts, and is strictly non-partisan and politically independent.

The *JO* is accessible to individuals and his office helps them prepare their complaints, frequently clearing up misunderstanding where official decisions were merely obscure, not wrong. His inspection tours reveal discrepancies between the letter and the spirit of laws being administered by courts, police, and other bodies, and as a result he may issue 'reminders' on what should be done. Only a few prosecutions occur each year, sometimes to insure that damages can be claimed. In other cases he appeals directly to the government for compensation, rather than using the blunter instrument of a prosecution. He can also recommend changes in the law or disclose shortcomings.

The administration of justice is the primary concern of the *JO*, and he looks not only for fair decisions but also for 'due process'. Many cases are concerned with licensing and with officials' discretion, where practice is more of a guide to the administrator than the law itself. He also makes certain that delays are minimized and that crime and punishment are only months, not years, apart. Prison administration, restraints on freedom of speech, and traffic ordinances have all been subject to investigations that have resulted in fines for individuals. But it is essentially the *JO*'s existence, not his activities, that maintains the high standards of Swedish public officials. By watching for errors and revealing their causes, by maintaining constant surveillance of new statutes, and by clearing up many invalid complaints, he protects not only the citizen's rights but the official's also.

The success in Sweden and neighbouring Finland of the concept of the Ombudsman led Denmark, Norway, and later New Zealand and Great Britain to adapt it to their needs. Similar institutions include military Ombudsmen in West Germany and Scandinavia, and the Inspector General in the US Army.

The first British Ombudsman, called the Parliamentary Commissioner, was appointed in 1967. His jurisdiction was limited to central government mal-administration. Where local government is involved, as with the police, National Health Service, or council housing, he cannot investigate. Local ombudsmen, as well as procedures for independent enquiries into complaints against the police and the health service, have been suggested, but successfully resisted. The first Parliamentary Commissioner, Sir Edmund Compton, established an undramatic but effective image for his job, with such criticism as there was being directed at the restricted jurisdiction, not the outcome, of his investigations.

In every country with such parliamentary commissioners, continuity of administration on the Scandinavian pattern is insured by neutral civil servants and ministers with responsibility to the legislature for their departments and cabinet control. Attempts to provide a 'Presidential Complaints and Action Committee' in the Philippines, whose constitutional system owes much to its old colonial ruler, the United States, do not augur well for a 'review board' within the executive branch. France has a system of administrative courts fulfilling a similar function – and so incidentally do the Swedes, who have an Ombudsman as well!

THE AMERICAN CONTEXT

The feasibility of an Ombudsman in America is less apparent than the need for one. The proposal has been put forward in different contexts on several occasions, but institutional barriers and, according to Ralph Nader, a 'hostile and suspicious political environment dominated by party professionals, state administrators, and representatives of organized pressure groups' have served to thwart the proposals.[1]

Representative Henry S. Reuss (Wisconsin) proposed a bill in 1963 to introduce an Administrative Counsel, appointed by the Speaker of the House and President of the Senate, who would look into delay or denial of benefit or improper penalty resulting from administrative action (or inaction). He was to have powers of inquiry similar to those of a Congressional committee, including power of subpoena. (Individual congressmen have less power.) Cases would reach him through congressmen, who would request him to investigate instances they could not handle satisfactorily through their own offices. Reports would go back to the congressman for action and/or publication.

Representative Reuss was concerned with the inevitable 'amateur' approach of even an experienced aide when it comes to determining the rights and wrongs of individual complaints. The Counsel would have an adequate staff for specializing in various 'agencies'. There would, of course, be the usual problem that the 'regulators' would have to come from the agency being regulated, since this is the only source of suitable talent.

Such staffing necessities always bring with them problems, as the Inspector General's Department in the Army has discovered. This institution, which dates back to the Revolutionary War, has always been manned by seconded line officers who, according to reports, 'evidence little commitment to the function of the IG, presumably because of their continuing loyalty to their previous status as line officers'. Soldiers usually consider it not only difficult but useless to petition the IG, and fear reprisals if they complain successfully. Citizens frequently view police review boards in the same light, and some of the disenchantment would no doubt rub off on an Ombudsman.

To date, American opinion has regarded the press and the ballot, the traditional checks on tyranny, as more useful than the strange foreign institution of Ombudsman. America is far too big for a highly personal operation on the Scandinavian pattern. Even if congressmen filtered the complaints, the office of the Ombudsman would be swamped if it were to consider all administrative processes. Attempts to introduce state- and city-level 'counsels' in Connecticut and Philadelphia have been no more successful despite the fact that on this smaller scale the problems of volume of business are less intractible.

During 1971, the Association for Computing Machinery appointed a dozen of its members 'computer ombudsmen' in major American cities. Their job, as advocates for people having trouble with computers, is to discover the real cause of the errors blamed on the computer, inform the public of their findings, and help put the situation right if possible. They work with Better Business Bureaux and other similar organizations but have no official status or power.

Traditionally, the separation of powers among legislature, judiciary, and executive is meant to constrain abuses. Unfortunately, however, not only has the volume of the executive's decisions increased, but also the distinction between administrative and policy decisions has blurred. As an appointed officer of the legislature, only the former would be within the purview of an Ombudsman.

Scale presents other problems besides volume of complaints. The different states lack the cohesion and homogeneity of Sweden or New Zealand. The Ombudsman's main weapon of publicity would be severely hampered in the absence of a nationwide daily press. Any proposal for a Congressional Administrative Counsel is therefore unlikely to succeed if his scope is to include the whole of the American administrative process. The very volume of injustices that make him so essential would overwhelm his capacity for effective action.

THE COMPUTERIZED ADMINISTRATION AND THE CITIZEN'S FRIEND

The objections outlined above are less powerful, however, if we consider just *new* administrative procedures. The congressman and his aides, even should they be as well informed as Rep. Gallagher, will find it difficult to handle and redress the numerous errors of those most perfect and heartless bureaucrats, the computers, and their more fallible human owners.

Most of the grievances, however, will be very similar: errors of fact; irrelevant information; improper disclosure; or petty tyranny through harassment. Consider how an Ombudsman could handle these.

Protection against tyranny by bureaucracy comes essentially from able and conscientious officials backed by procedural safeguards. But there is always a need for an outside check, since it is almost impossible to exclude patronage, politicking, and fixing in any human system.

Our Ombudsman must therefore be above party politics,

answerable to the legislature, not the executive, and be of personal stature comparable with the members of the house of Congress to whom he reports. He would be charged with protecting the individual citizen against abuse by the bureaucracy; with redressing the grievances of those crushed by administrative heavy-handedness; with insisting on the maintenance of high ethical standards; and with proposing further safeguards when inadequacies are revealed.

Clearly, to be effective he needs the support of the press, the courts, the legislature, and the executive; his role must be understood by everyone and his image must encourage their confidence. His powers would be comparable with his Scandinavian forebears: the option to prosecute, the ability to publicize, and the right to subpoena records. But these powers would apply *only* to the operations of the national and state data centres and to the regulatory commissions concerned with data utilities holding credit and other information on individuals.

This Ombudsman would operate by investigating not only cases referred by congressmen, but also those that come to him direct from the public, or are taken up by the mass media or by disinterested parties. Furthermore, he would inspect and audit the procedures and safeguards employed by the government data centres and watch over the administration of licensing and regulatory operations of the relevant commissions.

In particular, he would inspect the security and safety procedures, and analyse (by computer, naturally) the use made of the government data centres by individuals and agencies as duly recorded on their audit logs. His annual report would detail his activities and the types of problems and solutions he had discovered. He would insist on rapid redress of grievances and the avoidance of the unreasonable delays characteristic of so many judicial procedures. Equally important, his operations would be entirely free of charge to the plaintiff and there would be no filing fee.

As the servant of the legislature he would have no power to *direct* the action of a servant of the executive. He would not be able to over-rule a decision – only investigate and recommend; but he would comment not only on the legality but also the merits of administrative decisions. His objective would be not only to make officials responsible, but also to humanize the bureaucracy by digging to the roots of friction between it and the citizen. Only the humanity of man can preserve the bewildered citizen from the blind justice of ideal bureaucrats.

REFERENCES

1. Ralph Nader in Donald C. Rowat (ed.), *The Ombudsman – Citizen's Defender*, Allen & Unwin, 1965, p. 245.

28

THE SYSTEMS SCIENTIST'S ROLE

The world of the future will be an ever more demanding struggle
against the limitations of our intelligence, not a comfortable
hammock in which we can lie down to be waited upon by our
robot slaves. – *Norbert Wiener, God & Golem, Inc.*

ORDER AND CHAOS

Life is complex. To cope with it, we have to impose order on it.
By measurement we describe some of its more obvious attributes; and our measures, though of archaic origin, are against
a global standard of great precision. We can measure position,
time, mass, and electric current and thereby make a universally
intelligible description of any physical phenomenon.

The units of physics have their analogues in society: language allows communication, though imperfect because words
are inadequate measures of ideas; currency both permits and
distorts our measurement of value; custom dictates how
hands are shaken and ships shall pass; laws impose order on
the behaviour of citizens.

Systems are characterized by laws, natural or man-made, that
govern or explain the interactions of their components. The
solar system operates within the laws of gravitation; the telephone system is governed by the physical structure of its circuits which embody both the laws of physics and the logic of
the designers. The systems scientist seeks to determine the laws
necessary to direct a system to a desired goal. Every system
contains, and is contained by, other systems. The goals of the
inner system are set by those which contain it, and the rules
that govern its behaviour must allow achievement of these
goals. When the systems scientist engineers a system, he designs
or chooses his component subsystems for this purpose.

Systems design is comparatively easy when it can be based
on standard components. During the analysis of a system the

scientist seeks to determine the variety of components and their common characteristics.

The consistency of physical laws and the ability of scientists to isolate individual phenomena for persistent investigation led to early success in natural sciences. Systems which include men among their components are less easy to investigate because human behaviour exhibits such huge variety.

The human component is, however, capable of conforming to externally imposed rules provided these rules can be communicated to the individual and that he is willing to accept them. It is possible to impose order on traffic by instructing its component vehicles by use of traffic signals. An army of men can be manoeuvered as a system by the drilling of its subsystems.

Laws impose patterns of response on signals denoting interactions. They explain how the incoming information will be transformed into outgoing signals. If there are no interactions, there need be no laws to govern the system, for there can be no system. If the variety of transforming processes is too great, the performance of the system can neither be predicted nor controlled. So the first task of the systems engineer is to impose order – to 'lay down the law' – for those components whose performance would otherwise be unpredictable. This reduces the variety of responses of which the *components* are capable without necessarily reducing the variety of responses of which the *system* is capable, for it may markedly increase the versatility and controllability of the whole. (When we find such behaviour among men, we call it 'discipline' or 'team spirit'.) The standardization of the Gutenberg press, for instance, permitted the great variety of literature that followed.

The modern bureaucracy was an early achievement of the systems concept. The goals of consistency and impartiality were reached by restricting the ranges of responses and specifying a rigid set of procedures. The bureaucratic system, however, is only a component of a society which imposes the additional goals of sympathy and speed.

The systems scientist sees the world as a nest of systems, as in Fig. 28.1, the innermost systems simple and predictable,

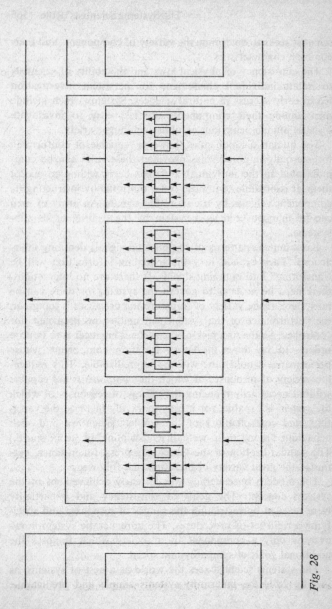

Fig. 28

either by nature or by imposed requirement, and the outermost providing the goals for its components and the environment in which they operate. When the system he views is a society of men, he looks for its goals and seeks to impose laws which maintain the flexible response of the whole while at the same time allowing *minimum* standardization of parts. Let us look at some of his attempts:

The History of Systems

Le Bourgeois Gentilhomme spoke prose without knowing it for forty years. There have been systems designers equally unknowing for as many centuries.

A system must have a purpose, and the earliest systems served the ends of political order and military success. The religious hierarchies were organized to maintain a status quo, acting thereby as homeostats for society. Military organization required both a command system and a planning capability from the earliest development of armies. The Trojan Horse resulted from a 'systems approach'. It began with a problem – an impregnable city; continued with an investigation of the response characteristics of the system's components – superstitious men; developed an input – the wooden horse – which these components would operate on in characteristic manner, i.e., pull it inside the gate; and finally produced an event – the opening of the gates from inside, for which the defence subsystem had no response.

Odysseus might have been a useful member of the team designing a missile early-warning system. He was adept at the art of combining elements into total systems whose performance was beyond anything hitherto devised. It is this preoccupation with the possible that distinguishes the systems scientist, and also explains why he is most often called an engineer.

Early human systems were hierarchical, with the law imposed from the top. With the development of communications, collective memory became embodied in the form of custom and law. Such systems grew, rather than being designed. Lacking tools of analysis, men invented gods to explain systems that had developed as a result of the characteristic response patterns of its components. Many of these patterns lost their original purpose and degenerated into hollow ritual and incantation, further confusing the cause-and-effect relationships within the system.

Similar preoccupations with the 'model of the transform' – the description of the operation of the component (as embodied in the above example in religious ritual) – are to be found in accounting. The double-entry book-keeping system is a set of procedures designed to maintain a description of the world it models. From this description, information could be more easily obtained than from the physical ships, warehouses, and vaults of the Venetian merchants who devised it. As the mortgagers of a great deal of salad oil were not alone in discovering, there is not necessarily a one-to-one correspondence between the model in the books and the real world it claims to portray.

Adam Smith was a systems analyst, but unlike Karl Marx he was not also a systems designer. He looked at the elements of the economic system and showed how the input of market information was transformed into an output of activity. The set of all system components communicating through the market resulted in a goal-seeking, self-optimizing system. Marx was not convinced, for he could see two types of human component in the system: capitalist and worker. Not liking the characteristic behaviour of the capitalist, he proposed his abolition and, simultaneously. modification by law of the characteristics of the worker.

The theory of economic systems has advanced a long way since Smith and Marx. Analytic techniques, however, are often still unable to treat components of a system as other than identical. The elements of social systems, be they economic, military, political, or industrial, are human. It is possible by law and discipline to force into them predictable behaviour – or in systems language, to give the components of the system identical 'transforms'.

The systems analyst begins by determining the similarities between the elements of the system, but he finishes by examining the differences and insuring that the goals are achieved despite them. Nothing is clearer than the essential differences between human elements of a system. In industrial and military organizations, these have to be overcome by discipline and reward.

Adam Smith recognized the value of division of labour, but he would not have chosen to describe it as a technique designed to minimize the variety of output of a human 'transform'. Division of effort works admirably up to the point at which the definition of inputs to one stage becomes governed solely by the potential outputs of the earlier stages. In a corporation the accounting function may provide the transformation of financial input data into the format required for public reporting, or into information for cost engineers, who require a totally different organization of the same figures.

The increase in the size of organizations led to the development of systems analysis to define the objectives of system components themselves. Sometimes the analysis must be followed by the design of a whole new set of compatible elements working towards the common goal.

Tools of the Trade

Systems analysis and design require a wide range of tools, many of which are borrowed from mathematics.

In this book we have discussed a variety of *models*. These are basic to systems science. A model without numbers cannot be manipulated, so *measurement* and *quantification* is a fundamental part of the description resulting from analysis, and the basis for the evaluation of systems design.

The models may be *analytic* and resemble the mathematical description of the solar system or the hydrogen atom. Such models are the basis of high-school science, and familiar in principle even to those to whom the theory of the lever was a traumatic experience. Analysis enables the scientist to abstract the essence of a system, formulate a model, and test it against reality. If the model is good, then its manipulation will closely parallel the physical manipulation of the system modelled. Our mathematical model of a lever tells us that we can balance a 2-kilogram weight 1 metre from the pivot by means of a 1-kilogram weight 2 metres from the pivot. We can check that model by experiment, and use the theory to design balances.

As systems become more complex, so do their mathematics. Analytic techniques have to give way here to *synthetic* techniques. The computer is ideal for such synthesis. The basic model governing the process can be supplied, followed by a large set of input components to be transformed individually and assembled into the output. To the computer it matters little whether the input components are the essential characteristics of radio-wave modulations or of invoices for insurance premiums, provided the programmer has constructed a model which represents the process correctly.

When the analyst investigates a system, his essential task is the determination of the process by which data are transformed. He uses an organized technique of investigation to find the objectives of the system and its components. In a human system he will be particularly concerned to see that the descriptions he gets from the people in it actually correspond to what does happen and what the system's users expect to happen. Frequently, the variety of behaviour is such that it can only be described statistically. The statistics so gathered will be used to test the model of the future system.

In designing a telephone system, for instance, the average number of calls each day gives little indication of the peak loads which govern the availability of trunk lines. So the system designer must build a model of the subscribers' telephone habits and use this as input to the model of his proposed system to see whether this will meet its objectives.

Simulation techniques of this kind can also be used to project the future behaviour of the system. The designer can investigate the impact of a continued rise in traffic, the development of computer-to-computer communications, the impact on revenue of a change in taxes, and so on.

These models are *analogs*, but they are not analogies. The essential difference lies in the fact that model and system have a common denominator in their mathematical structure.

Analogy may provide the insight necessary to discover the fundamental similarity later incorporated in the analog. But a true analog, or valid model, can be pushed to a logical conclusion without destroying the analogy. This is frequently the method by which models are tested. For example, the following model is absurd: the proportion of male Americans with arrest records was 20 per cent in 1965 and is rising at a constant 2 per cent per annum. What happens in 2006? Such 'saturation' tests, pushing the model to extremes, are used on more complex systems to test the validity of the model and improve it.

Analogy uses past experience as a guide to future happenings without proof of a basic similarity. Simulation, by contrast synthesizes experience based on fundamental relationships. This is why it is becoming increasingly a major tool of systems design, with applications in all spheres where systems are found.

A Way of Looking

We are claiming for the systems scientist neither unprecedented nor unique qualities, but we are claiming a viewpoint unusual in a world of ever-deepening specializations.

This viewpoint has two salient characteristics. First, the purpose of a system is extrinsic. The mountain is not climbed because it is there, nor is a phenomenon investigated for its intrinsic interest. The systems man has little intellectual rapport with the pure mathematician who committed suicide on discovering his work had proved *useful*. The determination of the proper set of goals for a system is the primary function of systems analysis.

Second, a system is described not in terms of what it *is*, but of what it *does*. Hence the prevalence of inputs, transforms, and outputs discussed in this book and others written by systems engineers. Ignorance of designing or repairing computers does not disqualify someone who knows the characteristics of its transform incorporating it into a system for bank accounting or controlling a moon shot.

When the systems scientist views systems science he sees, not a discipline, but an activity. That activity takes as its input the apparent chaos and complexity of an ongoing process and a set of partially identified problems, and transforms it, by an appropriate technique, into an output which includes a description of the system, a specification of its objectives, and a design for their better achievement.

MAN IN THE SYSTEM

Man as a system component provides so much variety that he is always difficult to analyse. The instantaneous output of a human 'transform' is theoretically determined by the totality of inputs in his lifetime and the immediate stimulus or 'current transaction'. For practical purposes – and systems engineering is nothing if not practical – this is irrelevant, since the calculation would take too long to be useful. The human element is therefore treated as unequal, random, wilful, obstinate, and very significant. In the mass, he may be described by statistical measures à la Gallup. Individually, his output will be governed not only by the current input but also by his memory and his goals externally set by other systems, such as his family or trade union, of which he is simultaneously a member.

Men are most valuable when their output is most unpredictable. If it were predictable, then, in principle, a machine could be built to imitate it. Keypunch operators are used because human handwriting takes on unpredictable forms. Again, as pilots of aircraft, men have unique advantages in their capacity to respond to the unexpected. A 100 per cent pre-take-off check, however, is much better left to a computer that will not forget a vital item.

Good systems exploit man's variety, accept the individual's inequality with other men, anticipate his failings as a processor of standardized input, and respect his membership in more than one system at a time. Bad systems – political, industrial, social, or military – assume that all men are equal and interchangeable, predictable, and dedicated to the sole purpose of achieving the goals of the system alone. Free men cannot be treated as automatons.

THE RESPONSIBILITIES OF SCIENTISTS

By any measure, the growth of knowledge is accelerating. There are more scientists, more reports, and greater investment every year in all sciences. The advance is inexorable and inevitable. Our capacity to cope with the changes in the train of discovery is less developed.

We have no way of preventing the advance of science, although we can slow or speed up its application by the differential allocation of resources. To avoid a fundamental discovery requires the prediction of what that discovery will be. The discovery of nuclear fission could not be avoided, but the Manhattan project need not have advanced its military exploitation so rapidly.

Increase in knowledge has necessarily brought greater depth of specialization. The view of the specialist is necessarily constrained to his own field, and he finds it difficult to see the future of his own discoveries. It is the engineer mopping up behind the battlefront of knowledge who exploits the breakthrough. If he, and the reinforcements of money and manpower, are not present, then development will be slow.

A discovery that is little publicized and less exploited will not influence the evolution of mankind. Man is the only species in a position to control his own evolution. By his selective patronage of development, he can determine the direction in which his environment changes.

The primary political problem raised by scientific advance is the direction of patronage. Our political organization tends to place power in the hands of those inadequately skilled to cope with this responsibility. The choice between alternative projects requires invidious distinctions to be drawn between and within medicine, space, food, construction, and every other field.

No country can claim to make such decisions entirely rationally, for none has even a scale of values or a technique for measuring return on investment. We do not know how to measure the ultimate value, in terms of society, of scientific research and development. Our measurement of the inputs evaluates only men and money, not the loss from the diversion of scarce resources from better projects. The cost of missed opportunity is not measured. There is probably a point at which the extra investment of money and talent begins to yield diminishing returns. It is sometimes said that the fruits of American efforts are disproportionate to the manpower involved, compared to smaller countries. Eventually, we must reach a talent ceiling beyond which there are no more potential scientists of suitable calibre. The main factor holding back the exploitation of computers will be a talent shortage.

At the moment, governments tend to be surprised by the consequences of scientific advances and fail to formulate the policies necessary to the smooth introduction of the consequent changes. Since governments lack the capacity to foresee the consequences of science, it is vital that those who do have the capacity make sure that their views of the future are widely known. This implies warning of the impending problems, not merely advertising the advantages connected with discoveries, large or small.

Rachel Carson's *Silent Spring* should be the model for the professional scientist. Among the distinctive characteristics of the 'professional' must be a loyalty to principles or ethics that go beyond concern for financial reward, a loyalty that if necessary results in opposition to the actions of his employer or government. There can be two opinions about the efficacy of defoliating chemicals, or about the ethics of chemical warfare, but there can be no argument about the need for the fullest information to be made available by those best able to discuss the matter. We are not proposing here the breaking of commercial secrecy; there is no need to explain how to make polyethylene film so as to warn of the dangers of electrostatic charges causing bags made from it to cling to children's faces and smother them.

The employer, on his part, like the government, must allow his employees freedom of speech on these matters. Nothing could be worse than if corporations were to prevent their employees from publishing their views on the implications of new technology. Such censorship could be overt, or it could be unwritten censorship implied in an 'organization man' structure. Either form is heinous in an age of ultra-rapid and potentially dangerous change.

The scientist must be responsible for publicizing the consequences of his ideas, and the engineer of the use he makes of them. He cannot abdicate his duty and let decisions be made in unnecessary ignorance.

DEFECTS IN THE INFORMATION SYSTEM

Equally, those who make decisions affecting the lives of others, be they businessmen, politicians, or mere voters, have a duty to keep themselves informed of significant scientific and technological developments.

Communication requires both transmitter and receiver. The scientist who tries – as we maintain he should – to communicate with his lay colleague has a right to expect that he will be both heard and understood. The scientist must achieve literacy, but it is no less necessary to communication for the non-scientist to achieve 'numeracy'. For without it, he has no way of determining the validity of the claims and warnings addressed to him.

We have already discussed the bewilderment caused by the advances in the field of computer technology and the changes in education that are needed to cope with it. The same scene can be reproduced mainfold in transport, food, medicine, and any other field.

Education must prepare the student for change. Unfortunately, it is not at present constituted to do so. Many schoolteachers still merely regurgitate their own education. The whole system is seized by a massive inertia due to the investment of past generations. Curricula tend to change slowly, and then only by accretion, not by substitution.

At the same time, there are two prevalent and dangerous attitudes inculcated by the schools. The first is an unnatural fear of logic, statistics, and mathematics, which are a part of the natural language of this century. The second is the appeal to authority rather than experiment in the determination of the truth. The result, in an age that has an overwhelming need for scepticism, is faith. Those who refer habitually to the past, be it embodied in methodology, textbook, folklore, or constitution, tend to perpetuate the errors of their predecessors. Because the past can be an untrustworthy guide in an era of such rapid change, such people are hardly fitted to alter the present to prepare for the future, for they are conditioned to believe what they are told.

SYNERGY, COOPERATION, AND FREEDOM

The recent increase in world population has been concentrated mainly in urban areas. The result has been an extreme increase in the frequency with which people interact with others.

> And this is good old Boston,
> The home of the bean and the cod,
> Where the Lowells talk to the Cabots,
> And the Cabots talk only to God.

Things have changed since that toast was proposed at Harvard half a century ago. Now we are obliged to talk and cooperate with each other, whether we like it or not. In the past, manners and customs lubricated personal contacts. But the characteristic time over which the conventions were established exceeded a generation. As new situations developed, laws had to be promulgated to regulate what threatened to turn contacts into collisions. There are rules for telegraph, roads and airlines, advertising, journalism, and television. Freedom has been necessarily sacrificed to orderliness, because custom could no longer restrain those who were tempted to indiscretion in the exercise of freedom. Indiscreet exercise of free will is coming to have more serious consequences, affecting greater numbers of people.

We are now finding that the rate of innovation has overtaken our power to legislate for its consequences. It may take five years from the time the consequences of a new development are first foreseen by a few until the enforcement of a law dealing with it is possible. Five years in the computer-business, for instance, is the length of a generation. Third-generation computers date from 1964, and have made possible the giant data banks and inter-computer communication with which we have been so concerned in this book. Public awareness has followed some way behind professional forecasts, and legislation has still to be derived from legislators' investigations.

Revolution is evolution in too much of a hurry. As presently constituted, the democratic system is ill-equipped to cope with change. Democracy provides a safeguard against the usurping of power necessary to bring about revolution. However, evolution has now reached revolutionary pace, and evolution is technological whereas government is political. Control of social change has been delegated – not by democratic choice, but by democratic default – to the researchers, the engineers, and the entrepreneurs. None of these is elected, and by the time the electorate comes to pass judgement on the representatives who might have protected the general interest against incursions, the damage has been done.

One alternative to this situation is *meritocracy*, the concentration of power in the hands of the able and intelligent. But is there any greater innate virtue in associating political dominance with brainpower, inherited or developed, than with social status, by birth or usurpation, or wealth, entailed or acquired? Each is in some sense an aristocracy, with a built-in sense of its own fitness to rule others and with all the arrogance this implies.

The other alternative is to make democracy fit to survive by developing the wisdom of the individual voter. This means improving not only the transmission of information, but also its reception and processing. The voters need to be educated to cope with unforeseen choices, not to respond traditionally to traditional cues. They may then be able to ensure that they choose representatives who are also capable of responding to changes not by opposition but by direction and control.

Innovation, of course, is the primary mission of applied science, but it seems unnatural for governing agencies to innovate. By their nature such agencies attempt to regularize behavior. By their dicta, for example, we all drive on the same side of the street and use the same methods of reporting income taxes. Those who have had some association with the federal government know that it is not easy for government agencies to sponsor innovation. All the funds that are wasted, the cutoffs, the letdowns are understandable because government activity does not lend itself to the kind of risk-taking, the quick decisions that innovation through science and technology best thrives on.[1]

Quick decisions are needed, not only to innovate, but to deal with the consequences of innovation. The impact is no longer local:

Today the typical American earns his livelihood working in a large impersonal organization that sells its products on a geographically large – often national – market, and spends his income on products produced in all corners of the country and the world. In contrast with a century and a half ago when a man's ability to earn a livelihood depended almost totally on the value of his effort in his local community, today his livelihood is in considerable part dependent upon the demand for goods and services – and hence for labor – in the economy as a whole. While formerly a local community had the economic wherewithal, and the economic autonomy, to take care of its own, this is no longer the case. The need for Social Security, unemployment compensation, compensatory fiscal and monetary policies, and similar programs follows directly from these developments, though the appropriate institutions and ways of thinking evolved quite slowly.[2]

This interdependence of citizens, the fruit of the interactions we have been stressing, is the primary social consequence of improved communications, the heart of technological advance.

The choice of priorities by reason, not emotion, is equivalent to cooperation, and in one sense it entails a sacrifice of freedom. Instead of making a 'free' decision about a course of action, the criteria by which the decision will be made are 'freely' chosen. The decision itself follows automatically, thus being essentially fixed, not free. This technique is not new. Edmund Burke, in his famous speech to the electorate of Bristol in 1774, said, 'Your representative owes you, not his industry only, but his judgement; and he betrays instead of serves you if he sacrifices it to your opinion.' The machinery of choice should be the informed judgement of representatives. Electors therefore decide, not by making all decisions, large and small, but by choosing decision-makers.

The modern Burke can quantify his judgement, and so test the consequences of his actions against the criteria his electors chose him to embody. He can state the impact on Humboldt County of extending the Redwoods national park, and thereby state the price to be paid for conservation. That price can be assessed against the value placed on the preservation of their environment by this electorate.

We are postulating the politician as a super-systems-scientist – or more properly, the employer thereof. For he deals in the totality of the effects of change, not in a limited subset brought to his notice by a lobby. He must abjure the temptation to oversimplify or to indulge in special pleading, and also, most difficult of all, he must apply to 'his' constituents the same criteria that he applies to others.

THE DUTY TO MOULD THE FUTURE

Now that it is within our power to mould the future, we have certain duties. Failure to carry these out will, in the long run, be catastrophic.

In the short run, scientific discovery is a random event, but technological development of discovery is within our control. We can choose whether or not to build supersonic-transport aircraft or manned moon-landing capsules, to control population, grow ten grains of rice where one grew before, house the world's people properly, or harvest the sea.

But there are things we must not do. We cannot abdicate responsibility for the use of knowledge, whether or not we made, sponsored, or exploited its discovery, or are merely its victims. The monstrous tragedy of thalidomide, the smog of Los Angeles, the cesspool called Lake Erie, the endless inanities of prime-time television, the subhuman victims of hallucinogenic drugs, and the corpses littering the highways could have been foreseen and should have been reduced to negligible proportions.

To do so requires foresight, not hindsight. Forewarned, we could have been forearmed with appropriate regulations. The advocates of innovation can be expected to advertise its benefits, but we need to know the potential for harm as well as good. We also need to consider the human response to the perturbations in a way of life. There are limits to the rate of change that can be imposed without destroying the social structure. The homeostatic system which maintains society in dynamic equilibrium with its environment needs time to develop its response to new situations. Part of that response must be the derivation of a cost-benefit equation covering the total impact of a proposal.

It is here that the systems scientist can offer a discipline and technique not available to his fellows dedicated to specialties requiring depth of vision and concentration of effort. He has the habit of looking at all aspects and consequences of a problem, unlike the pure scientist whose approach is to strip off and control irrelevant detail and to concentrate on the essence of the phenomenon he is studying. He also treats people, in all their variety, strengths, and weaknesses as vital. Systems exist for people, not vice versa.

REFERENCES

1. William O. Baker, 'The Dynamism of Technology' in Eli Grizberg (ed.), *Technology and Social Change*, Columbia University Press, New York, 1964, p. 88.

2. Richard R. Nelson, Melon J. Peck, and Edward D. Kalachek, *Technology, Economic Growth and Public Policy*, the RAND Corporation and the Brookings Institution, Washington DC, 1967, p. 146.

29

THE WORLD OF THE FUTURE

The future is a race between education and catastrophe.
– H. G. Wells.

The world can never be quite the same again. Perhaps on other planets in the depths of the universe, developing organisms reach a fundamental point in their evolution – the point at which they invent machines which can store, process, and transmit their intelligent thoughts. From that point on evolution will not simply be a matter of biological changes minutely altering the creatures as centuries pass. It will be evolution of the symbiotic group: intelligent-creatures-plus-their-machines. For at least a few centuries after the transition point the machines will develop at the breathless, reckless pace we see in our own computer industry today.

If man should tire of technological evolution it will probably still race onward because of the struggle for defence. For instance, SAGE was the father of many of the ideas in this book. The ABM and its successors will sire others. We deliberately omitted 'artificial-intelligence' techniques from our speculation in Section I, as they seem of dubious practicability as yet. They are likely to be needed, however, for military systems, and perhaps it is only from military requirements that the enormous budgets needed for their development could come. In the fifteen-year period *after* that which we have discussed, the computers will probably be much more clever.

To what extent will there be *separate* isolated computers in the future? Will they mostly be part of a vast interconnected network? A telephone would be a senseless device if it were isolated from the grid of switching and transmission equipment. Perhaps most computers of the future will equally need their 'grid' to obtain programs, data and the processing power of remote specialized machines. Perhaps much computing in the future will begin with the local machine interrogating a directory computer to find out where it can find the services it needs. We may think not so much in terms of separate machines but of one huge organism of linked computers which help each other.

Man will live out his life, dominated as before by emotions and human feelings, against this background of evolving interconnecting machines, increasingly omniscient, and as ubiquitous as the telephone. The rooms where he lives will initially have a telephone, then perhaps a computer terminal, then a multi-purpose screen used for television, picturephone, and computer data. As the years go by the screen will become bigger and more necessary. He will not be able to do without it any more than he can do without electricity now. Perhaps eventually the screen will occupy a whole wall of the rooms where he lives and works.

The latter is certainly beyond the fifteen-year time span which this book covers. Our time span is the beginning of this new period in evolution. During these years we must determine what laws, attitudes, and mechanisms of social control are needed, and lay their foundation.

If we look several decades ahead, rather than our mere fifteen years, we expect that many of the problems that we speak of will have been solved. Society will have new attitudes to help it survive in the new environment. The interim period will be chaotic because the machines are rapidly taking on their new functions before society is ready for them. The necessary changes in attitudes, ethics, laws, education, and employment will not have taken place in time.

Several decades in the future our ways will be different. Children will be brought up with the new machines. Perhaps they will learn to touch-type on terminals in kindergarten. Soon after they learn to speak to their parents they may learn to speak to computer terminals, in the precise, slow voice and limited vocabulary required. They may grow up to regard the machines as obtuse, although entertaining and staggeringly fast and powerful. They will learn to program as soon as they have learned arithmetic.

Briefly, let us paint a picture of a possible future within the lifetime of most readers of this book. Passing from the society of the 1960s to the fully computerized society is a little like passing through the sound barrier. The plane tends to vibrate. In the planes built after the Second World War, most violent juddering set in as they approached the sound barrier. Then, the transition past, they streak quietly into the blue skies, the immense speed becoming unnoticeable. In our transition to the computerized society the juddering is just beginning.

In the blue skies beyond the transition period, people are richer, with far more material goods and far more leisure time. We presume that they are better educated to use their leisure time. Perhaps persons of lesser IQ or lower aptitudes have found a valued place in the computerized society. Probably there is finer attention to food and other pleasures of life. Landscape gardening is taking place in the cities. There is sculpture on buildings, excellent television, and a respect for craftsmanship. Those who enrich the abundant leisure hours might become more highly regarded in this society than those who program and organize the machines (who were the high priests of the 1980s).

To the accepted stages of economic growth of a society, a new one will have been added. Societies will modify their behaviour patterns when a state of 'saturation of affluence' is reached. Eventually, as the GNP per capita rises through automation, the mass striving for more money and material goods must give way to other drives. A bigger electric can-opener does not give life a richer texture. Music, opera, gardens, informed dinner parties, informed discourse, and entertainment using the new media do. Education for leisure will have become of vital importance. The Protestant ethic will have given way, perhaps to attitudes closer to those of Athens in its prime, perhaps to new attitudes unclassifiable in terms of past history. Man will have the opportunity to once again become civilized.

Probably the public will have passed through the period of bewilderment with computers to a period of acceptance. The days when it was impossible to query an incorrect electricity bill will have gone. A bank statement, if needed, can appear on the home screen with alphabetic details, not cryptic digits. The terminals no longer produce unintelligible mnemonics. The programmer will be in trouble if he confuses his public, and the public can press a 'HELP' button if all is not clear.

Today's cherished attitudes towards privacy will have changed somewhat. Reluctantly, the public may have come to realize that the machines have information about them. They cannot prevent it. It is regarded as being for the general good. Most of it is reasonably safeguarded. Their relationship with the data networks is in some ways like that with their confessional. At least the data are accurate now. The earlier years, when embarrassing errors appeared on the files, have gone. Most of the public would now be upset if the machines did not have all their personal details. They might miss out on many of the beneficial aspects of the computerized society.

Only certain facts about individuals in the data banks are 'publicly' available; many are available to limited user groups. An employer can access certain facts; doctors and hospitals have access to personal information locked from other users; the police can look up only certain facts about people; and so on. The file locks now work very well, for much money was eventually spent on this aspect of technology. It would be extremely difficult for an unauthorized person to break in.

Nevertheless, all manner of very personal facts are stored. Somewhere in the machines are records giving one's past examination failures, broken engagements, arrests for drunkenness, failures to pay bills, driving convictions, psychiatric reports, and memberships of college political groups. Because in this society total information is stored, there has developed an ethic of almost total indifference to the information. Tolerance is now of great importance. Behaviour which in earlier decades would not have been tolerated by many employers and social groups is now smilingly ignored. While this is true in the older, richer countries, it is far from true in the newly developing nations who acquired their data networks before they had affluence, good education, or political maturity. In the older countries Orwell's *1984* now seems a strangely sick book, but in some of the developing countries truth has proved more subtle and insidious than such fiction.

For two decades crime continued the appalling growth rate that it had set in the 50s and 60s. The United States' murder rate doubled and then doubled again, and in some cities reached the staggering figure of 0·1 per cent of the population murdered annually. Crime-detection techniques of necessity became more and more elaborate. Some of the most complex computer systems are the police information and surveillance networks which now contain a detailed dossier on every citizen and alien. The computers predict the probability of each person committing crimes of various categories. The machines often know that a person is likely to commit murder, for example, before he actually does so. The police systems did much to change public attitudes toward 'privacy'. The majority of citizens are honest and no more worried by their police dossier than by the record of convictions on their driving licence. They *are* worried about the rise in crime and about their family's safety on the streets, and so the police computers are generally approved of. In the computerized society nobody fears hellfire and damnation any more; instead they have an electronic conscience.

Many of the facilities of the cities are controlled by computers. Air traffic has increased enormously and the airlanes are entirely under computer control. They are now far too crowded to be workable without such control. Landings are fully automatic and can be made in thick fog. Trains are also under computer control, as is city mass transportation, which has been extensively constructed following the prohibition of most cars in major city centres. The computers arrange for the various mass-transportation media to interconnect in an optimal fashion even when (rare) delays occur. Almost all journeys are booked in advance, with the computer arranging the interconnections. The highly successful 'dial-a-bus' services have been expanded to 'dial-a-journey', in which the computers plan and book the interconnections.

Working hours for all who travel to work have been staggered to minimize traffic peaks. Brief traffic jams are still occasionally caused by accidents, but the long evening crawl from the cities has been eliminated. Working days have been staggered also, and the weekend snarl of beach and country traffic has gone. The staggering of hours has been made possible by two factors. The first is efficient computerized storage and relay of messages (including spoken messages). The second is that many people work at home equipped with terminals and picturephone, and travel to work only once or twice a week, avoiding traffic peaks.

Many other aspects of city life are automated. Theatre bookings, dentists' appointments, health surveillance, and so on, are all monitored by the machines. The systems are entirely flexible. When the city-dwellers in their wanton way break appointments, fail to turn up, or change their mind at the last minute, the computers reschedule as well as possible. Many persons have developed an arrogant attitude towards this and think that it is their prerogative to change their mind whenever they want and that the computers should adjust accordingly. Many of the mechanisms of the city, then, have become part of an enormous programmed complex of machines. Far from being ruled by the computer, as some people feared in earlier decades, the citizen now expects to go his own sweet way and have the computers serve him, slavelike. When they fail, and he is kept waiting in a station or cannot obtain the movie he wants on his home screen, he soon protests that the level of programming or facility planning must be improved.

City-facility planning, in fact, has at last matured. Using computer techniques the requirements of the future can be assessed in time to allow for their construction. The mid twentieth century is now looked upon as the dark ages of city planning. Replacing despair over the plight of cities is the promise of an environment that is at once dramatic and beautiful, unpolluted, unjammed, quiet, and graced with trees and flowers. People have to walk somewhat further, into the traffic-free precincts, but that is becoming more acceptable in a society with more leisure time.

THE WAY AHEAD

If this brief description is anywhere near the truth, then a number of conclusions can be drawn from it about what we should be doing in the more immediate future.

First, one might say, we should decide whether or not we want such a world. We feel, however, that we are hardly free to make this decision. Automation is proceeding at a furious pace and it would take a cataclysmic force to stop it. The decision to computerize society is not one decision, but the amalgam of thousands of independent decisions by different groups to use the new technology. Most of these decisions are based upon economic factors, and so at the moment the way we use machines is determined by the economic laws of nature. There appears to be little that we are able to do beyond the confines of these laws today. These laws, as we saw in Chapter 1, are swinging ever more in favour of automation. Systems which are not economical today will be almost mandatory in ten years' time. We are in the grip of the machine.

Although it is a little alarming to consider the limited degree to which we appear to be able to choose our future environment, this is nothing new. Very few of today's cities were designed on a drawing board. They evolved; and the process of evolution was governed largely by the economics of the technologies available (for example, the costs of building, production, and transportation), the degree of farm mechanization, and the economical siting of factories and offices. The structure of today's environment was dependent to a major degree on the state of the art in engineering and science and the associated economics. So will it be with the future decades. Only when we become very rich will we be more free to choose our environment.

In practical terms it would be more reasonable to accept the massive use of computers and data banks as inevitable, and then to steer their usage into directions most beneficial to a society that is near 'saturation of affluence'.

If we can proceed with collective intelligence, the use of computers described on the previous few pages can make our lives much richer and our society more civilized. Today's world will probably seem crude and unpleasant by comparison. We look back a century now, and wonder at children working in coalmines and men walking eight miles to work and labouring an eighty-hour week. In the future people will look back in horror at our era of mental drudgery when intelligent men hardly used their intelligence and had no time for hobbies, scarcely saw their children on weekdays, and commuted twelve hours a week.

If we prefer the above image of a fully computerized society, then the temptation is to say: 'Let's get there as quickly as possible. We know we can work technical miracles – the moon-shot has taught us that. Why is it that a returning moon capsule arrives within seconds of the planned time, but a jet from Boston to New York has to wait in the sky for two hours before landing at Kennedy? Let us spend some of this enormous amount of money and technical capability on the problems which more directly affect our lives.'

There is much to be said for this viewpoint. Certain government-funded projects of immense technical complexity could have a very beneficial effect on our society, industry, and education. The moonshot development took eight years, and a similar lead-time may be necessary for some vast computer projects. The use of computers, however, affects so many aspects of our society that long-range planning is needed not just in technical areas, but in areas where there is no evidence yet that we are capable of adequate planning.

One of the main reasons for the juddering as we approach this particular 'sound barrier' will be the inability of today's institutions to react to the changes with sufficient speed. New laws will not be thought out and passed in time; management and administrations will mostly continue to operate in traditional ways; and the unions are concerned with only short-term advantages for their members. One can question whether the political machine is really adequate to handle the world we are building. Governments, influenced by the desire to return to power at the next election, cannot move far ahead of public opinion as they are only elected for terms of a few years, and thus find it difficult to do much planning far beyond the next election. They are mostly composed of laymen who represent their electorates by keeping up on the current situation, not future situations. The public is largely unaware of future situations and those they elect are unlikely to plan for these. When experts are used for forecasting requirements, time and again we see their advice rejected if it conflicts with the needs of the moment. Sociologists in general are equally concerned with the present. They formulate their views in the framework of a technology that is obsolete rather than one which looks into the future with an adequate lead-time.

Perhaps the most damage of all is done by the institutions that are concerned with education. It is appalling how most of these are failing to adapt to our rapid rate of change. The majority of dons at Oxford and Cambridge might as well be unaware that computers even exist. In India, where the number of *illiterates* is increasing at between 5 and 10 million people per year, Literacy House, an American organization for raising the level of literacy, uses *puppet shows* as its latest means of conveying information to people in the villages. Only one puppet show can be performed each night, as a large group of people is needed to run it; whereas inexpensively made film could be duplicated and shown in thousands of villages. The curriculum of schools throughout the rest of the world has barely changed to reflect the electronic era. There seems to be little or no attempt to assess the nature of the society in which a child will be living when he grows up in order to educate him accordingly. Instead, we have the attitude that 'what was good enough for me is good enough for him'.

In Chapter 23 we talked about the half-life of an education, and indicated that for many professional groups it is rapidly decreasing. The years ahead will bring a much greater obsolescence rate. The half-life of the technical training of computer personnel is about three years. As more and more jobs and functions in life become computerized, so the half-life of many persons' training will drop – in some cases to the three years of the computer man. We could contrast with this a half-life figure for the course content of a school subject, which might be defined as the time period in which one half of the material in the course is actually replaced. Some teachers hardly change their material from the time they first develop it to the end of their careers. Except for university science courses, a half-life of less than twenty years must be rare.

Lack of understanding of the new techniques brings bewilderment, and bewilderment breeds hostility. If we allow the bewilderment to increase, the hostility will increase. With the rapid introduction of terminals in the 1970s this could happen to a major degree if we are not careful. In some cases the clever will exploit the bewildered to their own ends.

A joint and massive effort is needed by our systems planners to produce man–machine interaction that is not bewildering, and by our educators to prepare the public for what is ahead.

As the rate of change within our society becomes faster, the 'generation gap' is going to widen, particularly if education for the new technology concentrates on the young and not the old. Children will learn to communicate with machines which their parents fail to comprehend. Children may be 'educated for leisure' while their parents stolidly maintain the Protestant ethic. Finding a way to increase the IQ of people at an early age would greatly help alleviate the imbalance of capabilities discussed in earlier chapters (see Fig. 19.3). If this becomes possible, as some biologists now think, this will only aggravate the generation gap. If we cannot raise IQ, we can almost certainly raise capability with machines by making children fluent with computer terminals at an early age. Young children learn very fast and find the terminals fun to use.

The trouble with any attempt to keep up with the technology is that it is now changing so fast that we are bound to cause major social disruptions. Human beings seem to need a high measure of familiarity in their environment and relationships. Perhaps future generations will be better able to cope with discontinuous change. They will be brought up to understand that re-education continues throughout life. Their decreased working hours will give them more time to adapt. Teachers will have learned to become adaptable. Institutions will be built with flexible structures which can better withstand social earthquakes.

Some nations today seem to be able to absorb a faster rate of change than others. They are more receptive to new ideas and cling to their traditions less tenaciously. The United States, for instance, is more adaptable than either Britain or France, whereas Japan is more adaptable than the United States. China might prove to be the most adaptable of all.

The ability and willingness to adapt rapidly may in the long run bring wealth to a society. It probably does not bring happiness. England, a myopic prisoner of its traditions, seems happier than the United States. The statistics suggest that a society in a state of rapid change is more likely to have a high divorce rate, a high proportion of mental illness, a high level of alcoholism, and an appalling crime rate. If the change is brought about with intense public idealism, this is less likely to be so, but in the affluent countries of the world idealism is increasingly hard to come by.

Because of this, some sociologists are suggesting that we must slow down the rate of change. If we can insulate ourselves to some degree from the turmoil of modern technology, we may build a society that is pleasant and satisfying to live in. However, what is good for the sociologist is not necessarily good for the economist. In a world of intense competition between nations, economic considerations dominate. Many Europeans, for example, want their countries to protect themselves from the powerful forces from the United States that are eroding their culture. They want to preserve their older values and be free to build a society of their own design. However, their countries cannot do this very effectively unless they largely isolate themselves economically. They would need to erect tariff barriers against American goods and to limit American investment. And this few countries would be prepared to do. Economic health depends upon modern technology, and so the technologist, not the sociologist, rules the roost.

The technologist, however, is becoming every year more disruptive. He is acquiring steadily more powerful tools. In the decades ahead some of his capabilities are going to become potentially disruptive in a most alarming way. If not yet, sooner or later as science progresses we shall have to ask ourselves: Can we give free rein to this particular aspect of technology? We have recognized the need for a test-ban treaty on nuclear weapons. Some of this century's other wonders will make less of a bang, but might in the long run be more dangerous. Is it too early to ask whether we are going to need a test-ban on other discoveries?

Consider, for example, the following developments which might appear in the next fifty years. Some experts say that every item in the following list is highly probable. To many of our readers we could say: This is your lifetime:

1. Electronic or chemical means of regulating moods, emotions, and desires, and perhaps for changing personality and character.

2. A substantial capability for controlling weather, which can be exercised by rich and powerful nations (such as China eventually), perhaps at the expense of the poorer nations.

3. Capability for determining the sex of unborn children.

4. Some measure of genetic control whereby characteristics of unborn children can be planned.

5. New and powerful techniques of mass persuasion and techniques for affecting human behaviour.

6. The continued upward rise of the curves in Chapter 1 at an exponential rate (as that in Fig. 1.6 has done for almost 100 years).

7. True 'artificial intelligence' in computers.

8. Computer-logic circuitry and mass-storage both mass-produced almost as cheaply as today's plastic goods.

9. A rise in the world's population to 8 billion (70 per cent African and Asian) and a rise in education so that most of these aspire to America's level of living space and affluence.

10. An increase in life expectancy of fifty years and consequently a higher proportion of elderly people than today.

11. A variety of means of overindulging the senses which are not physically harmful.

12. 'Morning-after' birth control and other methods which are completely safe.

13. Direct electronic communication with and stimulation of the brain.

14. A further vast improvement in electronic eavesdropping and cyberveillance devices.

15. Inexpensive biochemical warfare.

16. Vast satellite relays for direct pick-up of world television. Cheap wall-sized television screens which portray news to all nations with extreme vividness (riots, city guerilla warfare, affluence of living, results of biochemical-warfare attacks).

These are all predictions based upon *today's* understanding. If we had made a set of such predictions fifty years ago, we probably would have failed to include television, the computer, satellites, the moonshot, nuclear energy, radar, the laser, heart transplants, colour film, or 'the pill' – and the 'surprises' in technology in the next fifty years will almost certainly be greater than those of the past fifty.

The above list could no doubt be expanded at length. The further ahead into the future we look, the likelier science seems to be capable of wrecking our society, our ecology, and perhaps our individual personalities if it is used in an unplanned fashion. Entrepreneurial freedom to apply and market every scientific wonder cannot go on indefinitely, unchecked.

We have now reached the point where we have the *technical* capability to build almost any society we could wish for, but just as certainly still lack the *political* capability. Science gives rise to the greatest optimism as well as the greatest pessimism. Dickens's words on the French revolution might well apply to the computer revolution:

It was the best of times, it was the worst of times, it was the age of wisdom, it was the age of foolishness, it was the epoch of belief, it was the epoch of incredulity, it was the season of Light, it was the season of Darkness, it was the spring of hope, it was the winter of despair ...

If we can learn to control the terrible forces we are unlocking, then the computerized society could be rich and glorious and more civilized than any epoch known to man. If not, the decades ahead will be nightmare indeed.

GLOSSARY

The following definitions were adapted with permission from these sources:

1. United States of America Standards Institute, standard definitions.
2. United States of America Standards Institute, definitions proposed by Subcommittee x3.5 on Terminology and Glossary.
3. IBM Data Processing Glossary (form No. C20-1699-0).
4. *Automation: Its Anatomy and Physiology* by John Rose. Published by Oliver and Boyd, Edinburgh and London, 1967.

ACCESS. The process of obtaining data from or placing data in storage.[4]

ACCESS TIME. Time required to read out or write in data from a data storage system.[4]

ADDRESS. A label, name or number identifying the location where data are stored.[4]

ALGOL. A computer language designed mainly for programming scientific applications.

ALGORITHM. A fixed step-by-step procedure for solving problems.[4]

ANALOG. Pertaining to representation of data by means of continuously variable physical quantities. (Contrast with 'Digital'.)[1]

ANALOG COMPUTER. (1) A computer in which analog representation of data is mainly used. (2) A computer that operates on analog data by performing physical processes on these data. (Contrast with 'Digital computer'.)[2]

ANALYST. A person who defines problems and develops algorithms and procedures for their solution.[2]

APPLICATION PROGRAM. The working programs in a system may be classed as Application Programs and Supervisory Programs. The Application Programs are the main data-processing programs for that application. The Supervisory Programs control the components of the machine and schedule their work.

BATCH PROCESSING. A systems approach to processing where a number of similar input items are grouped for processing during the same machine run.[3]

BINARY. A system in which combinations of only two digits represent any number or quantity of units, i.e. a number representation system with a base of two.[4]

Decimal code	Binary code
0	0000
1	0001
2	0010
3	0011
4	0100
5	0101
6	0110
7	0111
8	1000
9	1001
10	1010

BIT. A coined word from *bi*nary digi*t*; this is one of the whole numbers, 0 or 1, in a single position, in the binary scale of notation.[1]

BIT RATE. The speed at which bits are transmitted, usually expressed in bits per second.[3]

BLOCK DIAGRAM. A diagram of a system, instrument or computer, in which the principal parts are represented by suitably associated geometrical figures to show both the basic functions and functional relationship between the parts. (Contrast with 'Flowchart'.)[2]

BROADBAND. Communication channel having a bandwidth greater than a voice-grade channel, and therefore capable of higher-speed data transmission.[3]

CHARACTER. A letter, digit, punctuation mark, or other sign used in the representation of information. Each character is uniquely represented by a group of bits.

CHARACTER SET. An agreed set of characters from which selections are made to denote data; the total number of a set is fixed, e.g. a 48-character set may contain the 26 letters of the alphabet, 10 numerals and 12 special characters.[4]

COBOL. A computer language designed mainly for programming commercial applications.

CONSOLE. That part of a computer used for communication between the operator or maintenance engineer and the computer.[1]

CORE STORAGE. The main or internal memory of a computer; it is usually expressed in terms of k (1,000), e.g. a 10k machine has 10,000 characters of core storage.[4]

CYBERNETICS. The study of control and communication in man and the machine (Wiener).

DATA BANK. A comprehensive collection of libraries of data. For example, one line of an invoice may form an item, a complete invoice may form a record, a complete set of such records may form a file, the collection of inventory control files may form a library, and the libraries used by an organization are known as its data bank. (Synonymous with 'Data base'.)[2]

DATA BASE. (See 'Data bank'.)[2]

DATA PROCESSING. The execution of a systematic sequence of operations performed upon data. (Synonymous with 'Information processing'.)[2]

DATA STORAGE. A device which accepts and retains units of information and which will produce them, unaltered, on command.

DEBUG. To remove all malfunctions or mistakes from a device, or, more usually, from a program.[4]

DIGITAL COMPUTER. (1) A computer in which discrete representation of data is mainly used. (2) A computer that operates on discrete data by performing arithmetic and logic processes on these data. (Contrast with 'analog computer'.)[2]

DIGITL DATA. Data represented in discrete, discontinuous form, as contrasted with analog data represented in continuous form. Digital data is usually represented by means of coded characters, for example, numbers, signs, symbols, etc.[2]

DIGITIZE. To render a continuous or analog representation of a variable into a discrete or digital form.[4]

DISK STORAGE. A storage device that uses magnetic recording on flat rotating discs.[3]

DISPLAY UNIT. A device that provides a visual representation of data.[3]

DUPLEXING. The use of duplicate computers, files or circuitry, so that in the event of one component failing an alternative one can enable the system to carry on its work.

ERROR-CORRECTING CODE. A code in which each acceptable expression conforms to specific rules of construction that also define one or more equivalent non-acceptable expressions, so that if certain errors occur in an acceptable expression, the result will be one of its equivalents, and thus the error can be corrected.[1]

ERROR-DETECTING CODE. A code in which each expression conforms to specific rules of construction, so that if certain errors occur in an expression, the resulting expression will not conform to the rules of construction, and thus the presence of the errors is detected. (Synonymous with 'Self-checking code'.)[1]

FACSIMILE (FAX). A system for the transmission of images. The image is scanned at the transmitter, reconstructed at the receiving station, and duplicated on some form of paper.[3]

FAIL SOFTLY. When a piece of equipment fails, the programs let the system fall back to a degraded mode of operation rather than let it fail catastrophically and give no response to its users.

FALLBACK PROCEDURES. When the equipment develops a fault the programs operate in such a way as to circumvent this fault. This may or may not give a degraded service. Procedures necessary for fallback may include those to switch over to an alternative computer or file, to change file addresses, to send output to a typewriter instead of a printer, and so on.

FEEDBACK. A means of automatic control in which the actual state of a process is measured and used to obtain a quantity that modifies the input in order to initiate the activity of the control system.[4]

FILE. A collection of related records treated as a unit. For example, one line of an invoice may form an item, a complete invoice may form a record, a complete set of such records may form a file, the collection of inventory control files may form a library, and the libraries used by an organization are known as its data bank.[2]

FILE ADDRESSING. A data or a file record have a key which uniquely identifies those data. Given this key, the programs must locate the address of a file record associated with these data. There is a variety of techniques for converting such a key into a machine file address.

FILE PACKING DENSITY. A ratio of amount of space (in words or characters) on a file, available for storing data to the total amount of data that are stored in the file.

FILE RECONSTRUCTION PROCEDURES. There is a remote possibility that vital data on the files will be accidentally destroyed by an equipment failure, or a program or operator error. The means of reconstructing the file must be devised should such an unfortunate circumstance occur. Vital data must be dumped onto tape or some other media, and programs must be written so that the file may be reconstructed from these data if necessary.

FORTRAN. A computer language designed mainly for programming scientific applications.

HARDWARE. The electrical, electronic, magnetic and mechanical devices or components of a computer.[4]

HEURISTIC. Trial-and-error method of tackling a problem, as opposed to the algorithmic approach.[4]

INFORMATION RETRIEVAL. A branch of computer sciences relating to the techniques for storing and searching large or specific quantities of information.[4]

INFORMATION SYSTEM. The network of all communication methods within an organization.[4]

INFORMATION THEORY. The mathematical theory concerned with channels, rates of transfer, noise, etc., relating to information.[4]

INPUT. Information which a control system's elements receive from outside.[4]

INSTABILITY. A condition of a feedback control system in which large sustained oscillations of the controlled variable occur, so that the latter is no longer controlled by input instructions.[4]

INSTRUCTION. A coded program step that tells the computer what to do for a single operation in a program.[4]

INTELLIGENCE. The developed capability of a device to perform functions that are normally associated with human intelligence, such as learning or reasoning.[4]

INTELLIGENCE (ARTIFICIAL). The study of computer and related techniques to supplement the intellectual capabilities of man.[4]

INTERFACE. The point of contact between different systems or parts of the same system; it may involve codes, speeds, sizes and formats.[4]

MACHINE LANGUAGE. Information recorded in a form directly understood by the computer.[4]

MAGNETIC CORE. A data storage device based on the use of a highly magnetic, low-loss material, capable of assuming two or more discrete states of magnetization.[4]

MAGNETIC DRUM. A data storage device using magnetized spots on a magnetic rotating drum; permits quasi-random medium-speed access to any part of its surface.[4]

MAGNETIC INK CHARACTER RECOGNITION. (See 'MICR'.)

MAGNETIC TAPE. A device for storing digital or analog data in the form of magnetized areas on a tape of plastic coated with magnetic iron oxide.[4]

MASTER AND SLAVE COMPUTERS. Where two or more computers are working jointly, one of these is sometimes a master computer and the others are slaves. The master can interrupt the slaves and send data to them when it needs to. When data pass from the slaves to the master it will be at the master's request.

MEAN TIME TO FAILURE. The average length of time for which the system, or a component of the system, works without fault.

MEAN TIME TO REPAIR. When the system, or a component of the system, develops a fault, this is the average time taken to correct the fault.

MEMORY PROTECTION. This is a hardware device which prevents a program from entering areas of memory that are beyond certain boundaries. It is useful in a multi-programmed system. Different programs in that system will be confined within different boundaries, and thus cannot do damage to each other.

MICR. Magnetic ink character recognition. The machine recognition of characters printed with magnetic ink. (Contrast with 'OCR'.)[1]

MICROSECOND. One-millionth of a second (μs).[4]

MILLISECOND. One-thousandth of a second (ms).[4]

MODULAR. The ability to increase a system in small steps (modules).[4]

NANOSECOND. One thousand-millionth of a second (ns).[4]

NUMERICAL CONTROL. A means of controlling machine tools through servo-mechanisms and control circuitry, so that the motions of the tool will respond to digital coded instructions on tape.[4]

OCR. Optical character recognition. Machine identification of printed characters through use of light-sensitive devices. (Contrast with 'MICR'.)

ON LINE. An on-line system may be defined as one in which the input data enter the computer directly from their point of origin and/or output data are transmitted directly to where they are used. The intermediate stages such as punching data into cards or paper tape, writing magnetic tape, or off-line printing, are largely avoided.

OUTPUT. Information which a control system transmits as a result of its input.[4]

PARALLEL. Simultaneous processing of the individual parts of a whole, such as bits on characters.[4]

PARAMETER. A variable corresponding to a given condition; an arbitrary constant as distinguished from a fixed constant.[4]

PERIPHERAL EQUIPMENT. Ancillary devices under the control of the central processor, e.g., magnetic tape units, printers or card readers.[4]

PL/I. A computer language designed for programming both scientific and commercial applications.

PROCESS CONTROL. Pertaining to systems whose purpose is to provide automation of continuous operations. This is contrasted with numerical control, which provides automation of discrete operations.[3]

PROCESSOR. (1) In hardware, a data processor. (2) In software, a computer program that includes the compiling, assembling, translating, and related functions for a specific programming language, for example, COBOL processor, FORTRAN processor.[1]

PROGRAM. A set of coded instructions to direct a computer to perform a desired operation or solve a predefined problem.[4]

PROGRAM TAPE. A magnetic or punched paper tape which contains the sequence of instructions required for solving a problem on a digital computer and coded in language which may be read by the computer.[4]

PUNCH CARD. Thin cards on which digits are represented by holes in selected locations for storing data.[4]

PUNCHED TAPE. A paper or plastic tape in which holes are punched to serve as a digital storage device.[4]

PUNCHED CARD. (1) A card punched with a pattern of holes to represent data. (2) A card as in (1) before being punched.[1]

RANDOM ACCESS. Access to data storage in which the position from which information is to be obtained is not dependent on the location of the previous information, e.g. as on magnetic drums, disks, or cores.[4]

RANDOM-ACCESS FILES. These are storage media holding a large amount of information in such a way that any item may be read or written at random with a short access time, i.e., usually less than one second. Example of random-access files are disk storages, drums, and magnetic strip files.

REAL TIME. A real-time computer system may be defined as one that controls an environment by receiving data, processing them and returning the results sufficiently quickly to affect the functioning of the environment at that time.

REASONABLENESS CHECKS. Tests made on information reaching a system or being outputted from it to ensure that the data in question lie within a given reasonable range.

RESPONSE TIME. This is the time the system takes to react to a given input. If a message is keyed into a terminal by an operator and the reply from the computer, when it comes, is typed at the same terminal, response times may be defined as the time interval between the operator pressing the last key and the terminal typing the first letter of the reply. For different types of terminal, response time may be defined similarly. It is the interval between an event and the system's response to the event.

SEEK. A mechanical movement involved in locating a record in a random-access file. This may, for example, be the movement of an arm and head mechanism that is necessary before a read instruction can be given to read data in a certain location on the file.

SELF-CHECKING NUMBERS. Numbers which contain redundant information so that an error in them, caused, for example, by noise on a transmission line, may be detected. A number may, for example, contain two additional digits which are produced from the other digits in the number by means of an arithmetical process. If these two digits are not correct it will indicate that the number has in some way been garbled. The two additional digits may be checked by the computer as a safeguard against this.

SERIAL ACCESS. (1) Pertaining to the sequential or consecutive transmission of data to or from storage. (2) Pertaining to the process of obtaining data from, or placing data into, storage where the time required for such access is dependent upon the location of the data most recently obtained or placed in storage. (Contrast with 'Random access'.)[2]

SYSTEM. An assembly of components united by some form of regulated interaction to form an organized whole. Also a collation of operations and procedures, men and machines by which an industrial or business activity is carried on. In the realm of computers a system is defined as an organization of hardware, software and people for cooperative operation to complete a set of tasks for desired purposes.[4]

SYSTEMS ANALYSIS. The organized step-by-step study of the detailed procedure for collection, manipulation and evaluation of data about an organization, for the purpose of determining what must be accomplished and the best method of accomplishing it in order to improve control of a system.

SOFTWARE. General-purpose programs used to extend the capabilities of computers, including compilers, assemblers, monitors, executive routines, etc.[4]

SUPERVISORY PROGRAMS. Those computer programs designed to coordinate service and augment the machine components of the system, and coordinate and service Application Programs. They handle work scheduling, input/output operations, error actions, and other functions.

TELEPROCESSING. A form of information handling in which a data processing system utilizes telecommunication facilities.

TERMINALS. The means by which data are entered into the system and by which the decisions of the system are communicated to the environment it affects. A wide variety of terminal devices have been built, including teleprinters, special keyboards, light displays, cathode tubes, thermocouples, pressure gauges and other instrumentation, radar units, telephones, and so on.

TIME SHARING. Participation in available computer time by multiple users, via terminals. Characteristically, the response time is such that the computer seems dedicated to each user.[3]

WORD. A set of characters or bits which is handled by the computer circuits as a unit; word lengths are fixed or variable, depending on the computer.[4]

INDEX

PENGUINEWS *AND*
PENGUINS IN PRINT

Every month we issue an illustrated magazine, *Penguinews*.
It's a lively guide to all the latest Penguins, Pelicans and
Puffins, and always contains an article on a major Penguin
author, plus other features of contemporary interest.

Penguinews is supplemented by *Penguins in Print*, a
complete list of all the available Penguin titles – there are
now over four thousand!

The cost is no more than the postage; so why not write
for a free copy of this month's *Penguinews*? And if you'd
like both publications sent for a year, just send us a
cheque or a postal order for 30p (if you live in the United
Kingdom) or 60p (if you live elsewhere), and we'll put you
on our mailing list.

Dept EP, Penguin Books Ltd,
Harmondsworth, Middlesex

Note: *Penguinews* and *Penguins in Print* are not
available in the U.S.A. or Canada

*A Volume in the Pelican Library
of Business and Management*

COMPUTERS, MANAGERS
AND SOCIETY

Michael Rose

'Here is a book on computers and computer
technology written by a sociologist, and it is one of the
very few which sets out clearly, simply and, even more
important, objectively what computers are all about. It is
also amusing . . .' – *New Society*

After a general survey of the development of computer-
controlled data processing, Michael Rose examines the
complex effects of the computer upon the clerical worker –
the new opportunities, the dangers of alienation, the
threat of technological unemployment. He then focuses
upon the fast-developing problems of managers. Many of
the standard managerial functions can already be
programmed. But should executives delegate qualitative
decisions to a machine? And if so, how far can and should
these changes go?

'Computerization' presents managers with new
opportunities on a structural scale unmatched since the
Industrial Revolution. Do they really understand the new
situation? Can they, when it is transforming itself so
rapidly? And are we enough aware of the effects of the
computer upon an even larger social group – society itself –
now faced with the need to clarify its whole attitude to
technological change?

'Knowledgeable, intelligent and clearly written' – *The Times
Literary Supplement*

ELECTRONIC COMPUTERS

S. H. Hollingdale and G. C. Tootill

Although little more than twenty years old, electronic computers are reshaping our technological society. This Pelican explains how computers work, how problems are presented to them, and what sort of jobs they can tackle. Analog and digital computers are compared and contrasted and recent syntheses of the two techniques described.

To survey a difficult subject so throughly necessitated the collaboration of two authors, both of whom hold senior posts directly connected with computers. With the general reader in mind they have taken particular care with the specialist jargon of their subject, explaining each term as it occurs. At the same time the technique of programming is given in sufficient depth to prepare a novice to cope with a manufacturer's handbook, and the computer, in its varying embodiments, is described in enough detail to give him confidence in learning to use one.

In addition the authors have devoted two chapters to the history of computers and the fascinating story of such pioneers of calculating machines as Charles Babbage.